Two Manuals Combined into One Book

AF800511

Carpentry

And

Combat Engineer

Department of Defense

Pentagon
Publishing

Field Manual
No. 5-426

Headquarters
Department of the Army
Washington, DC, 3 October 1995

CARPENTRY

TABLE OF CONTENTS

DISTRIBUTION RESTRICTION: Approved for public release; distribution is unlimited.
*This publication supersedes FM 5-551, 23 February 1984.

Page

Page

Page

PREFACE

This manual is intended for use as a training guide and reference text for engineer personnel responsible for planning and executing theater of operations (TO) construction. It provides techniques and procedures for frame construction, preparation and use of bills of materials (BOMs), building layout, forming for concrete slabs and foundations, framing and finish carpentry, roof framing and coverings, bridge and wharf construction, and the materials used for these operations.

The proponent for this publication is the United States Army Engineer School (USAES). Submit changes for improvement on Department of the Army (DA) Form 2028 (Recommended Changes to Publications and Blank Forms) to Commandant, United States (US) Army Engineer School, ATTN: ATSE-TD-D, Fort Leonard Wood, Missouri 65473-6650.

Unless this publication states otherwise, masculine nouns and pronouns do not refer exclusively to men.

CONSTRUCTION DRAWINGS — CHAPTER 1

Before the carpenter can begin his work, he must have a complete set of construction drawings or prints. He must be familiar with how the drawings are prepared. He should also know how to read the drawings so that he will understand what he must do to comply with their requirements. This chapter will help the carpenter to reach that understanding

ARCHITECTURAL SYMBOLS. LINE CONVENTIONS, AND MATERIAL CONVENTIONS

The plan for a building must give all the details necessary to construct the building. Therefore, it usually consists of a collection of sheets, called a *set of plans.* Each sheet shows the details of a different phase or part of the construction. Several of the sheets will be devoted to floor plans. Other sheets in the set will show construction details (such as wiring, plumbing, and air-conditioning details and types and quantities of materials). Refer to Appendix A for conversion tables.

Architectural symbols on construction drawings show the type and location of windows (Figure 1-1), doors (Figure 1-2), and other features. They show the general shape of an actual architectural feature and show any motion that is supposed to occur.

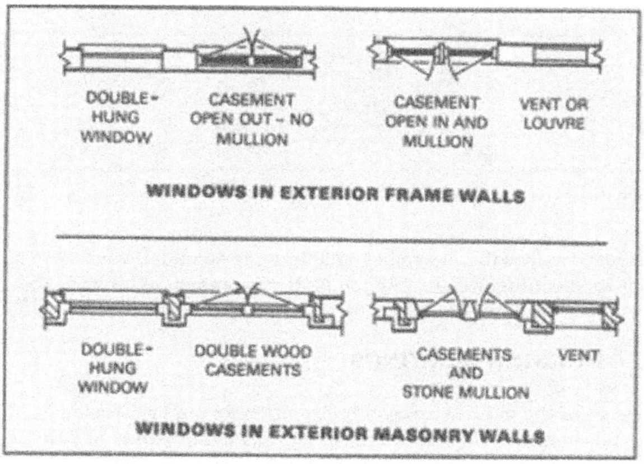

Figure 1-1. Window symbols

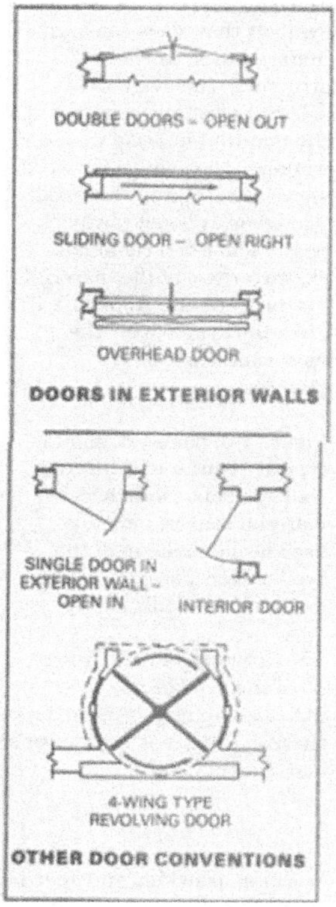

Figure 1-2. Door symbols

Line conventions are different types and weights of lines used to represent the features of an object. The meaning of a line with certain characteristics has been standardized and will be the same on any drawing. These line conventions must be understood in order to read drawings. The line conventions most often encountered in construction prints are shown and described in Figure 1-3.

NAME	CON-VENT-ION	DESCRIPTION AND APPLICATION	EXAMPLE
VISIBLE LINES		HEAVY, UNBROKEN LINES. USED TO INDICATE VISIBLE EDGES OF AN OBJECT.	
HIDDEN LINES		MEDIUM LINES WITH SHORT, EVENLY SPACED DASHES. USED TO INDICATE CONCEALED EDGES.	
CENTER LINES		THIN LINES MADE UP OF LONG AND SHORT DASHES ALTERNATELY SPACED AND CONSISTENT IN LENGTH. USED TO INDICATE SYMMETRY ABOUT AN AXIS AND LOCATION OF CENTERS.	
DIMENSION LINES		THIN LINES TERMINATED WITH ARROWHEADS AT EACH END. USED TO INDICATE DISTANCE MEASURED.	
EXTENSION LINES		THIN, UNBROKEN LINE. USED TO INDICATE EXTENT OF DIMENSIONS.	
LEADER		THIN LINE TERMINATED WITH ARROWHEAD OR DOT AT ONE END. USED TO INDICATE A PART, DIMENSION, OR OTHER REFERENCE.	

Figure 1-3. Line conventions

Material conventions are symbols that show the type of material used in the structure. The symbol selected normally represents the material in some way; for example, the symbol for wood shows the grain in the wood. (However, it is not always possible to use a common characteristic of the material for the symbol.) (Appendix B gives the symbols for the most common types of materials.)

Figure 1-4, page 1-4, shows typical exterior and interior wall symbols. (Note how the material conventions are used in the makeup of the symbols for masonry, brick, and concrete walls.)

The carpenter should know all of the symbols for materials to help him read a construction drawing. A symbol on a drawing should always be checked if there is any doubt about its meaning. Refer to Appendix B for common abbreviations and symbols.

WORKING DRAWINGS

Working drawings and specifications are the main sources of information for supervisors and technicians responsible for the actual construction. The construction working drawing gives a complete graphic description of the structure to be erected and the construction method to be followed.

A set of working drawings includes both general and detail drawings. General drawings consist of plans and elevations; detail drawings consist of sections and detail views.

SITE PLANS

A site plan (also called *a plot plan)* shows—

- Property lines and locations.
- Contours and profiles.
- Building lines.
- Locations of structures to be built.
- Existing structures.
- Approaches.
- Finished grades.
- Existing and new utilities (such as sewer, water, and gas).

NAME	CON-VENT-ION	DESCRIPTION AND APPLICATION	EXAMPLE
PHANTOM OR DATUM LINE		MEDIUM SERIES OF ONE LONG DASH AND TWO SHORT DASHES EVENLY SPACED ENDING WITH LONG DASH. USED TO INDICATE: ALTERNATE POSITION OF PARTS, REPEATED DETAIL, OR A DATUM PLANE.	
STITCH LINE		MEDIUM LINE OF SHORT DASHES. EVENLY SPACED AND LABELED. USED TO INDICATE STITCHING OR SEWING.	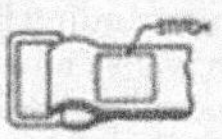
BREAK (LONG)	 (WOOD)	THIN, SOLID RULED LINE WITH FREE-HAND ZIGZAGS. USED TO REDUCE SIZE OF DRAWING REQUIRED TO DELINEATE OBJECT AND REDUCE DETAIL.	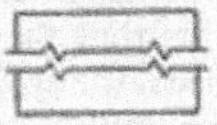
BREAK (SHORT)		THICK, SOLID FREE-HAND LINES. USED TO INDICATE A SHORT BREAK.	
CUTTING OR VIEWING PLANE— VIEWING PLANE OPTIONAL	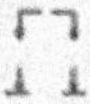	THICK, SHORT DASHES. USED TO SHOW OFFSET WITH ARROWHEADS TO SHOW DIRECTION IN WHICH SECTION OR PLANE IS VIEWED OR TAKEN.	
CUTTING PLANE FOR COMPLEX OR OFFSET VIEWS		THICK, SHORT DASHES. USED TO SHOW OFFSET WITH ARROWHEADS TO SHOW DIRECTION VIEWED.	

Figure 1-3. Line conventions (continued)

Figure 1-5, page 1-5, shows a typical site plan. Appropriate outlines show the location of each building. The new facility can be located by referring to the schedule of facilities on the plan. The site plan has a north-pointing arrow to indicate site north—not magnetic north. Each facility has a number (or code letter) to identify it in the schedule of facilities. The contour lines show the elevation of the earth surfaces. (All points on a contour have the same elevation.)

Distances are given between principal details and reference lines. (The coordinate reference lines on the figure are centerlines of the roads surrounding the area.) All distances in a plan view simply give the horizontal measurement between two points; they do not show terrain irregularities. (The sizes of proposed facilities are given in the schedule of facilities.)

Examine the site plan shown in Figure 1-5 to see what information can be obtained from it. For example, the contour lines show that the ground surface of the site area slopes. The location and identification of each facility are given. Most of the facilities are spaced at least 60 feet apart, but the library (facility No. 3) and the recreation building (facility No. 4) are only 15 feet apart. The library is the smallest of the four buildings and is closest to the road- the east wall of the library is 20 feet from the centerline of the road, while the other buildings are 30 or 60 feet from the centerline.

ELEVATIONS

Elevations are drawings that show the front, rear, or side view of a building or structure. Sample elevation views are given in Figure 1-6, page 1-6. Construction materials may be shown on the elevation. The ground level (called the grade) surrounding the structure may also be shown. When more than one view is shown on a drawing sheet, each view is given a title. If any view has a scale different from that shown in the title block, the scale is given beneath the title of that view.

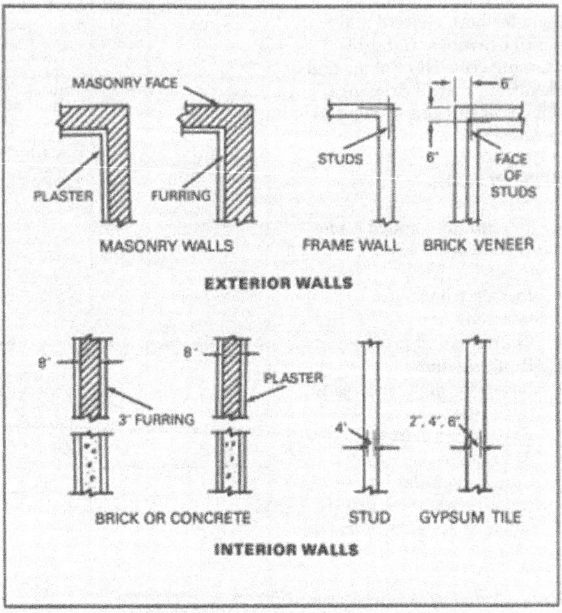

Figure 1-4. Typical wall symbols

The centerline symbol of alternate long and short dashes shows finished floor lines The hidden line symbol of short, evenly spaced dashes shows foundations below the grade line.
Note that Figure 1-6 shows the footings are below grade.

Elevations show the locations and types of doors and windows. Each different type of window shown in the elevations is marked; the three types of windows shown here are marked W-1, W-2, and W-3. These identifying marks refer to a particular size of window whose dimensions are given in a table known as the window schedule. In some cases, rough opening dimensions of windows are given on the drawing. Note that the building shown here has two double doors on each side and a double door at each end. The elevation also shows that at the end of the building with the loading platform, the door is at the level of the stage floor; all the other doors are at grade level.

FLOOR PLANS

A floor plan is a cross-sectional view of a building. The horizontal cut crosses all openings regardless of their height from the floor. The development of a floor plan is shown in Figure 1-7, page 1-7.

A floor plan may show, among other things, the outside shape of the building; the arrangement, size, and shape of the rooms; the type of material and the length, thickness, and character of the building walls at a particular floor; the type, width, and location of the doors and windows; the types and locations of utility installations; and the location of stairways. A typical floor plan is shown in Figure 1-8, pages 1-8 and 1-9.

As you read the floor plan in Figure 1-8, note the features of the recreation building. The lines with small circles show wiring for electrical outlets; appropriate symbols show the plumbing fixtures. These features are important to the carpenter from the standpoint of coordination. He

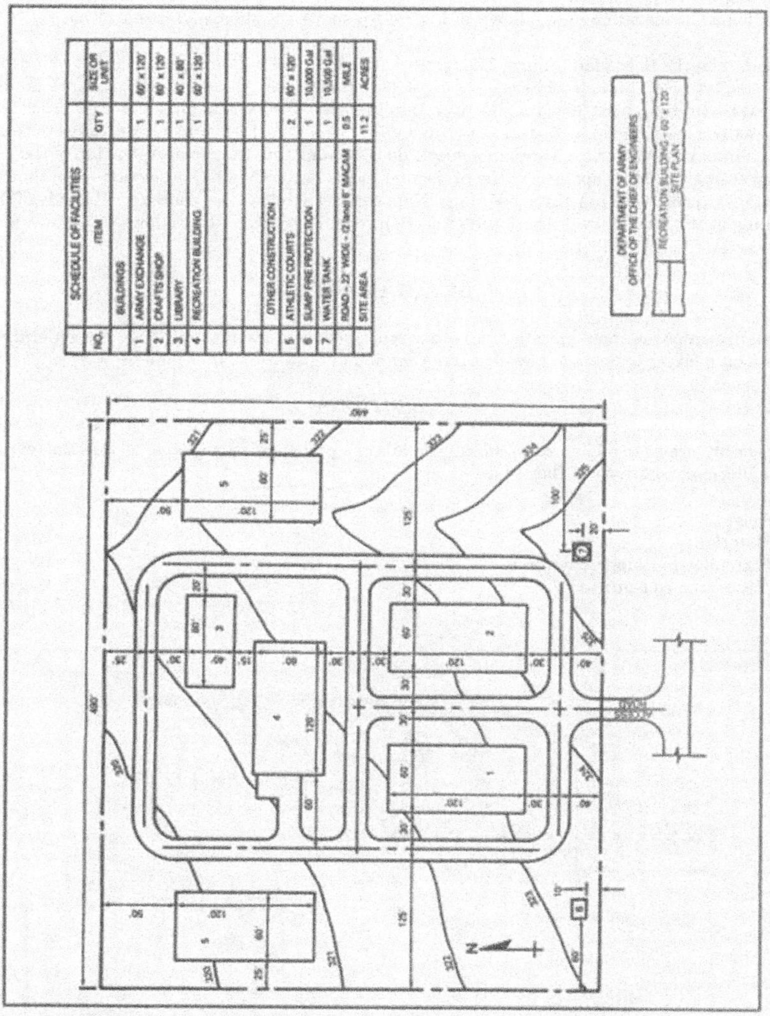

Figure 1-5. Typical site plan

may have to make special provisions, at various stages of construction, for the placement of electrical or plumbing fixtures. Installation of these fixtures should be coordinated at the appropriate time with the electrician, plumber, and foreman.

As you examine the floor plan, note that the interior of the building will consist of an auditorium, a lobby with a post exchange (PX) counter, a men's toilet, a women's toilet, a projection room on a second level above the lobby, two dressing rooms, and a stage. The stage may not be apparent but, by noting the steps adjacent to each dressing room, it can be seen that there is a change in elevation. (The elevation view, shown in Figure 1-6 shows the stage and its elevation.)

Note that on the floor plan (Figure 1-8, pages 1-8 and 1-9) all building entrance and exit doors are the same type (1D) and all windows are the double-hung type. All interior single doors are the same (2D), and two double doors (3D) open into the lobby from the auditorium. The projection room will be reached via a 15-riser stairway located in a 12- x 18-foot room. Entrance to this room will be from the auditorium through a single door opening into the room. At the top of the stairway, a single door opens into the projection room. The wall of the projection room that faces the stage (inside wall) has three openings. Note that no windows are shown for the side of the building at the second level where the projection room is located, but windows are shown at the main level.

DETAIL DRAWINGS

Detail drawings are more specific than other types of construction plans. They are generally drawn on a larger scale and show features that do not appear on other plans.

SECTIONS

Sections are drawn to a large scale showing details of a particular construction feature that cannot be given in a general drawing. They show—

- Height.
- Materials.
- Fastening and support systems.
- Any concealed features.

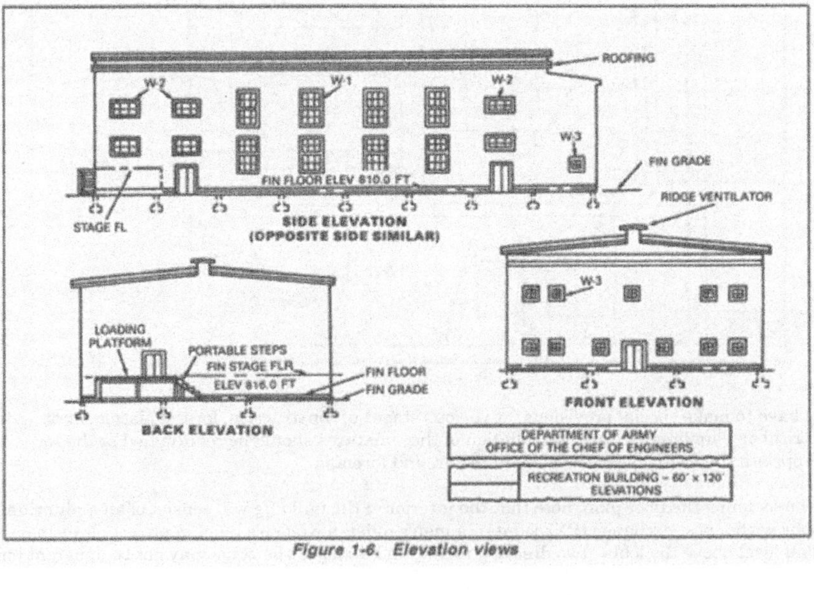

Figure 1-6. Elevation views

A typical wall section, with parts identified by name and/or size, is illustrated in Figure 1-9, page 1-10. This figure shows how a structure looks when cut vertically by a cutting plane.

Wall sections are very important to construction supervisors and to the craftsmen who do the actual building. They show the construction of the wall, as well as the way in which structural members and other features are joined to it. Wall sections extend vertically from the foundation bed to the roof. Sections are classified as typical and specific.

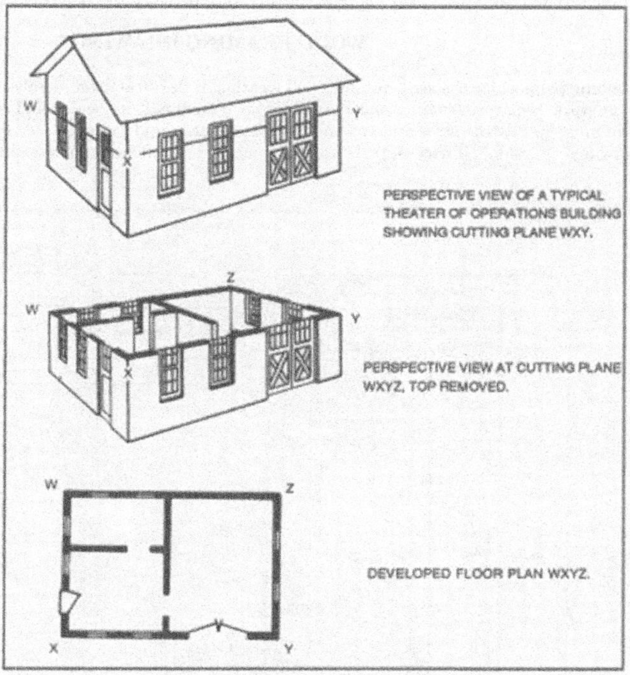

Figure 1-7. Floor-plan development

Typical Sections

Typical sections are used to show construction features that are repeated many times throughout a structure.

Specific Sections

When a particular construction feature occurs only once and is not shown clearly in the general drawing, a cutting plane is passed through that portion.

DETAILS

Details are large-scale drawings which show features that do not appear (or appear on too small a scale) on the plans, elevations, and sections. Sections show the builder how various parts are connected and placed. Details do not have a cutting-plane indication, but are simply noted by a code. The construction of doors, windows, and eaves is usually shown in detail drawings. Figure 1-10 shows some typical door framing details, window wood-framing details, and an eave detail for a simple type of cornice. Other details which are customarily shown are sills, girder and joint connections, and stairways.

Figure 1-11, page 1-12, shows how a stairway is drawn in a plan and how riser-tread information is given. For example, on the plan, DOVVN 17 RISERS followed by an arrow means that there are 17 risers in the run of stairs going to the firs floor from the floor above, in the direction indicated by the arrow. The riser-tread diagram provides height and width information. The standard for the riser, or height from the bottom of the tread to the bottom of the next tread, ranges from 6 1/2 to 7 1/2 inches.

The tread width is usually such that the sum of riser and tread is about 18 inches (a 7-inch riser and 11-inch tread is standard). On the plan, the distance between the riser lines is the width of the tread.

WOOD FRAMING DRAWINGS

Framing plans show the size, number, and location of the structural members constituting the building framework. Separate framing plans may be drawn for the floors, walls, and roof. The floor framing plan must specify the sizes and spacing of joists, girders, and columns used to support the floor. Detail drawings are added, if necessary, to show the methods of anchoring joists and girders to the columns

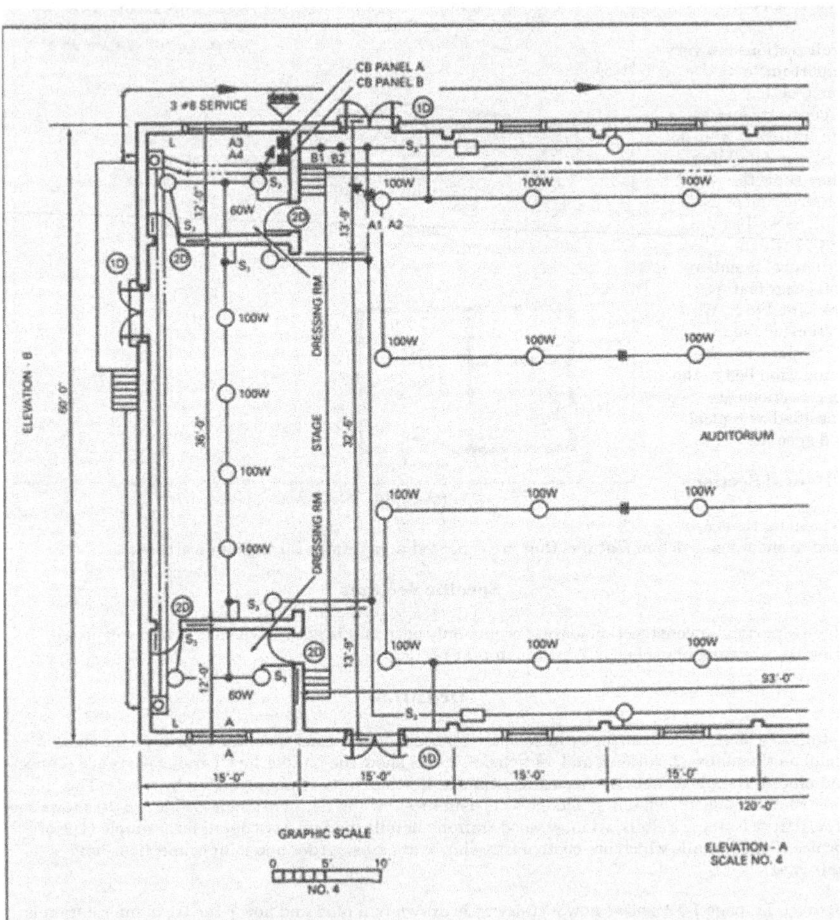

Figure 1-8. Typical

and foundation walls or footings. Wall framing plans show the location and method of framing openings and ceiling heights so that studs and posts can be cut. Roof framing plans show the construction of the rafters used to span the building and support the roof. Size, spacing, roof slope, and all necessary details are shown. Working prints for TO buildings usually show details of all framing.

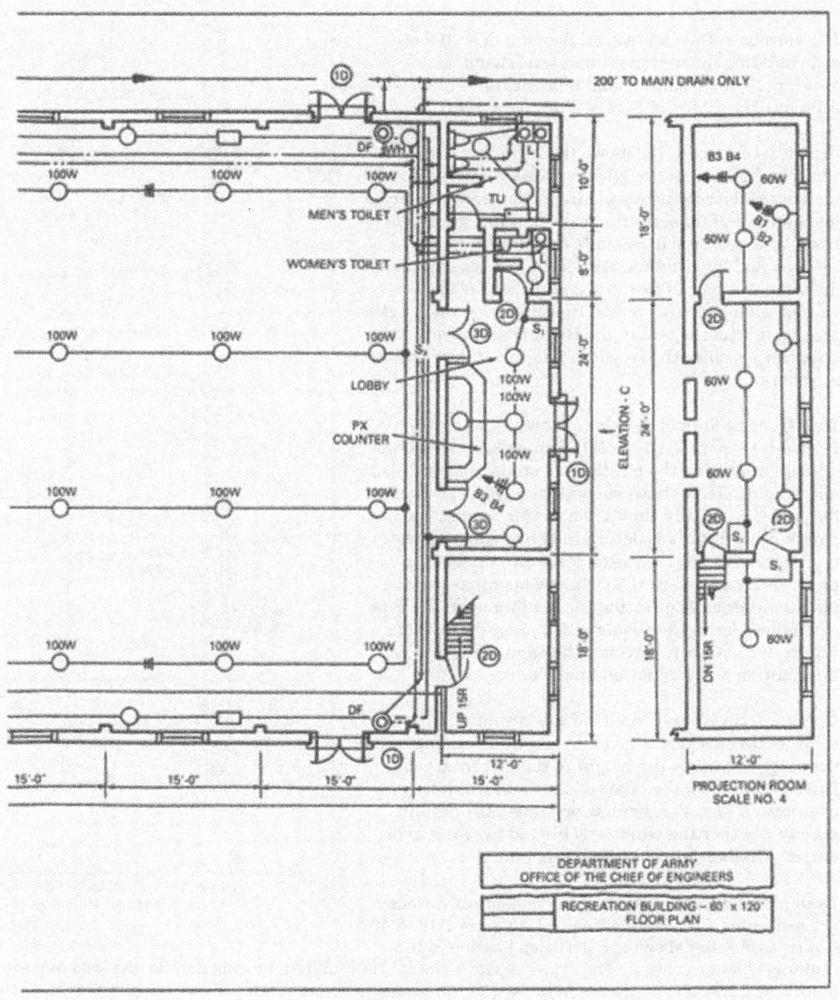

floor plan

LIGHT WOOD FRAMING

Light framing is used in barracks, bathhouses, administration buildings, light shops, hospitals, and similar structures.

Detailed drawings of foundation walls, footings, posts, and girder details normally used in standard TO construction are shown in Figure 1-12.

The various details for overall framing of a 20-foot-wide building (including ground level, window openings, braces, splices, and nomenclature of framing) are shown in Figure 1-13, page 1-14.

Figure 1-14, page 1-15, shows floor framing details showing footings, posts, girders, joists, reinforced sections of floor for heavy loads, section views covering makeup of certain sections, scabs for joint girders to posts, and post-bracing details as placed for cross sections and longitudinal sections. On a construction drawing, the type of footings and the size of the various members are shown. In some cases the lengths are given, while in others the BOM that accompanies the print specifies the required lengths of the various members.

Wall framing for end panels is shown in view A in Figure 1-15, page 1-16. Wall framing plans are detail drawings showing the locations of studs, plates, sills, and bracing. They show one wall at a time. The height for panels is usually shown. From this height, the length of wall studs is determined by deducting the thickness of the top or rafter plate and the bottom plate. Studs placed next to window openings may be placed either on edge or flat, depending upon the type of windows used. Details for side panels (view B of Figure 1-15, page 1-16) cover the same type of information as listed for end panels.

Chapter 6 covers the details of wall framing. The space between studs is given in the wall framing detail drawing, as well as the height of the girt from the bottom plate, and the types of door and window openings, if any. For window openings, the details specify whether the window is hinged to swing in or out, or whether it is to be a sliding panel.

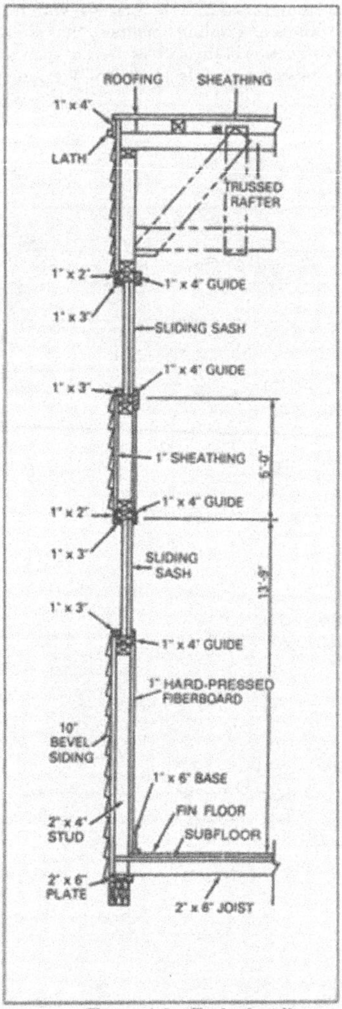

Figure 1-9. Typical wall section

Examples of drawings showing the makeup of various trussed rafters are given in Figure 1-16, page 1-18. A 40-foot trussed rafter showing a partition bearing in the center is shown in view A. The drawing shows the splices required, bracing details, the stud and top plate at one end of the rafter, and the size of the members.

A typical detail drawing of a 20-foot trussed rafter is shown in view B of Figure 1-16. Filler blocks are used to keep the brace members in a vertical plane, since the rafter and bottom chord are nailed together rather than spliced.. The drawing shows placement of the rafter tie on the opposite side from the vertical brace. Usually the splice plate for the bottom chord (if one is needed) is placed on the side where the rafters are to be nailed so that it can serve also as a filler block.

A modified trussed rafter, shown in view C of Figure 1-16, page 1-18, is used only when specified in plans for certain construction. It will not be used in areas subject to high wind velocities or moderate to heavy snowfall. In this type of trussed rafter, the bottom chord is placed on the rafters above the top plate.

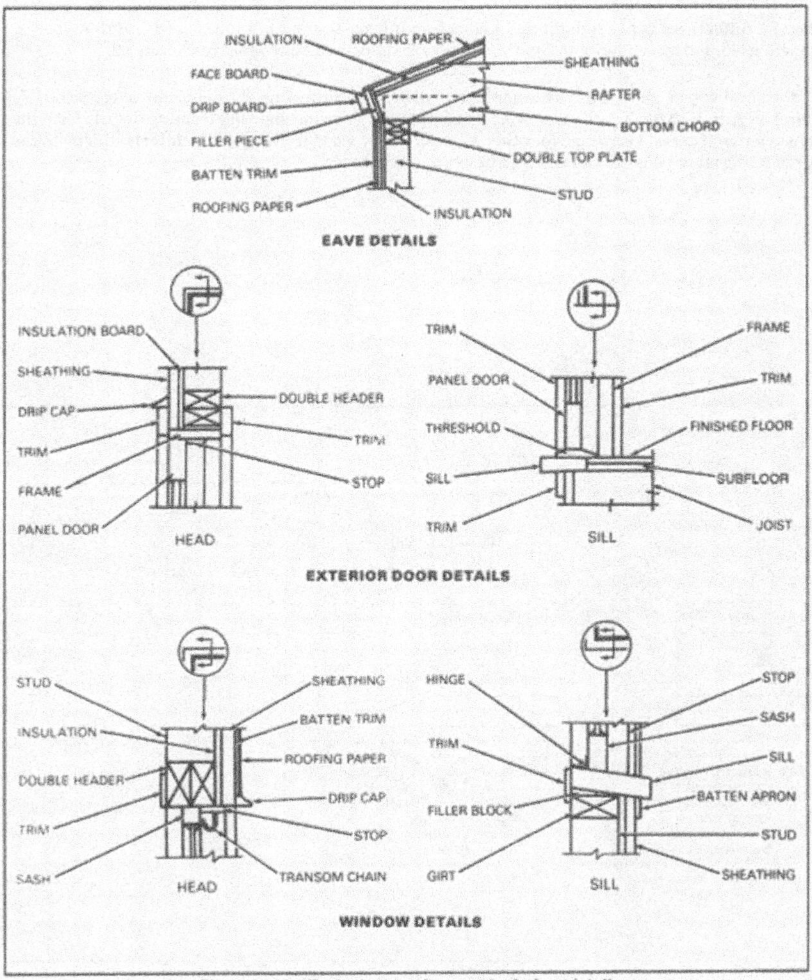

Figure 1-10. Typical eave, door, and window details

The construction plans will specify the best type of trussed rafter for the purpose. The drawings must show, in detail, the construction features of the rafter selected.

Another type of truss is the W-truss, shown in Figure 1-17, page 1-19. It may be used in either TO or residential construction, time permitting.

HEAVY WOOD FRAMING

Heavy wood framing consists of framing members (timber construction) at least 6 inches in dimension (for example, 2 x 6 inches or 4 x 12 inches). Examples of this type of framing can be found in heavy roof trusses, timber-trestle bridges, and wharves.

The major differences between light and heavy framing are the size of timber used and the types of fasteners used. Fasteners for both light and heavy framing will be covered in Chapter 2.

Note the similarities, as well as the differences, between the drawings for light and heavy wood framing. Figure 1-18, page 1-20, is a typical example of a drawing showing framing details for light and heavy roof trusses. Drawings for other types of heavy wood framing will similarly illustrate the kinds of material to be used and the way in which it is joined.

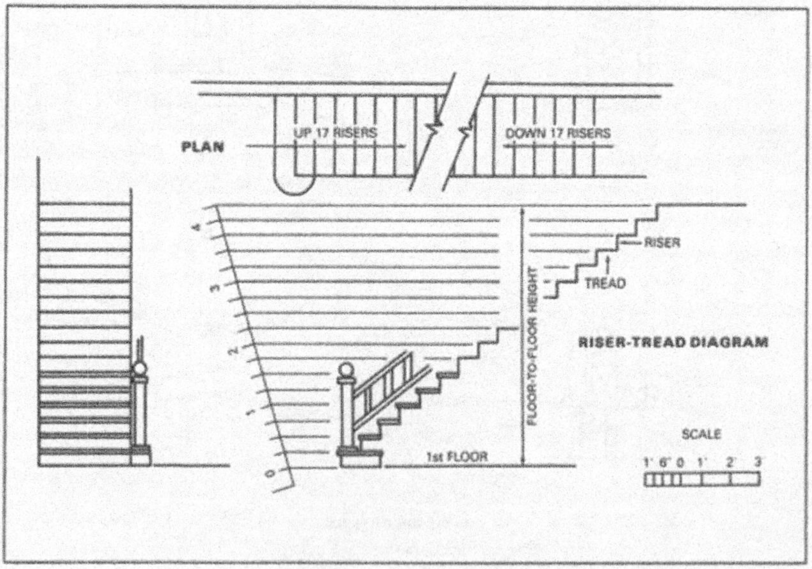

Figure 1-11. Stairway and steps

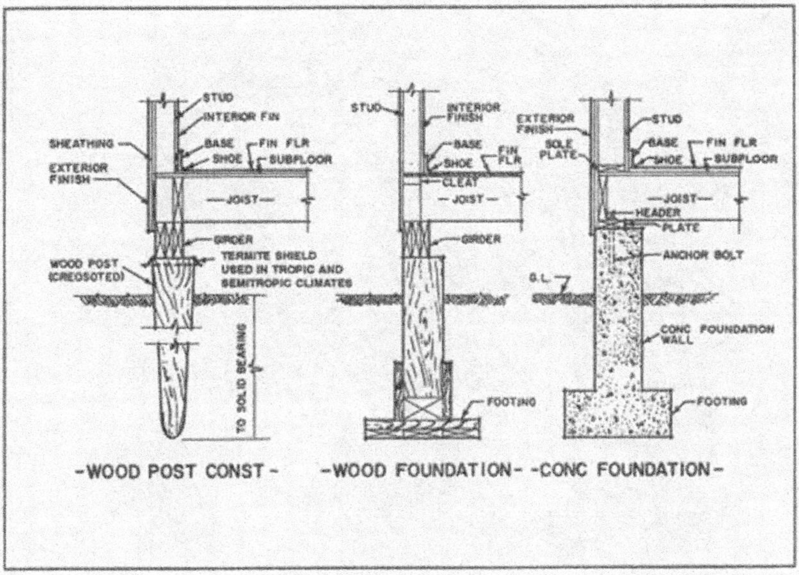

Figure 1-12. Typical foundation wall, post, footing, and girder details

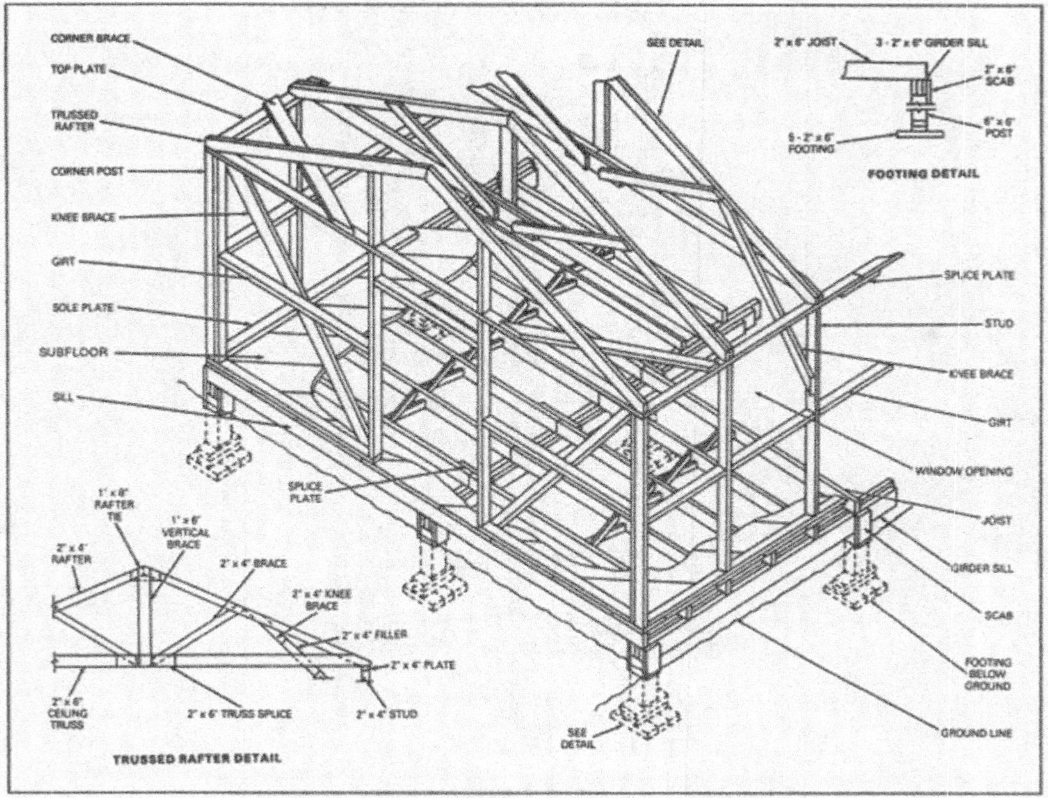

Figure 1-13. Light framing details (20-foot-wide building)

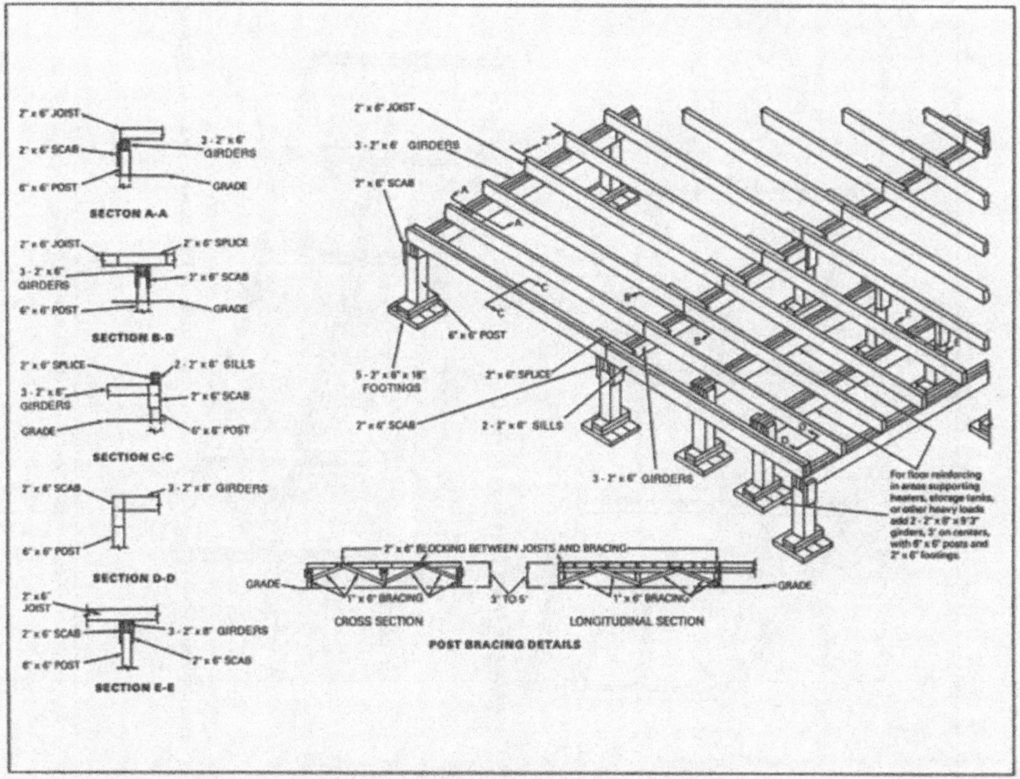

Figure 1-14. Floor framing details (20-foot-wide building)

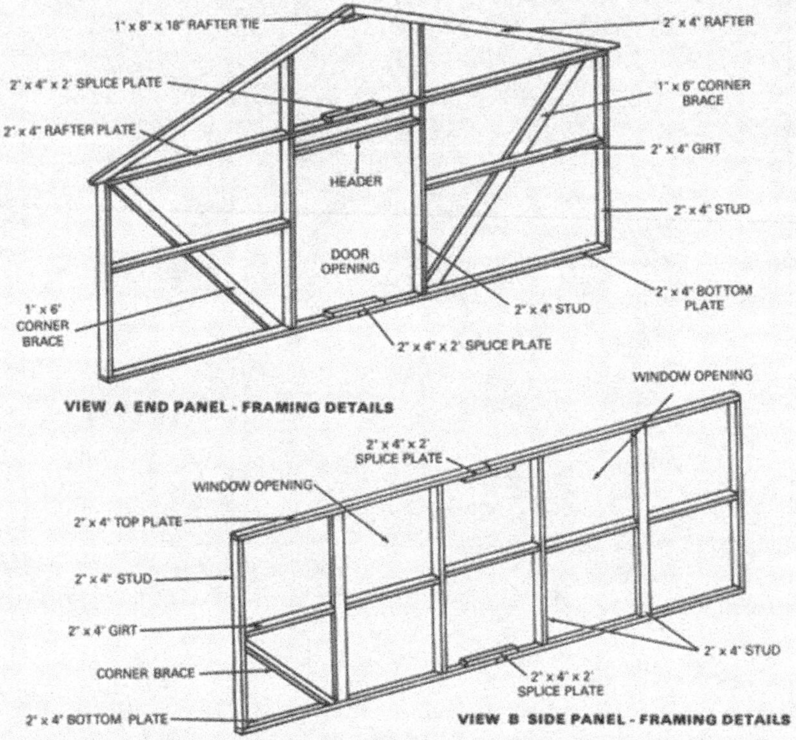

Figure 1-15. Typical wall-panel framing details

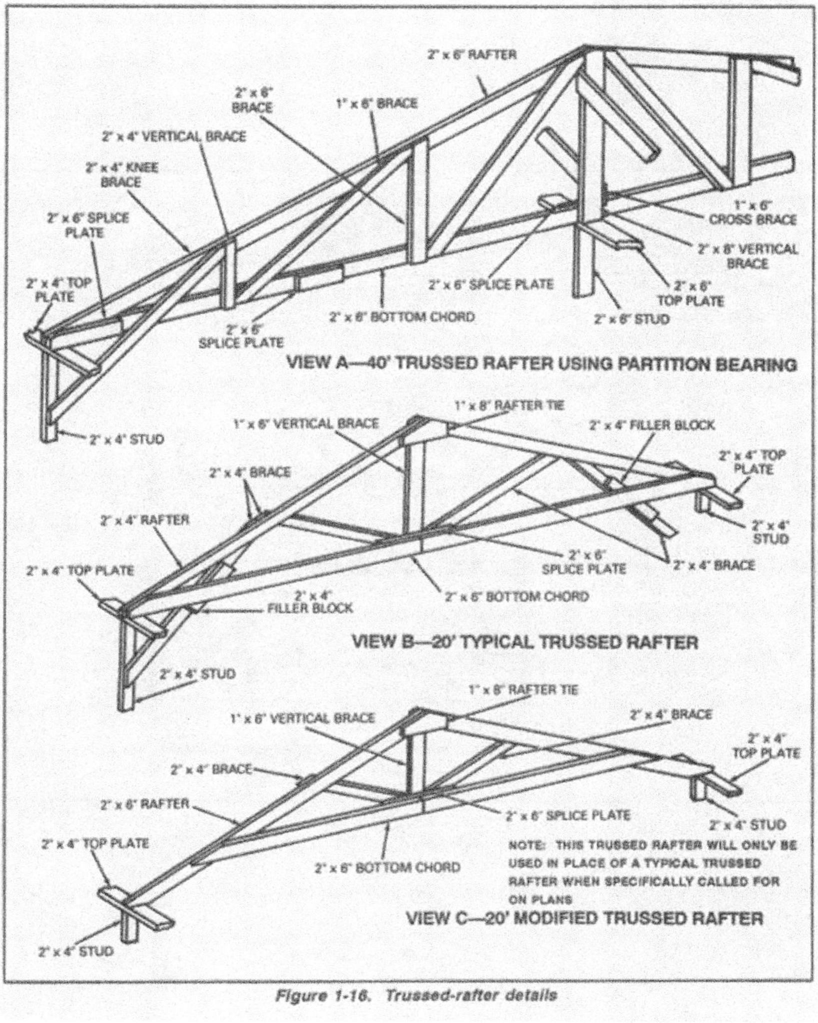

Figure 1-16. Trussed-rafter details

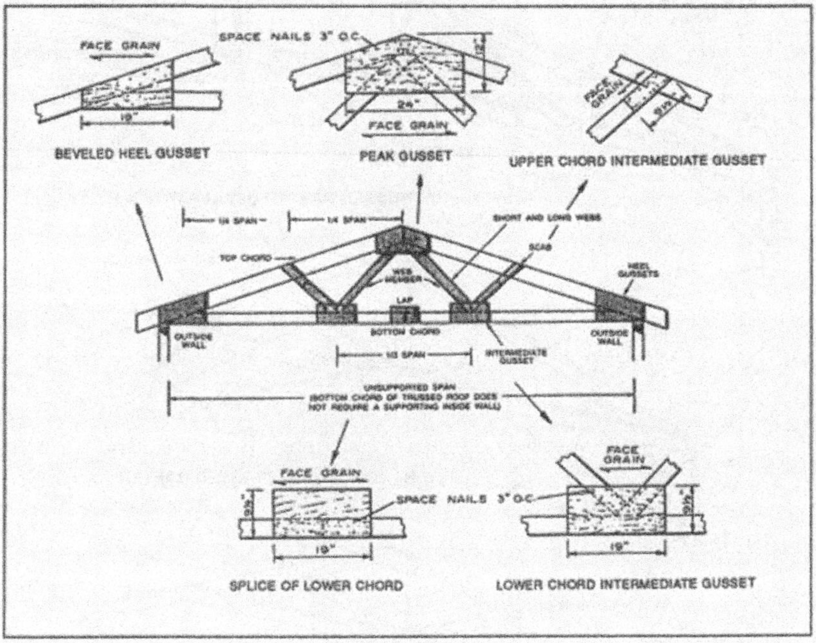

Figure 1-17. W-truss

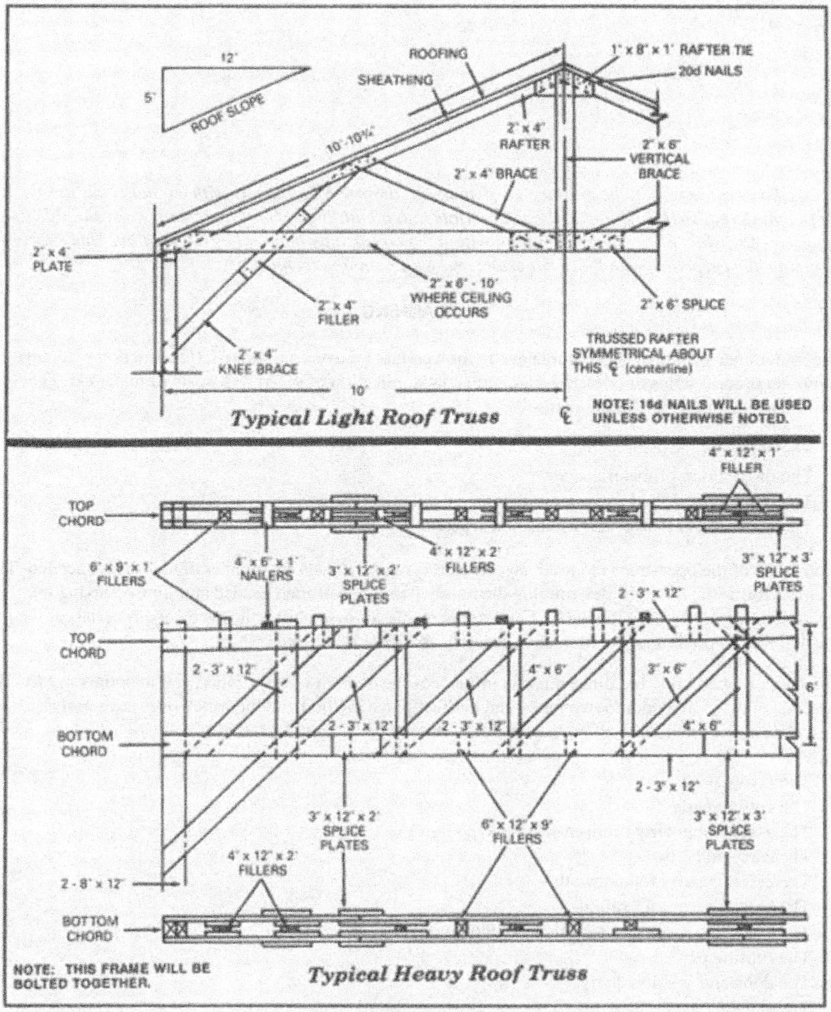

Figure 1-18. Roof trusses

CONSTRUCTION PLANNING AND MATERIALS

CHAPTER 2

Successful construction in the military TO demands a thorough familiarity with the materials used and the planning required to assemble these materials into a building—a building that is most suitable for the requirements, while most effectively fitting within the constraints imposed by the locality. This chapter acquaints the carpenter with those materials and with planning techniques.

PLANNING

The importance of properly planning construction cannot be overemphasized. Construction planning allows an orderly series of operations and prevents duplication of effort and waste of materials. The major considerations in planning are—

- The construction-plant layout.
- The distribution of materials.
- The number of skilled and semiskilled men available.
- The number and type of units to be constructed.

From a list of the operations required, an estimate is made of the total number of man-hours needed. This estimate forms the basis for determining the number and type of men needed and for organizing the erection crew(s). (Refer to Appendix C.) Arrangements for assembling the necessary materials at the job site and for the preliminary cutting and assembly are made in advance.

The method of erecting buildings directly influences the amount of time, labor, and materials needed. The methods may be divided into two types: the built-in-place method and the panel, *or preassembly (prefabricated),* method. Working parties for both methods should be set up as follows:

- The layout party.
- The cutting party.
- The assembling party (built-in-place method only).
- The carrying party.
- The erecting party for sidewalls.
- The erecting party for rafters.
- The sheathing party (built-in-place method only).
- The roofing party.
- The door-and-window party.
- The finishing party.

BUILT-IN-PLACE METHOD

In this method, each piece is separately erected in *its* proper place. When *using* the built-in-place method, the noncommissioned officer in charge (NCOIC) of construction divides the men into working parties, whose duties may be as follows:

- Laying out the foundation.
- Grading and excavating.
- Laying out and cutting various sizes of material.
- Carrying material to the cutting and erecting parties.

If a party finishes its task before the building is completed, it is assigned a new task. For example, if the party laying out the foundation completes its work before erection of the building has begun, it is assigned a new duty (such as cutting rafters).

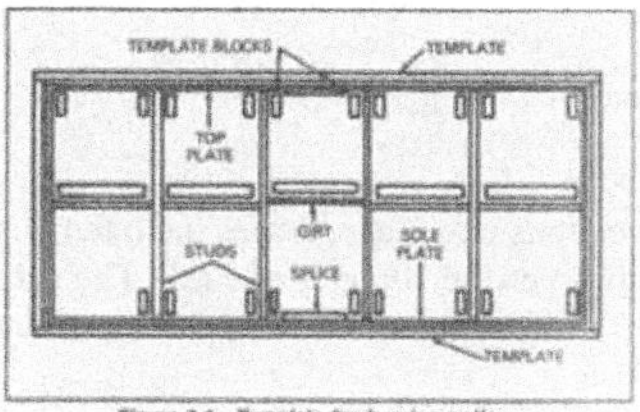

Figure 2-1. Template for framing walls

Parts of the building are built in the following order: footings, posts, sills, joists, floor, soles, studs, plates, girts, rafters, bracing, siding, sheathing, roofing, doors, windows, steps, and inside finish (if used).

PANEL METHOD

In this method (also called preassembly (prefabricated)), a complete section is built as a unit and then set in the building in its proper place. It is used extensively because it makes for greater speed, better control over working parties, and better use of manpower. For example, it allows the use of a standard list of sizes for each similar section. Standard plans shown in Technical Manuals (TMs) 5-02-1 and 5-302-2 further simplify construction. The panel method requires careful planning before the actual construction. Before measuring and cutting lumber, the number and size of sections that are alike should be determined from the construction drawing. This ensures availability of the correct numbers of each piece.

The carpenter assigns a crew to cut and assemble one section. (Several cutting and assembling parties may be used at one time on different types of sections.)

In most cases, a template (Figure 2-1) is built as a guide for assembling the section. The template should be built square and should be correct in size.

The number and size of each piece in a section are given to the man in charge of the cutting party. The cutting party cuts the timber to the correct length with a handsaw or power saw. The length is measured with a square and a tape. After one piece has been cut, it may be used as a pattern for marking the remaining pieces (Figure 2-2). The pattern is set up by nailing two blocks to the piece of correct size, one near each end, as shown in the figure. These blocks act as stops to hold the pattern in place on the timber to be marked.

The plate and sole are placed in the template with the studs and girts between them; then the door and window posts, if any, are placed as shown in Figure 2-1. The girts, sole, and plate are nailed to the studs with 16d or 20d nails. If insulation board is used, it and the wall sheathing are put on the section before it is taken out of the template. Applying the wall finish before raising the section makes the use of scaffolds or ladders unnecessary.

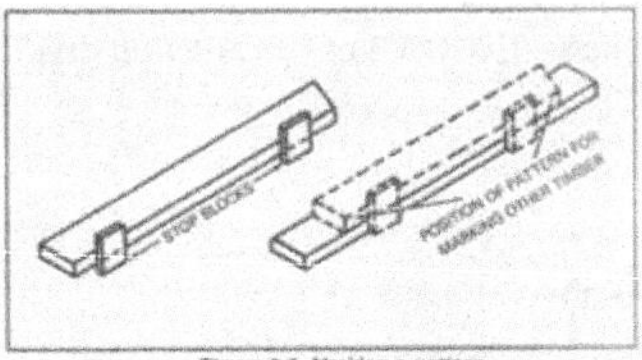

Figure 2-2. Marking a pattern

The erecting party sets the sections into place, braces them temporarily, and nails them together. The end section should be first; it may be erected on graded earth. The sidewall sections are next; they should be erected so as to keep the two walls even. The rafter party can then place the rafters on the walls.

The panel method of erection may be used for all types of small buildings and large warehouses. When this method is used for large buildings, cranes are used to place sections too heavy to be handled by hand.

> **WARNING**
>
> **When machinery is used, use caution in fastening the lifting cable or rope to avoid damaging the section.**

MATERIALS

The primary components used in frame construction are lumber and hardware. This section includes information on the types and sizes of lumber as well as a description of various metal fasteners.

LUMBER

Lumber varies greatly in structural characteristics. A carpenter must learn about lumber so that he can choose the most suitable material for each job. This section covers the types, standard sizes, and uses of lumber for construction carpentry. It also covers the methods of measuring lumber quantities in terms of board feet, which is the unit by which it is ordered.

Grades

Lumber, as it comes from the sawmill, is divided into three main classes: yard lumber, structural material, and factory and shop lumber. However, only yard lumber will be considered here. It is classified on the basis of quality. The carpenter must choose a quality that is suitable for the intended purpose. At the same time, he must exercise economy by not choosing a better (and therefore more expensive) grade than is required. Lumber is subdivided into classifications of select lumber and common lumber.

Select Lumber. Select lumber is of good appearance and finishing. It is identified by the following grade names for comparison of quality:

- *Grade A is* suitable for natural finishes and is practically clear.
- *Grade B is* suitable for natural finishes, is of high quality, and is generally clear.
- *Grade C is* suitable for high-quality paint finishes.
- *Grade D is* suitable for paint finishes between high-finishing grades and common grades and has somewhat the nature of both.

Common Lumber. Common lumber is suitable for general construction and utility purposes. It is identified by the following grade names for comparison of quality:

- *No. 1 common is* suitable for use without waste, it is sound and tight knotted, and it may be considered *watertight* lumber.
- *No. 2 common is* less restricted in quality than No. 1, but of the same general quality. It is used for framing, sheathing, and other structural forms where the stress or strain is not too great.
- *No. 3 common* permits some waste, and it is lower in quality than No. 2. It is used for such rough work as footing, guardrails, and rough flooring.
- *No. 4 common* permits waste, is of low quality, and may have coarse features such as decay and holes. It is used for sheathing, subfloors, and roof boards in the cheaper types of construction, but its most important industrial outlet is for boxes and crates.
- *No. 5 common is* not produced in some kinds of lumber. It is used for boxes, crates, and dunnage, for which the quality requirement is very low.

Uses

In frame construction, lumber is used primarily for the frame and walls.

Frames. Building frames are the wood forms constructed to support the finished members of a structure. These include posts, girders (beams), scabs, joists, subfloors, sole plates, girts, knee braces, top plates, and rafters. Softwoods are usually used for wood framing and all other construction carpentry covered in this manual.

No. 2 common lumber is used for framing. Heavy frame components, such as beams and girders, are made by combining several pieces of framing material.

Walls. The exterior wall of a frame structure usually has three layers: sheathing, building paper, and siding.

Sheathing and siding lumber are normally grade No. 2 common softwood, which is with solid knots, no voids.

Siding is either vertically or horizontally applied. Theater construction may limit available material to lap siding for both horizontal and vertical surfaces. For local procurement, there are several types of drop and bevel siding, which is applied horizontally. (See page 6-32, siding.)

Sizes

Lumber is usually sawed into standard dimensions (length, width, and thickness). This allows uniformity in planning structures and in ordering materials. Table 2-1 lists the common widths and thicknesses of wood in rough and in dressed dimensions in the US. Standards have been established for dimension differences between the quoted size of lumber and its standard sizes when dressed. ***Quoted size*** refers to dimensions prior to surfacing. These dimension differences must be taken into consideration. A good example of the dimension difference is the common 2 x 4. As shown in Table 2-1, the familiar quoted size 2 x 4 is the rough or nominal dimension, but the actual dressed size is 1 1/2 x 3 1/2 inches. Lumber is sawn in standard sizes used for light framing

- Thickness: 1, 2, and 4 inches.
- Width: 2, 4, 6, 8, 10, and 12 inches.
- Length: 8, 10, 12, 14, 16, 18, and 20 feet.

The actual dimensions of dressed lumber are less than the sawn dimensions because of drying and planing (or finishing). For the relative difference between swan (standard or nominal) dimensions and actual sizes of construction lumber, see Table 2-1.

Table 2-1. Nominal and dressed sizes of lumber

Nominal Size (In Inches)	Dressed Size (In Inches)
1 x 3	3/4 x 2 1/2
1 x 4	3/4 x 3 1/2
1 x 6	3/4 x 5 1/2
1 x 8	3/4 x 7 1/4
1 x 10	3/4 x 9 1/4
1 x 12	3/4 x 11 1/4
2 x 4	1 1/2 x 3 1/2
2 x 6	1 1/2 x 5 1/2
2 x 8	1 1/2 x 7 1/4
2 x 10	1 1/2 x 9 1/4
2 x 12	1 1/2 x 11 1/4
3 x 8	2 1/2 x 7 1/4
3 x 12	2 1/2 x 11 1/4
4 x 12	3 1/2 x 11 1/4
4 x 16	3 1/2 x 15 1/4
6 x 12	5 1/2 x 11 1/2
6 x 16	5 1/2 x 15 1/2
6 x 18	5 1/2 x 17 1/2
8 x 16	7 1/2 x 15 1/2
8 x 20	7 1/2 x 19 1/2
8 x 24	7 1/2 x 23 1/2

Plywood is usually 4 x 8 feet and varies from 1/8 to 1 inch in thickness.

The amount of lumber required is measured in board feet. A board foot is a unit measure representing an area of 1 foot by 1 foot, 1 inch thick. Thus, a board that is 1 inch thick, 1 foot wide, and 1 foot long measures 1 board foot. A board that is 1 inch thick, 1 foot wide, and 12 feet long measures 12 board feet.

To determine the number of board feet in one or more pieces of lumber, use the following formula:

$$\textit{Board feet} = \frac{N \times T(in) \times W(in) \times L(ft)}{12} \text{ or } \frac{N \times T(in) \times W(in) \times L(in)}{144}$$

where—
N = number of lumber pieces
T = thickness
W = width
L= length

Examples of board feet computations are shown in Figure 2-3.

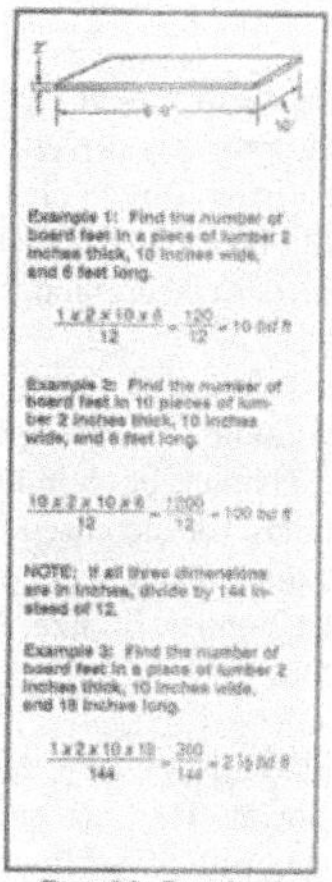

Figure 2-3. Examples of board-feet computations

HARDWARE

A wide variety of fasteners are used for frame construction in the TO. These fasteners are all made of metal. They are classified as nails, screws, bolts, driftpins, corrugated fasteners, and timber connectors.

Nails

Nails, the most common type of metal fasteners, are available in a wide range of types and sizes.

Some basic nail types are shown in Figure 2-4, page 2-6. The common nail is designed for rough framing. The box nail is used for toenailing and light work in frame construction. The casing nail is used in finished carpentry work to fasten doors and window casings and other wood trim. The finishing nail and brad are used for light, wood-trim material and are easy to drive below the surface of lumber with a nail set.

The size of a nail is measured in a unit known as a penny. Penny is abbreviated with the lowercase letter *d.* It indicates the length of the nail. A 6d nail is 2 inches long; a 10d nail is 3 inches long (Figure 2-5, page 2-6). These measurements apply to *common, box, casing,* and *finish* nails only. *Brads* and small *box* nails are identified by their actual length and gauge number.

A nail, whatever the type, should be at least three times as long as the thickness of the wood it is intended to hold. Two thirds of the length of the nail is driven into the other piece of wood for proper anchorage. The other one-third of the length provides the necessary anchorage of the piece being fastened. Protruding nails should be bent over to prevent damage to materials and injury to personnel.

There are a few general rules to be followed in the use of nails in building. Nails should be driven at an angle slightly toward each other to improve their holding power. You should be careful in placing nails to provide the greatest holding power. Nails driven with the grain do not hold as well as nails driven across the grain. A few nails of proper type and size, properly placed an properly driven, will hold better than a great many driven close together. Nails are generally considered the cheapest and easiest fasteners to be applied.

Figure 2-6 shows a few of the many specialized nails. Some nails are specially coated with zinc, cement, or resin materials. Some have threading for increased holding power. Nails are made from many materials, such as iron, steel, copper, bronze, aluminum, and stainless steel.

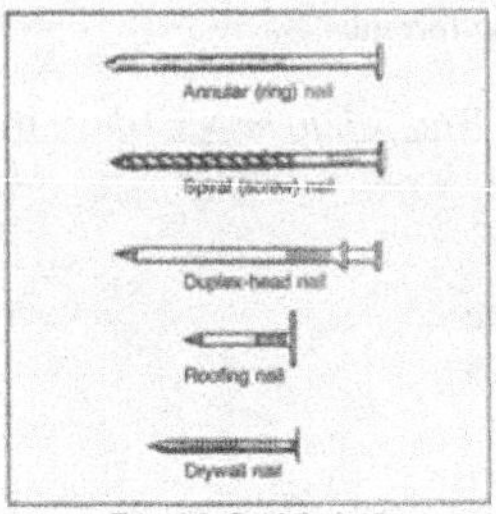

Figure 2-6. Specialized nails

Annular and spiral nails are threaded for greater holding power. They are good for fastening paneling or plywood flooring. The drywall nail is used for hanging drywall and has a special coating to prevent rust. Roofing nails are not specified by the penny system;

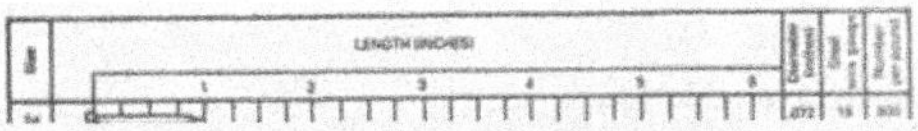

Table 2-2. Sizes, types, and uses of nails

SIZE	LGTH(IN)	DIAM(IN)	REMARKS	WHERE USED
2d	1	.072	Small head	Finish work, shop work
2d	1	.072	Large flathead	Small timber, wood shingles, lathes
3d	1¼	.08	Small head	Finish work, shop work
3d	1¼	.08	Large flathead	Small timber, wood shingles, lathes
4d	1½	.098	Small head	Finish work, shop work
4d	1½	.098	Large flathead	Small timber, lathes, shop work
5d	1¾	.098	Small head	Finish work, shop work
5d	1¾	.098	Large flathead	Small timber, lathes, shop work
6d	2	.113	Small head	Finish work, casing, stops, and so forth, shop work
6d	2	.113	Large flathead	Small timber, siding, sheathing, and so forth, shop work
7d	2¼	.113	Small head	Casing, base, ceiling, stops, and so forth
7d	2¼	.113	Large flathead	Sheathing, siding, subflooring, light framing
8d	2½	.131	Small head	Casing, base, ceiling, wainscot, and so forth, shop work
8d	2½	.131	Large flathead	Sheathing, siding, subflooring, light framing, shop work
8d	1¼	.131	Extra-large flathead	Roll roofing, composition shingles
9d	2¾	.131	Small head	Casing, base, ceiling, and so forth
9d	2¾	.131	Large flathead	Sheathing, siding, subflooring, framing, shop work
10d	3	.148	Small head	Casing, base, ceiling, and so forth, shop work
10d	3	.148	Large flathead	Sheathing, siding, subflooring, framing, shop work
12d	3¼	.148	Large flathead	Sheathing, subflooring, framing
16d	3½	.162	Large flathead	Framing, bridges, and so forth
20d	4	.192	Large flathead	Framing, bridges, and so forth
30d	4½	.207	Large flathead	Heavy framing, bridges, and so forth
40d	5	.225	Large flathead	Heavy framing, bridges, and so forth
50d	5½	.244	Large flathead	Extra-heavy framing, bridges, and so forth
60d	6	.262	Large flathead	Extra-heavy framing, bridges, and so forth

This chart applies to wire nails, although it may be used to determine the length of cut nails.

they are referred to by length. They are available in lengths from 3/4 to 2 inches and have large heads. The double-headed nail, or duplex-head nail, is used for temporary construction, such as form work or scaffolding. The double head on this nail makes it easy to pull out when forms or scaffolding are torn down. Nails for power nailing come in rolls or clips for easy loading into a nailer. They are coated for easier driving and greater holding power. Table 2-2 gives the general sizes and types of nails preferred for specific applications.

Screws

Screws are more expensive than nails in both time and money, but are sometimes required for superior results. They provide more holding power than nails and can be easily tightened to draw material securely together. Screws are neater in appearance and may be withdrawn without damaging the material. The common wood screw is usually made of unhardened steel, stainless steel, aluminum, or brass. The steel may be bright-finished or blue, or it may be zinc, cadmium, or chromeplated. Wood screws are threaded for approximately 2/3 of the length of the screw from a gimlet point and have a slotted head.

Screws vary in length and size of shaft. Each length is made in a number of shaft sizes identified by a number that shows relative differences in the diameter of the screws. Proper screw size number indicates the wire gauge of the body, the drill or bit size for the body hold, and the drill or bit size for the starter hole. Table 2-3 shows screw sizes and dimensions. Table 2-4 shows applicable drill and auger bit sizes for screws.

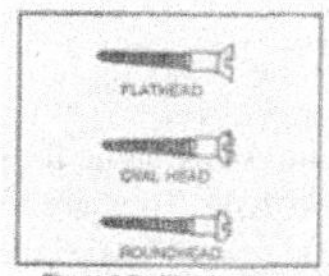

Figure 2-7. Wood screws

Both slotted and Phillips (cross-slotted) flathead and oval-head screws are countersunk enough to allow a covering material to be used. Slotted roundhead and Phillips roundhead screws are not countersunk, but are driven firmly flush with the surface.

Table 2-3. Screw sizes and

LENGTH (IN)	SIZE 0	1	2	3	4	5	6	7	8	9	10	11	12	1[illegible]
1/4		x	x	x										
3/8	x	x	x	x	x	x	x	x	x	x				
1/2		x	x	x	x	x	x	x	x	x	x	x	x	
5/8		x	x	x	x	x	x	x	x	x	x	x	x	x
3/4			x	x	x	x	x	x	x	x	x	x	x	x
7/8			x	x	x	x	x	x	x	x	x	x	x	x
1				x	x	x	x	x	x	x	x	x	x	x
1 1/4					x	x	x	x	x	x	x	x	x	x
1 1/2					x	x	x	x	x	x	x	x	x	x
1 3/4					x	x	x	x	x	x	x	x	x	x
2						x	x	x	x	x	x	x	x	x
2 1/4						x	x	x	x	x	x	x	x	x
2 1/2						x	x	x	x	x	x	x	x	x
2 3/4						x	x	x	x	x	x	x	x	x
3							x	x	x	x	x	x	x	x
3 1/2							x	x	x	x	x	x	x	x
4									x	x	x	x	x	x
4 1/2									x	x	x	x	x	
5													x	
6													x	

GAUGE AND DIAMETER													
STEEL WIRE GAUGE DIAMETER (INCHES)	17 .054	15 .072	14 .080	13 .091	12 .105	11 .120	10 .135	9 .148	8 .162	7 .177	6½ .184	6 .192	5 .207
STEEL WIRE GAUGE DIAMETER (INCHES)	4½ .215	4 .225	3 .243	2½ .253	2 .262	1 .283	½ —	0 .306	00 .331	00½ —	000 .362	0000 .393	

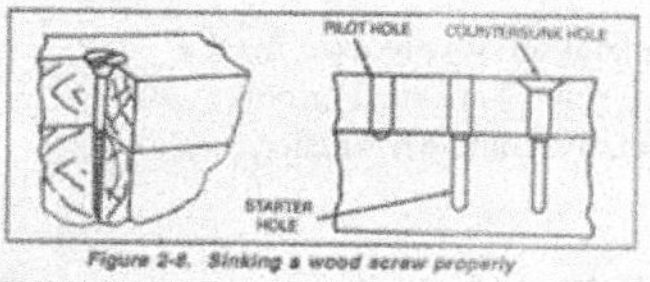

Figure 2-6. Sinking a wood screw properly

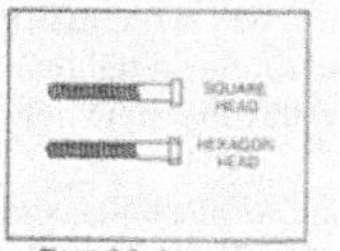

Figure 2-8. Lag screws

The most common types of screws, along with their uses, are discussed in the following paragraphs.

Wood Screws. Wood screws are designated according to head style. The most common types are flathead, oval head, and roundhead (Figure 2-7) with either slotted or Phillips heads. Their sizes vary from 1/4 to 6 inches. Screw sizes up to 1 inch increase by eighths, screws from 1 to 3 inches increase by quarters, and screws from 3 to 6 inches increase by half inches.

Table 2-4. Drill and auger bit sizes

SCREW SIZE		1	2	3	4	5	6	7	8	9	10	12	14	16	18
NOMINAL SCREW BODY DIAMETER		.073	.086	.099	.112	.125	.138	.151	.164	.177	.190	.216	.242	.268	.294
		5/64	3/32	3/32	7/64	1/8	9/64	5/32	11/64	11/64	3/16	7/32	15/64	17/64	19/64
PILOT HOLE	DRILL SIZE	5/64	3/32	7/64	7/64	1/8	9/64	5/32	11/64	3/16	3/16	7/32	1/4	17/64	19/64
	BIT SIZE	—	—	—	—	—	—	—	—	—	—	4	4	5	5
STARTER HOLE	DRILL SIZE	—	1/16	1/16	5/64	5/64	3/32	7/64	7/64	1/8	1/8	9/64	5/32	3/16	13/64
	BIT SIZE	—	—	—	—	—	—	—	—	—	—	—	—	—	4

To prepare wood for receiving the screws (Figure 2-8), a pilot hole the diameter of the screw is bored into the piece of wood to be fastened. A smaller starter hole is then bored into the piece of wood that is to act as anchor or to hold the threads of the screws. The starter hole has a diameter less than that of the screw threads and is drilled to a depth of 1/2 or 2/3 the length of the threads to be anchored. This method assures accuracy in placing the screws, reduces the possibility of splitting the wood, and reduces the time required.

Lag Screws. The Army name for *lag screws* (Figure 2-9) is *lag bolt, wood-screw type.* They are longer and heavier than the common wood screw and have coarser threads, which extend from a cone or gimlet point slightly more than half the length of the screw. Square-head and hexagon-head lag screws are usually placed with a wrench. They are used when ordinary wood screws would be too short or too light and spikes would not be strong enough. Lag screw sizes are given in Table 2-5, page 2-10. Combined with expansion anchors, lag screws are used to frame timbers to existing masonry.

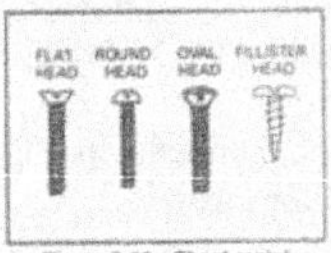

Figure 2-10. Sheet-metal screws

Sheet-Metal Screws. Sheet-metal screws are used for the assembly of metal parts. These screws are steel or brass with four types of heads: flat, round, oval, and McAllister, as shown in Figure 2-10, page 2-10.

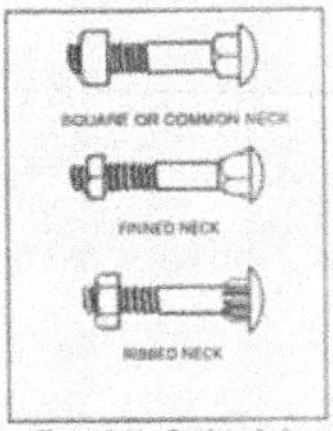

Figure 2-11. Carriage bolts

Bolts

Bolts are used when great strength is required or when the work must be disassembled frequently. Nuts are usually used for fastening bolts. The use of washers between the nut and wood surfaces, or between both the nut and the head and their opposing surfaces, will avoid marring the surfaces and will permit additional torque in tightening. Bolts are selected by length, diameter, threads, style of head, and type.

Carriage Bolts. Carriage bolts come in three types: square neck, finned neck, and ribbed neck, as shown in Figure 2-11.

In each type of carriage bolt, the part of the shank immediately below the head grips the materials into which the bolt is inserted. This keeps the bolt from turning when a nut is tightened down on it or removed. The finned carriage bolt has two or more fins extending from the head to the shank. The ribbed type has longitudinal ribs, splines, or serrations on all or part of a shoulder, located immediately beneath the head.

Holes bored to receive carriage bolts are bored to a tight fit for the body of the bolt and counterbored to permit the head of the bolt to fit flush with or below the surface of the material being fastened. The bolt is then driven into the hole with a hammer.

Table 2-5. Lag-screw sizes

LENGTHS (INCHES)	DIAMETERS (INCHES)			
	1/4	3/8, 7/16, 1/2	5/8, 3/4	7/8, 1
1	X	X		
1 1/2	X	X	X	
2, 2 1/2, 3, 3 1/2, so forth, 7 1/2, 8 to 10	X	X	X	X
11 to 12		X	X	X
13 to 16			X	X

Carriage bolts are chiefly for wood-to-wood use, but may also be used for wood-to-metal. If used for wood-to-metal application, the bolt head should be fitted to the wood item. Metal surfaces are sometimes predrilled and countersunk to allow the use of carriage bolts for metal-to-metal fastening. Carriage bolts can be obtained from 1/4 to 1 inch in diameter and from 3/4 to 20 inches long. (Table 2-6 lists carriage-bolt sizes.) A common flat washer should be used between the nut and the wood surface with carriage bolts.

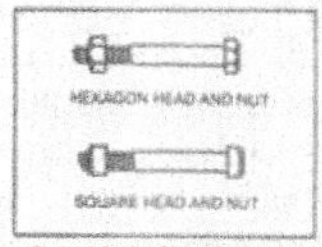

Figure 2-12. Machine bolts

Machine Bolts.
Machine bolts (or *cap screws)* (Figure 2-12) are made with cut national fine or national coarse threads. These threads extend from twice the diameter of the bolt plus 1/4 inch (for bolts less than 6 inches in length) to twice the diameter of the bolt plus 1/2 inch (for bolts over 6 inches in length). Machine bolts are precision made and generally applied metal to metal where close tolerance is needed. The head may be square, hexagon, double hexagon, rounded, or flat countersunk. The nut usually corresponds in shape to the head of the bolt with which it is used. (Machine bolts are externally driven only.)

Table 2-7. Machine-bolt sizes

LENGTHS (INCHES)	DIAMETERS (INCHES)				
	1/4, 3/8	7/16	1/2, 9/16, 5/8	3/4, 7/8, 1	1 1/8, 1 1/4
3/4	x				
1, 1 1/4	x	x	x		
1 1/2, 2, 2 1/2	x	x	x	x	
3, 3 1/2, 4, 4 1/2, so forth, 9 1/2, 10 to 20	x	x	x	x	x
21 to 25			x	x	x
26 to 30				x	x

A machine bolt is selected on the basis of head style, length, diameter, and type of thread. The hole through which the bolt is to pass is bored to the same diameter as the bolt. Machine bolts are made in diameters from 1/4 to 3 inches and may be obtained in any length desired. (Table 2-7 lists machine-bolt sizes.)

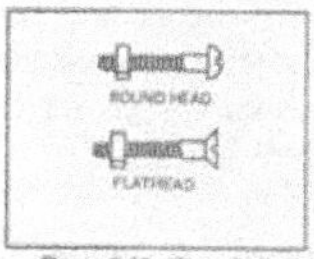

Figure 2-13. Stove bolts

Stove Bolts. Stove bolts (Figure 2-13) are less precisely made than machine bolts. They have either flat or round slotted heads and have threads extending almost the full length of the body. Stove bolts are generally used with square nuts and may be applied metal to metal, wood to wood, or wood to metal. If flatheaded, they are countersunk; if roundheaded, they are used flush with the surface.

Table 2-6. Carriage-bolt sizes

LENGTHS (INCHES)	DIAMETERS (INCHES)		
	3/16, 1/4, 5/16, 3/8	7/16, 1/2	9/16, 5/8, 3/4
3/4	x		
1	x	x	
1 1/4	x	x	x
1 1/2, 2, 2 1/2, so forth, 9 1/2, 10 to 20	x	x	x x

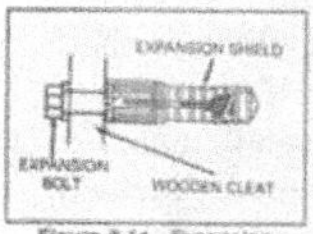

Figure 2-14. Expansion bolt and shield

Expansion Bolts. An expansion bolt is a bolt used together with an expansion shield (Figure 2-14), which is usually made of lead or plastic, to provide anchorage in substances in which a threaded fastener is useless. The shield (or *expansion anchor) is* inserted in a predrilled hole and expands when the bolt is driven into it. Wedged firmly in the hole, the shield provides a secure base for the grip of the fastener.

The expansion shield can be used with a nail, screw, or bolt. The shield may be obtained separately or may include the nail, screw, or bolt.

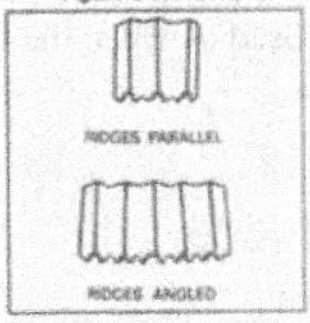

Figure 2-16. Corrugated fasteners

Driftpins

Driftpins (called *driftbolts* for supply purposes) (Figure 2-15) are long, heavy, threadless bolts used to hold heavy pieces of timber together. Driftpins have heads, and they vary in diameter from 1/2 to 1 inch, and in length from 18 to 26 inches.

To use the driftpin, make a hole in the timber slightly smaller than the diameter of the pin. Drive the pin into the hole. It is held in place by the compression action of the wood.

Corrugated Fasteners

Corrugated fasteners are used to fasten joints and splices in small boards, particularly in miter joints and butt joints. Corrugated fasteners are made of 18- to 22-gauge sheet metal with alternate ridges and grooves; the ridges vary from 3/16 to 5/16 inch, center to center. One end is cut square; the other end is sharpened, with beveled edges.

There are two types of corrugated fasteners: one with ridges running parallel, the other with ridges running at a slight angle to one another (Figure 2-16). The latter type tends to compress the material, since the ridges and grooves are closer at the top than at the bottom. Corrugated fasteners vary from 5/8 to 1 1/8 inches wide and from 1/4 to 3/4 inch long. Ridges on the fasteners range from three to six ridges per fastener.

Corrugated fasteners are a great advantage when used to fasten parallel boards together (as in fastening tabletops), to make any type of joint, and to substitute for nails where nails may split the lumber. The fasteners have a greater holding power in small lumber than nails do. The proper method of using the fasteners is shown in Figure 2-17.

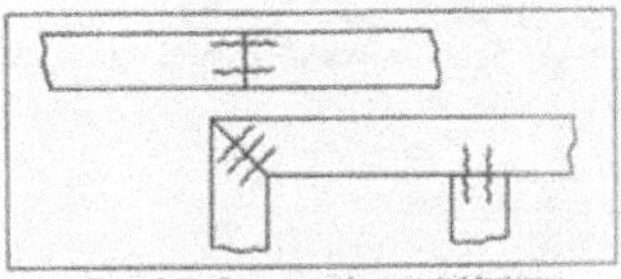

Figure 2-17. Proper use of corrugated fasteners

Timber Connectors

Timber connectors are metal devices for increasing the joint strength in timber structures. Efficient connections for either timber-to-timber joints or timber-to-steel joints are provided by the several types of timber connectors. The right type is determined by the kind of joint to be made and the load to be carried. The connectors—

- Eliminate much of the complicated framing of joints.

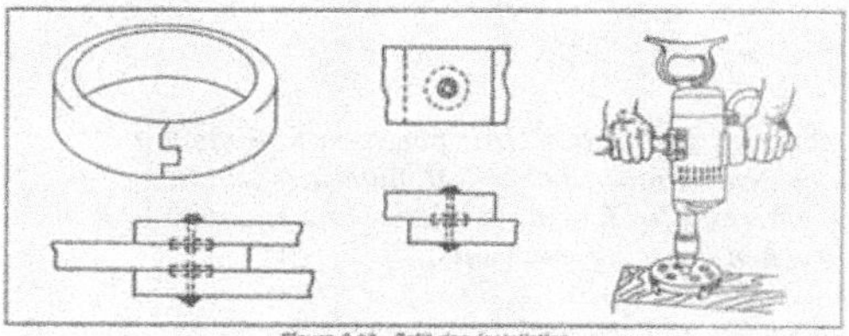

Figure 2-18. Split-ring installation

- Simplify the design of heavy construction.
- Provide greater efficiency in the use of material.
- Reduce the amount of timber and hardware used.
- Save time and labor.

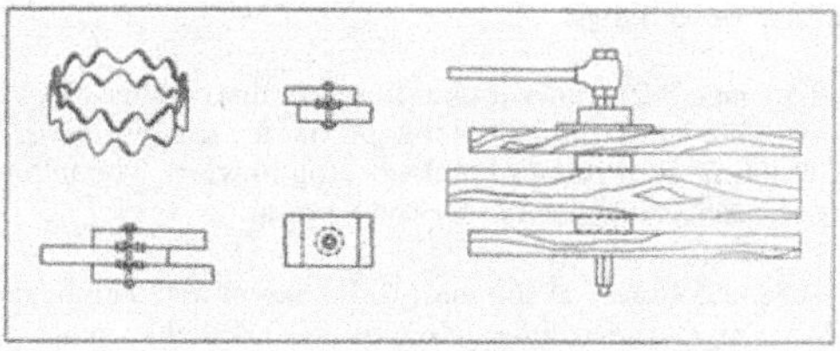

Figure 2-19. Toothed-ring installation

Split rings (Figure 2-18) are made of low-carbon steel and have 21/2- and Winch diameters. Split rings are used between two timber faces for heavy construction. They fit into grooves that are cut half the depth of the ring into each of the timber faces. The grooves are made with a special bit used in an electric, air, or hand drill. The tongue-and-groove split in the ring permits the ring to bear equally against the cone wall and the outer wall of the groove into which it is placed. The inside bevel and mill edge make installation into, and removal from, the groove easier.

Toothed rings (Figure 2-19) are corrugated and toothed and are made from 16-gauge plate, low-carbon steel. They are used between two timber frames for comparatively light construction and are embedded into the contact faces of the joint members by pressure.

BILLS OF MATERIALS CHAPTER 3

A BOM is a list of all materials needed to complete a structure. It is based on takeoffs and estimates of the materials needed. It includes item number (parts and materials), name, description, unit of measure, quantity and, where called for, the stock size and number, and sometimes the weight. The carpenter uses it when ordering materials.

A BOM is usually made up by the draftsman when the original drawings are prepared. However, when no BOM accompanies field prints, it must be developed up by the construction crew. For this reason, a carpenter should be able to develop a BOM, as well as work with one. Accuracy can best be obtained by having a separate bill prepared by at least two estimators. The bills may then be compared and one copy corrected or both used to make up a final BOM.

Before a BOM can be prepared, a materials takeoff list and a materials estimate list are prepared.

MATERIALS TAKEOFF LIST

The first step leading to preparation of a BOM is a materials takeoff list. This is a list of all parts of the building, taken from the plans, usually by tallying and checking off the items indicated on the drawings and specifications. Both architectural and engineering plans provide the names and sizes of the items that are to be listed.

For example, Figure 3-1 (page 3-2) shows a plan for the substructure of a 20- x 40-foot TO building. Table 3-1, page 3-3, is the materials takeoff list for this building. This list identifies all parts of the building, starting with its base and working upward. The following paragraphs are an example of computing the materials needed for the footers.

Look at the first and second columns of the materials takeoff list. The first column gives the *item* (footers); the next column gives the *number of pieces (46)* needed to make up the item.

The 20- x 40-foot building shown on the plan requires 15 foundation posts. Since three pieces are needed for each footer, a total of 45 pieces is needed.

The *length in place (1* foot 5 inches) is the actual length of the member after it has been cut and is ready to be nailed in place. The *size* (2 x 6) refers to the nominal size of the lumber. The *length* refers to the standard lengths available from the lumberyard or depot, such as 8-, 10-, and 12-foot pieces of stock.

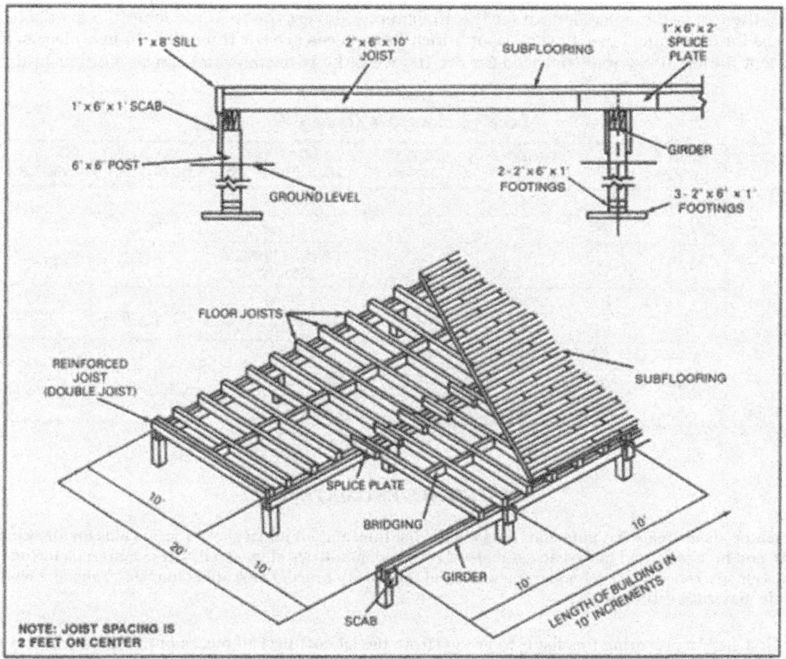

Figure 3-1. 20- x 40-foot TO building substructure

Table 3-1. Materials takeoff list for a 20- x 40-foot building

ITEM NAME OR USE OF PIECE	NO. OF PIECES	UNIT	LENGTH IN PLACE	SIZE	LENGTH	NO. PER LENGTH	QUANTITY
1. Footers	45	PC	1'- 5"	2x6	10'	7	7
2. Spreaders	30	PC	1'- 4"	2x6	8'	6	5
3. Foundation Post	15	PC	3'- 0"	6x6	12'	4	4
4. Scabs	20	PC	1'- 0"	1x6	8'	8	3
5. Girders	36	PC	10'- 0"	2x6	10'	1	36
6. Joists	46	PC	10'- 0"	2x6	10	1	46
7. Joist Splices	21	PC	2'- 0"	1x6	8'	4	6
8. Block Bridging	40	PC	1' - 10⅜"	2x6	8'	4	10
9. Closers	12	PC	10'- 0"	1x8	10'	1	12
10. Flooring	800	BF	RL	1x6	RL	—	—

Select the most economical length for the 15 footers. Convert the required length to available lengths for economical use. Seven 1-foot 5-inch-long pieces are cut from each 10-foot piece of stock; 45 1-foot 5-inch pieces require seven 2 x 6 x 10 pieces. Leftover material can be used for bridging.

Table 3-2. Materials estimate list

ITEM	SIZE & LENGTH	UNIT	TAKEOFF QUANTITY	WASTE ALLOWANCE	ADDITIONAL REQUIREMENTS	TOTAL QUANTITY	BD FT MEASURE
1.	6 x 6 x 12	PC	4	1	None	5	180
2.	2 x 6 x 10	PC	89	9	None	98	980
3.	2 x 6 x 8	PC	15	2	3 For Temporary Bracing	20	160
4.	1 x 8 x 10	PC	12	2	None	14	91
5.	1 x 6 x 8	PC	9	2	2 For Batter Boards	13	52
6.	1 x 6 x RL	BF	800	160	None	960	960
7.	16d	LB	—	—	36 Nails, Framing	36	—
8.	8d	LB	—	—	23 Nails, Flooring	23	—

MATERIALS ESTIMATE LIST

The materials estimate list puts materials takeoff list information into a shorter form, adds an allowance for waste and breakage, and makes an estimate of required quantities of material. These materials include nails, cement, concrete form lumber, tie wire, and temporary bracing or scaffold lumber. (Table 3-2 is a sample materials estimate list.)

The first step in preparing this list is to group (from the takeoff list) all pieces of the same size and length. For example, start with the largest size lumber that can be found on the materials takeoff list. Group all the pieces of that same size and length that appear on the list. This groups the total number of each size and length that will be needed. Continue in this way with each size of lumber, working down to the smallest size and length of material.

The sizes and pounds of nails needed should also he added to the list. Two formulas used for estimating the number of pounds of nails needed are shown below:

Estimating quantity of nails required

- For flooring, sheathing, and other 1-inch material, use the following formula:

Number of pounds (2 penny through 8 penny) =

$$\frac{penny}{4} \times \frac{board\ measure}{100}$$

- For framing materials that are 2 inches or more, use the following formula:

Number of pounds (10 penny through 60 penny) =

$$\frac{penny}{6} \times \frac{board\ measure}{100}$$

For each material size, a waste factor must be added. For flooring, sheathing, and other 1-inch material, add a waste allowance of 2 percent; for all other materials 2 inches and larger, add 10 percent to the total number.

Next, estimate the amount of additional requirements for materials (such as bracing) that are not shown on the plans. Add the total quantity for each size and length of material. If it is used for siding, sheathing, or flooring, convert it to board feet using the method given in Chapter 2 (page 2-5).

BILL OF MATERIALS FORMAT

The information for the BOM (DA Form 2702) is taken from the materials estimate list. The BOM is used to requisition these materials. The materials estimate list contains information on materials needed for a project, but it also contains much information on materials needed for a project, but it also contains much information of little interest to depot personnel. Therefore, use the simplified BOM (DA Form 2702, Figure 3-2) to requisition materials.

The rest of the building would be analyzed in the same way.

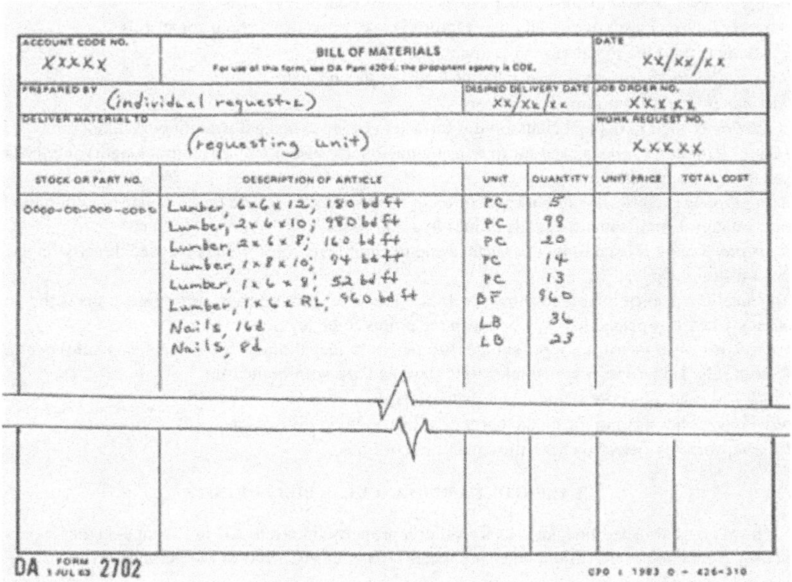

ACCOUNT CODE NO. XXXXX

BILL OF MATERIALS
For use of this form, see DA Pam 420-6; the proponent agency is COE.

DATE xx/xx/xx

PREPARED BY (individual requester)

DESIRED DELIVERY DATE xx/xx/xx

JOB ORDER NO. XXXXX

DELIVER MATERIAL TO (requesting unit)

WORK REQUEST NO. XXXXX

STOCK OR PART NO.	DESCRIPTION OF ARTICLE	UNIT	QUANTITY	UNIT PRICE	TOTAL COST
0000-00-000-0000	Lumber, 6x6x12; 180 bd ft	PC	5		
	Lumber, 2x6x10; 980 bd ft	PC	98		
	Lumber, 2x6x8; 160 bd ft	PC	20		
	Lumber, 1x8x10; 94 bd ft	PC	14		
	Lumber, 1x6x8; 52 bd ft	PC	13		
	Lumber, 1x6xRL; 960 bd ft	BF	960		
	Nails, 16d	LB	36		
	Nails, 8d	LB	23		

DA FORM 1 JUL 63 2702

GPO : 1983 0 - 426-310

Figure 3-2. Sample DA Form 2702

BUILDING LAYOUT AND FOUNDATION

CHAPTER 4

The details of layout and planning are essential to proper construction of a building. Layout prepares the site for the foundation which must be planned and completed for each building being constructed. This chapter introduces the carpenter to the tools, materials, and techniques used in the effective accomplishment of these vital layout and planning functions.

LAYOUT

Layout techniques are described in the following paragraphs. The following are the most commonly used layout tools and materials:

- *A string line is used* to distinguish the dimensions of the building layout.
- *A sledgehammer is used* to sink corner stakes or batter boards and posts.
- *A posthole auger is* used to dig the holes required to set posts properly in some soils.
- *A handsaw is* used to cut batter boards and posts.
- *An ax or a hatchet is* used to sharpen batter-board posts and stakes.
- *A hammer is* used for building batter boards.
- *A chalk line is* used to deposit chalk on the surface in order to make a straight guideline.
- *A 100-foot/30-meter tape is* used for measuring diagonally (usually in a 100 foot length) and for laying out excavation or foundation lines.
- *Tracing tape is* used for laying out excavation or foundation lines. The tape is made of cotton cloth approximately 1 inch wide. It usually comes in a 200 foot length.
- *A carpenter's level is* used to level a surface and to sight level lines. It may be used directly on the surface or with a straightedge.
- *A line level* has a spirit bubble to show levelness. The level is hung from a taut line. It gives the greatest accuracy when it is placed halfway between *the* points to be leveled.
- *An automatic level* measures approximate differences in elevation *and can* establish *grades* over *limited* distances. The landscape, level bubble, and index line *are* seen *in* the tube.
- *8d nails are* used to secure string line to batter boards.
- *A plumbing bob is* used to locate the corners of the building dimensions.
- *A framing square is used* to check the squareness of lines.

LAYING OUT A RECTANGULAR BUILDING SITE

Working from an established line, such as a road or a property line (line AB in Figure 4-1) that is parallel to construction, establish the maximum outer perimeter (AB, CD, AC, BD) of the building area.

Measure away from the front line (AB) along the side lines (AC and BD) the distances (AO and BO) desired to the dimension of the project that is to run parallel to the front line.

Stretch a line tightly from point O to O. This line will mark the project's frontage.

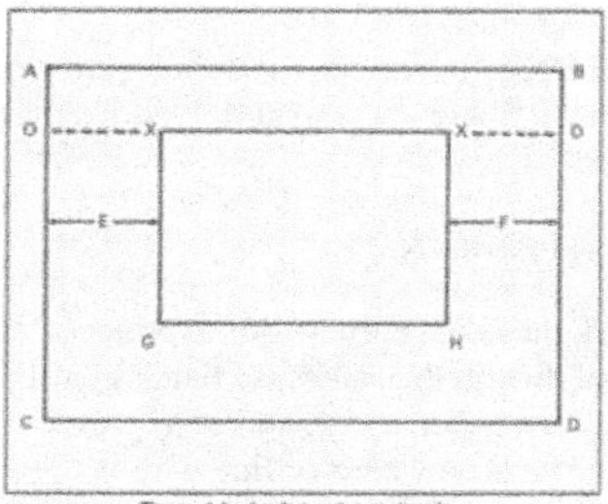

Figure 4-1. Laying out a rectangle

Measure in from lines AC and BD along line OO one-half the difference between the length of line OO and the desired length of the project. The points (X and X) will constitute the front corners of the project.

The two distances OX and XO establish the distances E and F. Extend lines from the two front corners, X and X, parallel to AC and BD at the distances established as E and F for the required depth of the project. This provides the side lines of the project (XG and XH).

Joining the extreme ends of side 111 XH will provide the rear line (GH) of the project.

After the four corners (X, X, G. and H) have been located, drive stakes at each corner. Batter boards may be erected at these points either after the stakes have been set or while they are being set. Dimensions are determined accurately during each step.

LAYING OUT AN IRREGULAR BUILDING SITE

Where the outline of the building is other than a rectangle, the procedure in establishing each point is the sam as described for laying out a simple rectangle. However, more points have to be located, and the final provin of the work is more likely to reveal a small error. When the building is an irregular shape, it is advisable to first lay out a large rectangle which will comprise the entire building or the greater part of it. This is shown in Figure 4-2 as HOPQ When this is established, the remaining portion of the layout will consist of small rectangles, each of which can be laid out and proved separately. These rectangles are shown as LMNP ABCQ, DEFG, and IJKO in Figure 4-2.

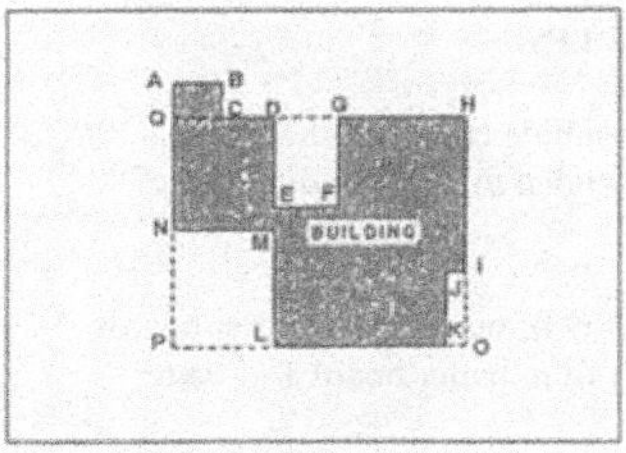

Figure 4-2. Laying out irregular projects

SETTING BATTER BOARDS

Batter boards are a temporary framework used to assist in locating corners when laying out a foundation. Batter-board posts are made from 2 x 4 or 4 x 4 material; corner stakes are made from 2 x 2s. Batter boards a made from 1 x 4 or 1 x 6 pieces.

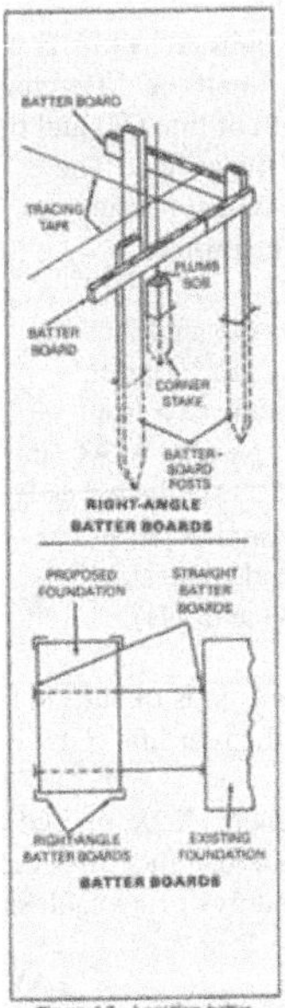

Figure 4-3. Locating batter boards

Staking Procedures

Corner stakes are driven to mark the exact corners of the project. Excavating for a foundation will disturb the stakes, so batter boards are set up outside the boundary established by the stakes to preserve definite and accurate building lines. Heavy cord or fine wire is stretched from one batter board to another to mark these lines.

Location of Batter Boards

Figure 4-3 shows how to locate batter boards. Right-angle batter boards are erected 3 or 4 feet outside of each corner stake. Straight batter boards are erected 3 or 4 feet outside of the line stakes.

Construction of Batter Boards

Right-angle batter boards should be fastened to the posts after the posts are sunk. Since the boards should be at the exact height of the top of the foundation, it may be desirable to adjust the height by nailing the boards to the stakes after the stakes have been sunk. Right-angle batter boards may be nailed close to perpendicular by using a framing square and should be leveled by means of a carpenter's level before they are secured. Then, angle saw cuts may be made or nails driven into the tops of the boards to hold the lines in place. Separate cuts or nails may be used for the building line, the foundation line, the footing line, and excavation lines. These grooves permit the removal and replacement of the lines in the correct position.

EXTENDING LINES

The following procedure applies to a simple layout as shown in Figure 4-4, page 4-4, and must be amended to apply to different or more complex layout problems:

Step 1. After locating and sinking stakes A and B. erect batter boards 1, 2, 3, and 4. Extend a chalk line (X) from batter board 1 to batter board 3, over stakes A and B.

Step 2. After locating and sinking stake C, erect batter boards 5 and 6. Extend chalk line Y from batter board 2 over stakes A and C to batter board 6.

Step 3. After locating and sinking stake D, erect batter boards 7 and 8. Extend chalk line Z from batter board 5 to batter board 7, over stakes C and D.

Step 4. Extend line O from batter board 8 to batter board 4, over stakes D and B.

Where foundation walls are wide at the bottom and extend beyond the outside dimensions of the building, th excavation must be larger than the laid-out size. To lay out dimensions of this excavation, measure out as far as required from the building line on each batter board and stretch lines between these points, outside the firs layout.

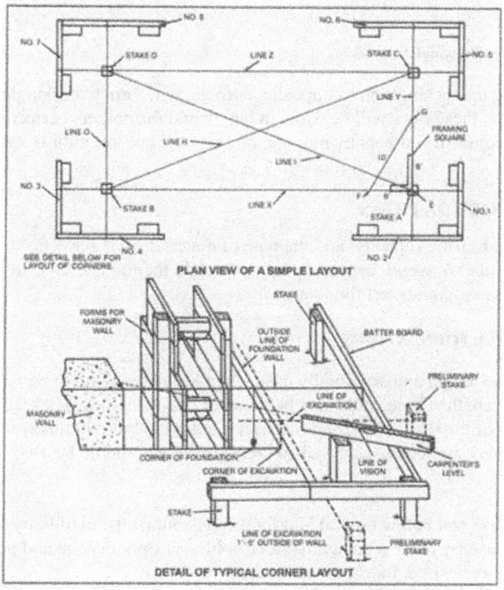

Figure 4-4. Laying out building lines from batter boards

The lines may be at a right angle where they cross the corner layout stakes, found by holding a plumb bob ov the corner layout stakes and adjusting the lines until they touch the plumb-bob line. All lines should be checked with a line level or a carpenter's level.

SQUARING LINES

The two methods commonly used for squaring extended lines are the 6-8-10 method and the diagonal methoc

The 6-8-10 Method

After extended lines are in place, measure line EF for a distance of 6 feet (Figure 4-4). Measure line EG for a distance of 8 feet. Adjust the lines (Y and X) until FG equals 10 feet. Multiples of 6-8-10 may be used for large layouts; for example, 12-16-20 for a layout 50 feet by 100 feet. For accuracy, never start with a measurement of less than 6 feet.

The Diagonal Method

If the layout is rectangular, lines H and I, cutting the rectangle from opposing corners, will form two triangles as shown in Figure 4-4. If the rectangle is perfect, these lines will be equal in length and the corners perfectly square. If lines H and I are not equal in length, adjust the corners by moving the lines right or left until H and are equal.

FOUNDATIONS

Foundations vary according to their use, the soil-bearing capacity, and the type of material available. The material may be cut stone, rock, brick, concrete, tile, or wood, depending on the weight the foundation is to support. Foundations may be classified as wall or column (pier) foundations.

WALL FOUNDATIONS

Wall foundations (Figure 4-5) are solid their total length and are usually used when heavy loads are to be carried or where the earth has low supporting strength. These walls may be made of concrete, rock, brick, or cut stone, with a footing at the bottom. Because of the time, labor, and material required to build it, this type wall will be used in the TO only when other types cannot be used. Steel-rod reinforcements should be used ir all concrete walls.

Rubble stone masonry is used for walls both above and below ground and for bridge abutments. In military construction, it is used when form lumber for masonry units is not available. Rubble masonry may be laid up with or without mortar; if strength and stability are desired, mortar must be used.

Coursed rubble is assembled of roughly squared stones in such a manner as to produce approximately continuous horizontal bed joints.

Random rubble is the crudest of all types of stonework. Little attention is paid to laying the stone in courses. Each layer must contain bonding stones that extend through the wall. This produces a wall that is well tied together.

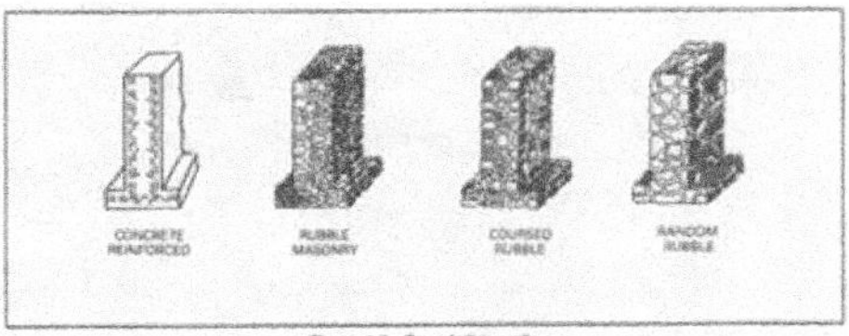

Figure 4-5. Foundation walls

COLUMN OR POST FOUNDATIONS

The use of column or post foundations constructed from masonry or wood saves time and labor. The posts or columns are spaced according to the weight to be carried. In most cases, the spacing is 6 to 10 feet apart. The sketches in Figure 4-6 show the different types of posts with appropriate types of footing. Wood posts are generally used, since they are installed with the least time and labor. When wood posts extend 3 feet or more above the ground, braces are necessary (Figure 4-7).

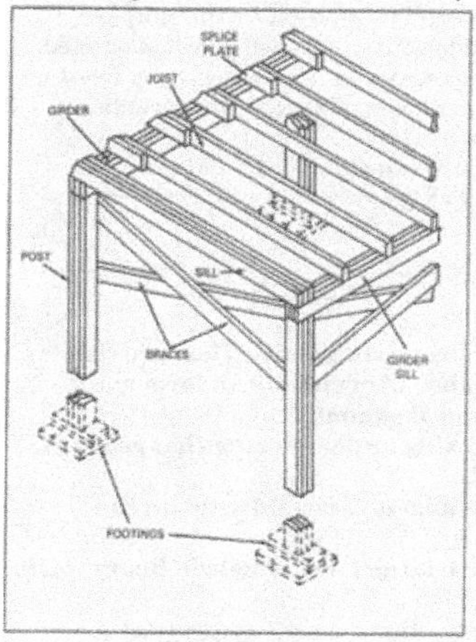

Figure 4-7. Posts with braces

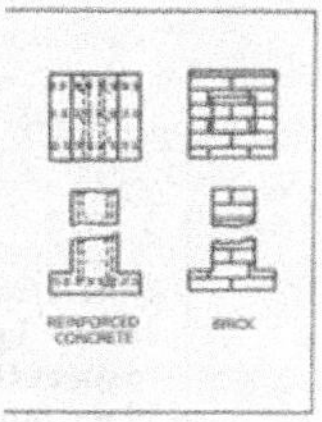

FORMS FOR CONCRETE

CHAPTER 5

Forms play a major role in concrete construction. They give the plastic concrete its shape and hold it until it hardens. Forms protect the concrete, assist in curing it, and support any reinforcing rods or conduits embedded in the concrete. This chapter familiarizes the carpenter with the design and construction of various types of forms.

FORM DESIGN

Forms for concrete must be tight, rigid, and strong. If forms are not tight, loss of concrete may cause a honeycomb effect, or loss of water may cause sand-streaking. The forms must be braced enough to stay in alignment. Special care is needed when bracing and tying down forms used in applications such as retainer walls, where the mass of concrete is large at the bottom and tapers toward the top. In this type of construction and in the first pour for walls and columns, the concrete tends to lift the form above its proper elevation. (Field Manual 5-742 gives formulas and tables for designing forms of proper strength.)

FORM CONSTRUCTION

Although forms are generally constructed from wood, fiber, earth, or metal, the TO carpenter usually constructs wood or fiber forms.

Wood forms are the most common in building construction. They are economical, easy to handle, and easy to produce, and they adapt to many shapes. Form lumber can be reused for roofing, bracing, or similar purposes.

Lumber should be straight, strong, and only partially seasoned. Kiln-dried lumber tends to swell when soaked with water. Swelling may cause bulging and distortion. If green lumber is used, allow for shrinkage, or keep it wet until the concrete is in place. Softwoods (pine, fir, and spruce) are the most economical; they are light, easy to work with, and generally available.

Wood coming in contact with concrete should be surfaced (smooth) on the side towards the concrete and on both edges. The edges may be square, shiplap, or tongue-and-groove. Tongue-and-groove lumber makes a more watertight joint, which reduces warping.

Plywood is economical to use for wall and floor forms; however, plywood used for this purpose should be made with waterproof glue and marked for use in concrete forms. Plywood is warp-resistant and can be used more often than other lumber.

An advantage of using plywood for forms is the great number of sizes available. It is made in thicknesses

CAUTION

- **Watch for protruding nails. They are the principal cause of accidents on form work.**
- **Inspect tools frequently.**
- **Place mud sills under shoring that rests on the ground.**
- **Protect all men on scaffolds and on the ground.**
- **Do not raise large form panels in heavy gusts of wind.**
- **Brace all shoring securely to prevent collapse of form work.**

of 1/4, 3/8, 1/2, 5/8, and 3/4 inch, and in widths up to 48 inches. The 8-foot lengths are most commonly used. The 6/8- and 3/4-inch thicknesses are most economical. Thinner plywood requires solid backing to prevent deflection. The 1/4-inch thickness is useful for curved surfaces.

Waterproof cardboard and other fiber materials are used for round concrete columns and other preformed shapes. Forms are made by gluing layers of fiber together and molding them to the right shape. The advantage is that fabrication at the job site is not necessary.

FOUNDATION AND FOOTING FORMS

When possible, earth is excavated to form a mold for concrete-wall footings. If wood forms are needed, the four sides are built in panels.

Panels for two opposite sides are made at exact footing width; the other pair has two end cleats on the inside spaced the length of the footing plus twice the sheathing thickness. One-inch-thick sheathing is nailed to vertical cleats spaced on 2-foot centers. Two-inch dressed lumber should be used for the cleats.

NOTE: Panels are held in place with form nails until the tie wire is installed. Nails should be driven only part way from the outside so that they can be easily removed.

Tie wires are wrapped around the center cleats. Wire holes on each side of the cleat should be less than 1 inch in diameter to prevent mortar leaks. Reinforcing bars must be placed before the wire is installed.

For forms 4 feet square or larger, stakes are driven as shown in Figure 5-1. These stakes, and 1 x 6 boards nailed across the top of the form, prevent spreading. Panels may be higher than the required depth of footing, since they can be marked on the inside to show the top of the footing. If the footings are less than 1 foot deep and 2 feet square, forms can be constructed of 1-inch sheathing without cleats (Figure 5-2).

When placing a footing and a small pier at the same time, the form is built as shown in Figure 5-3. To ensure that support for the upper form does not interfere with the placement of concrete in the lower form, 2 x 4 or 4 x 4 pieces

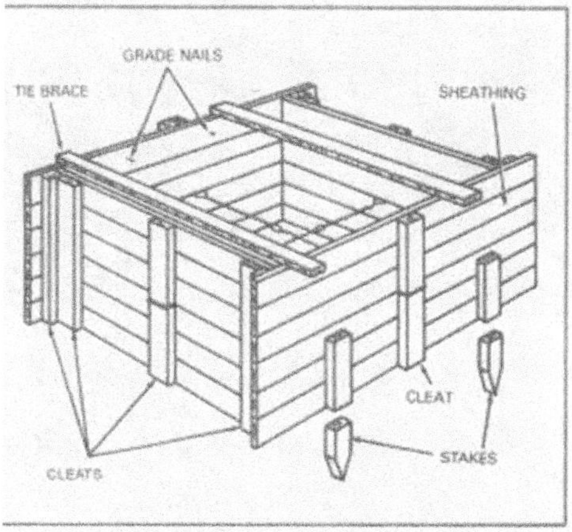

Figure 5-1. Typical large footing form

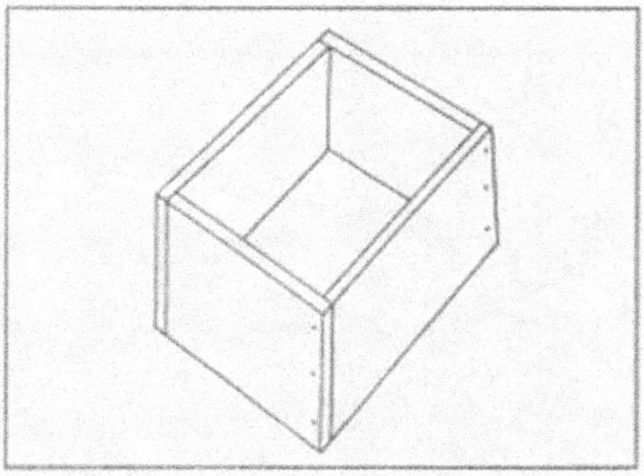

Figure 5-2. Typical small footing form

are nailed to the top of the lower form (as shown). The top form is then nailed to these pieces.

Construction and bracing of forms for wall footings are shown in Figure 5-4. The sides are 2-inch lumber held in place by stakes and held apart by spreaders. The short brace shown at each stake holds the form in line.

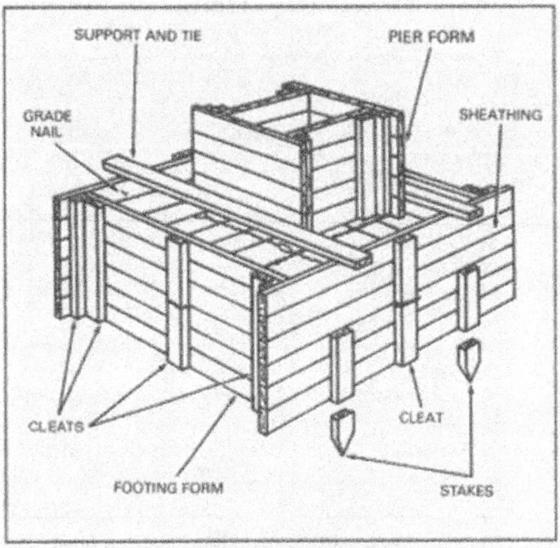

Figure 5-3. Footing and pier form

WALL FORMS

Wall forms are made of wall "panels" and other parts shown in Figure 5-5 page 5-4. These Darts are described as follows:

Wall Panels

Wall panels are made by nailing the *sheathing* to the *studs* and can be built in place or prefabricated elsewhere. Prefabricated wall panels should be no more than 10 feet long so that they can be easily handled. Figure 5-6, page 5-4, shows how wall panels are connected and how wall corners are constructed.

Sheathing

Sheathing forms the surface of the concrete. It should be smooth, especially if the finished surface is to be exposed. It is normally 1-inch (3/4-inch dressed) tongue- and-groove lumber or 3/4-inch plywood. Concrete is plastic when placed in the form, so sheathing should be watertight. tongue-and-groove lumber or plywood gives a watertight surface. Reinforce sheathing to prevent bulging from the weight of the concrete.

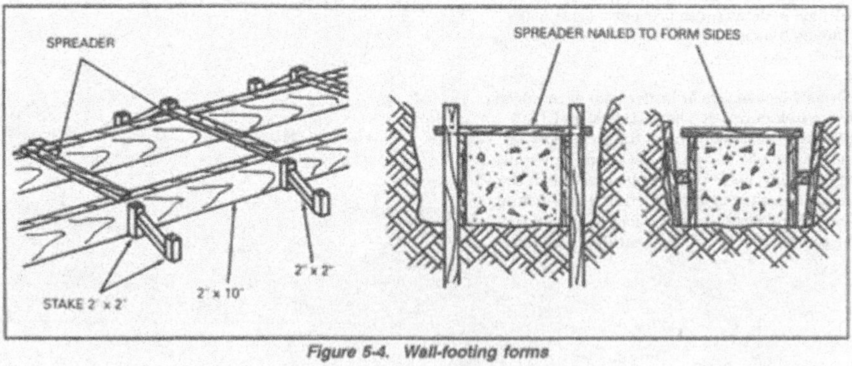

Figure 5-4. Wall-footing forms

Studs

Vertical studs make the sheathing rigid. These studs are generally made from 2 x 4 lumber. Studs also require reinforcing when they extend more than 4 feet.

Wales

Double wales reinforce the wall form. They also tie wall panels together and keep them in a straight line. They run horizontally and are lapped at the corners.

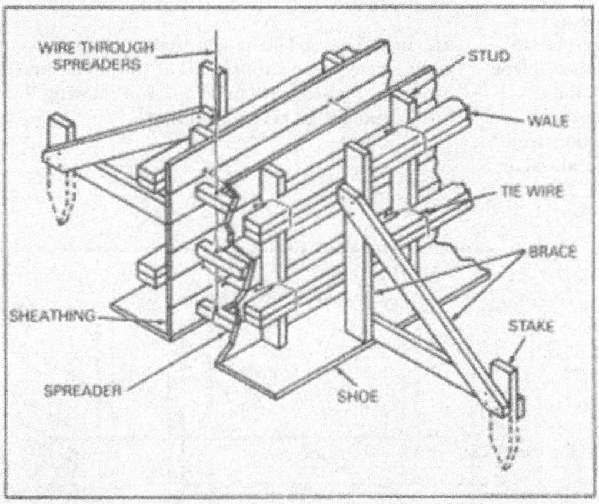

Figure 5-5. Concrete wall form

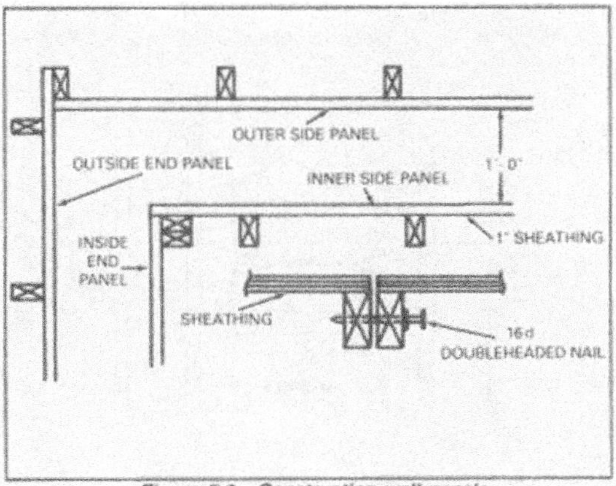

Figure 5-6. Constructing wall panels

Braces

Braces give the forms stability. The most common brace uses a horizontal member and a diagonal member nailed to a stake and to the stud or wale. The diagonal member of the brace should make a 30° angle with the horizontal member. Additional bracing may be provided by strongbacks (vertical members) behind the wales or in the corner formed by intersecting wales. (Braces are not part of the form design and are not considered as providing additional strength.)

Shoe Plates

Shoe plates are nailed into the foundation or footing and must be carefully placed to maintain the wall dimensions and alignment. Studs are tied into the shoe plate.

Spreaders

Spreaders must be placed near each *tie wire*. Spreaders are cut to the same length as the thickness of the wall and placed between the two sheathing surfaces of the forms. They are not nailed, but are held in place by friction.

Spreaders are removed as the forms are filled (Figure 5-7), so that they will not become embedded as the concrete hardens. A wire is attached to the spreaders to allow them to be pulled out of the form after the concrete has put enough pressure on the walls to allow easy removal. The wire fastened to the bottom spreader passes through a hole drilled off center in each spreader above it.

Pulling on the wire will remove the spreaders one after another as the concrete level rises in the forms.

Tie Wires

Tie wires hold the forms secure against the lateral pressures of unhardened concrete. Double strands are always used. Ties keep wall forms together as the concrete is positioned; Figure 5-8 shows two ways of doing this. The wire should be No. 8 or 9 gauge, black, annealed iron wire. Barbed wire may be used in an emergency. Tie spacing should be the same as the stud placing, but never more than 3 feet. Each tie is formed by looping the wire around a wale, bringing it through the form, and looping it around the wale on the opposite side. The tie wire is made taut by twisting it with a smooth metal rod or a spike.

NOTE: Wire ties should be used only for low walls or when *tie rods* are not available.

Tie Rods

An alternate to tie wires and spreaders, the tie rod and spreader combination is shown in Figure 5-9, page 5-6. After the form is removed, each rod is broken off at the notch. If appearance is important, the holes should be filled with a mortar mix.

The use of a wood strip as a wedge when curtain walls and columns are placed at the same time is shown in Figure 5-10, page 5-6. In removing the forms, the wedge is removed first.

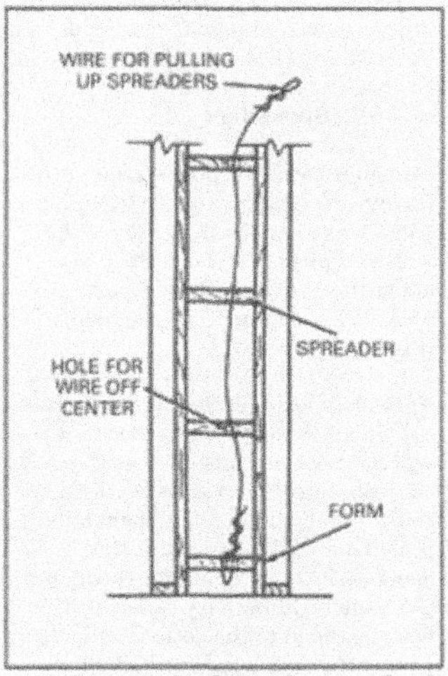

Figure 5-7. Removing wood spreaders

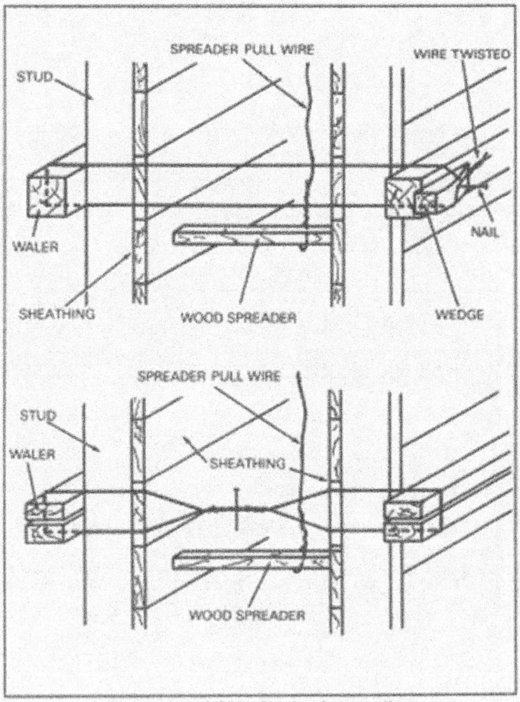

Figure 5-8. Wire ties for form walls

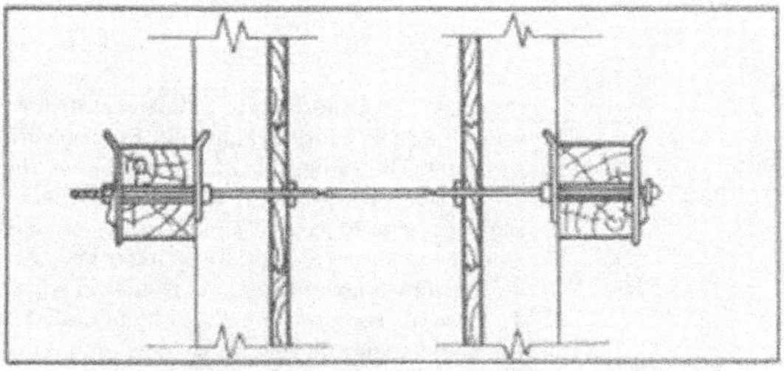

whether the bottom is to be left until the concrete is strong enough for shoring to be removed. Beam forms receive little bursting pressure but must be shored at close intervals to prevent sagging.

The bottom of the form is the same width as the beam; it is in one piece for its full width. Form sides are 1-inch tongue-and-groove material; they lap over the bottom (as shown). The sheath is nailed to 2 x 4 struts placed on 3-foot centers. A 1 x 4 piece is nailed along the struts to support the joists for the floor panel. The sides of the form are not nailed to the bottom but are held in position by continuous strips. Crosspieces nailed on top serve as spreaders. After erection, the slab panel joints hold the beam in place.

A beam and girder assembly is shown in Figure 5-13, page 5-8. The beam bottom butts tightly against the side of the girder and rests on a 2 x 4 nailed to the girder side. Details in Figure 5-13 show the clearances for stripping and the allowances for movement caused by the concrete's weight. The 4 x 4 posts are spaced to support the concrete and are wedged at the bottom or top for easy removal.

COLUMN FORMS

Sheathing runs vertically in column forms to save saw cuts. Corner joints are firmly nailed to ensure watertight construction. Battens or narrow strips of boards (cleats) are placed directly over the joints to fasten the several pieces of vertical sheathing together.

A column and footing form is shown in Figure 5-11. The column form is erected after the steel reinforcement is assembled and is tied to dowels in the footing. The form should have a **cleanout** hole in the bottom to help remove debris. The lumber removed to make the cleanout holes should be nailed to the form so that it can be replaced before the concrete is positioned.

BEAM AND GIRDER FORMS

Figure 5-12 shows both beam and girder forms. The type of construction of these forms depends on whether the form is to be removed in one piece or

FLOOR FORMS

Floor panels are built as shown in Figure 5-14, page 5-9. The 1-inch tongue-and-groove sheathing or 3/4-inch plywood is nailed to 1 x 4 cleats

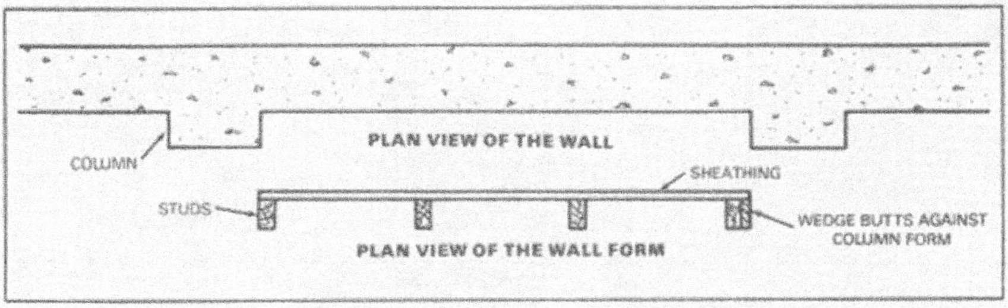

Figure 5-10. Wall form for curtain walls

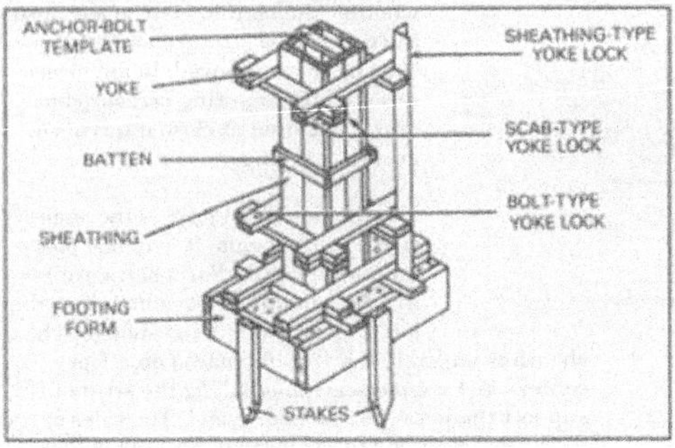

on 3-foot centers. These panels are supported by 2 x 6 joists. Spacing of joists depends on the thickness of the concrete slab and the span of the beams. If the slab spans the distance between two walls, the panels are used in the same manner as when beams support the floor slab.

STAIR FORMS

A method for building stair forms up to 3 feet in width is shown in Figure 5-15, page 5-9. The underside of the steps should be 1-inch tongue-and-groove sheathing. This platform should extend 12 inches beyond each side of the stairs to support stringer bracing blocks. The back of the panel is shored with 4 x 4 pieces (as shown). The 2 x 6 cleats nailed to the shoring should rest on wedges to make both adjustments and removal of the posts easy. The side stringers are 2 x 12 pieces cut as required for the treads and risers. The face of the riser should be 2-inch material, beveled (as shown).

FORM REMOVAL

Forms should be built to allow easy removal without danger to the concrete. Before concrete is placed, forms are treated with oil or other coating material to prevent the concrete from sticking. The oil should penetrate the wood to prevent water absorption. A light

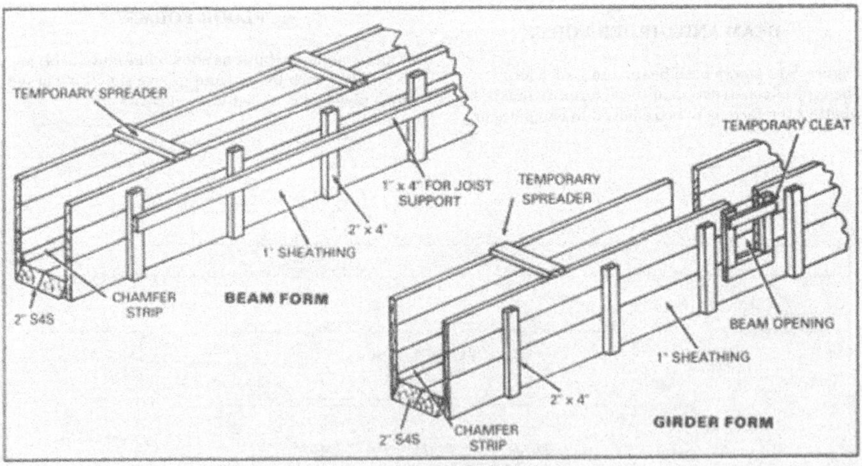

Figure 5-12. Beam and girder forms

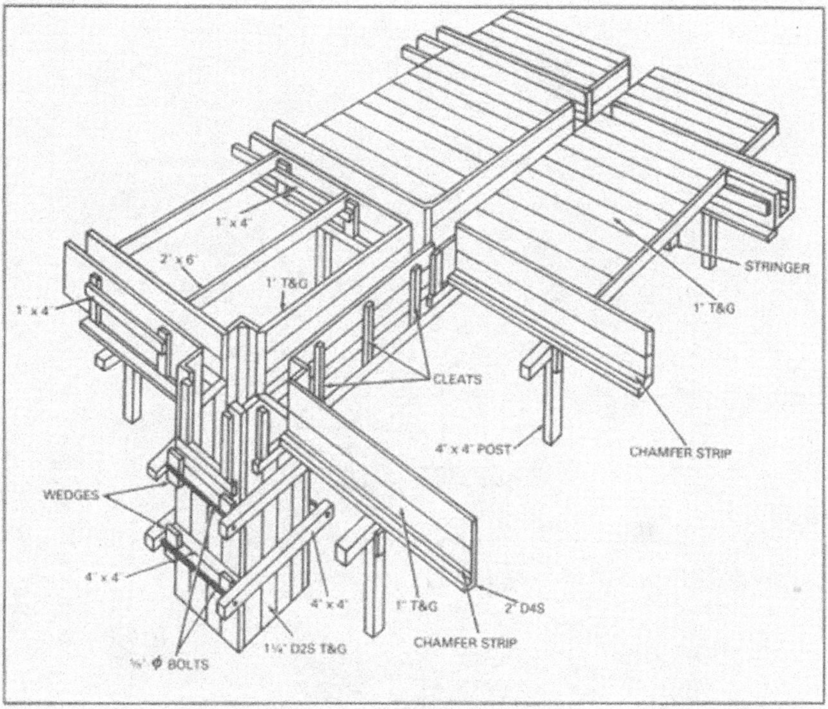

Figure 5-13. Beam, column and floor form

bodied petroleum oil will do. On plywood, shellac is more effective than oil. If forms are to be reused, painting helps preserve the wood.

If form oil is not available, wetting with water may be substituted in an emergency to prevent sticking.

Wood wedges should be used to wedge forms against concrete, rather than a pinchbar or other metal tool. To avoid breaking the edges of concrete, forms should not be jerked off after wedging has been started at one end. Forms to be reused should be cleaned and oiled immediately. Nails should be removed as forms are stripped.

CAUTION

--Permit only workmen doing the stripping in the immediate area.
--Do not remove forms until the concrete has set.
--Pile stripped forms immediately to avoid congestion, exposed nails, and other hazards.

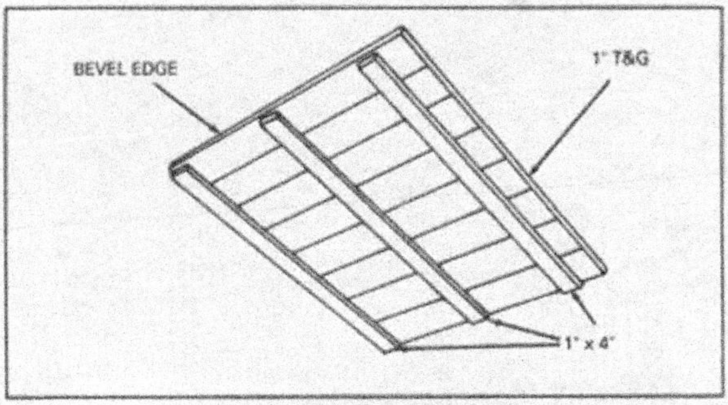

Figure 5-14. Floor slab form

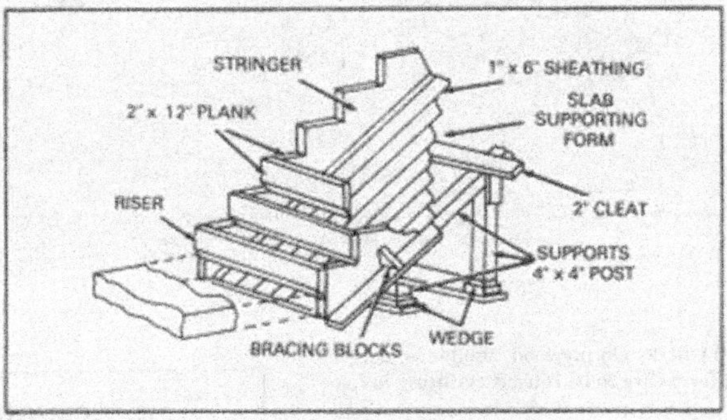

Figure 5-15. Stairway slab form

ROUGH FRAMING — CHAPTER 6

After the foundation is built and the batter boards are removed, the carpenter builds the framework. The framework consists of beams, trusses, walls and partitions, flooring, ceilings, sheathing and siding, stairways, roof framing and coverings (Chapter 7), and doors and windows (Chapter 8). This chapter familiarizes the carpenter with materials, tools, and techniques used to build the framework.

TYPES OF FRAMING

Framing consists of *light, heavy,* and *expedient* framing.

LIGHT FRAMING

There are three principal types of framing for light structures: western, balloon, and braced. Figure 6-1, page 6-2, illustrates these types of framing and specifies the nomenclature and location of the various members.

Light framing is used in barracks, bathhouses, and administration buildings. Figure 6-2, page 6-3, shows some details of a 20-foot wide building (such as ground level, window openings, braces, and splices) and labels the framing parts.

Much of light framing can be done in staging areas while staking out, squaring, and floor framing is being done. Subflooring can begin when a portion of the floor joists has been laid. The better-skilled men should construct the frame, and with good coordination, a large force of men can be kept busy during framing.

Western Frame

The western or platform frame (Figure 6-1, 1) is used extensively in military construction. It is similar to the braced frame, but has boxed-sill construction at each floor line. Also note that cross bridging is used between the joists and bridging is used between the studs. The platform frame is preferred for one-story structures since it permits both the bearing and nonbearing walls (which are supported by the joist) to settle uniformly.

Balloon Frame

The balloon frame (Figure 6-1, 2) is a widely used type of light framing. The major difference between balloon and braced framing in a multistory building is that in balloon framing the studs run the full length, from sill to rafters. It is customary for second-floor joists to rest on a 1- x 4-inch ribbon that has been set into the studs. The balloon frame is less rigid than a braced frame.

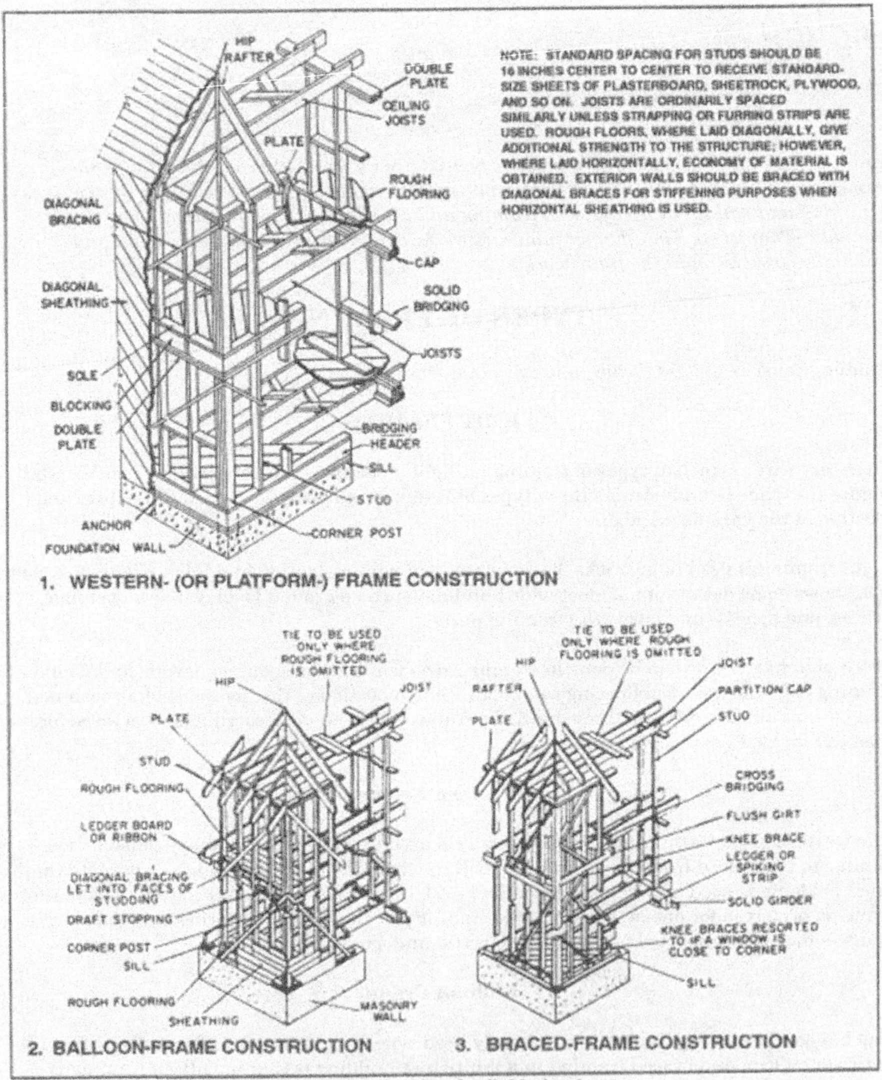

Figure 6-1. Framing for light structures

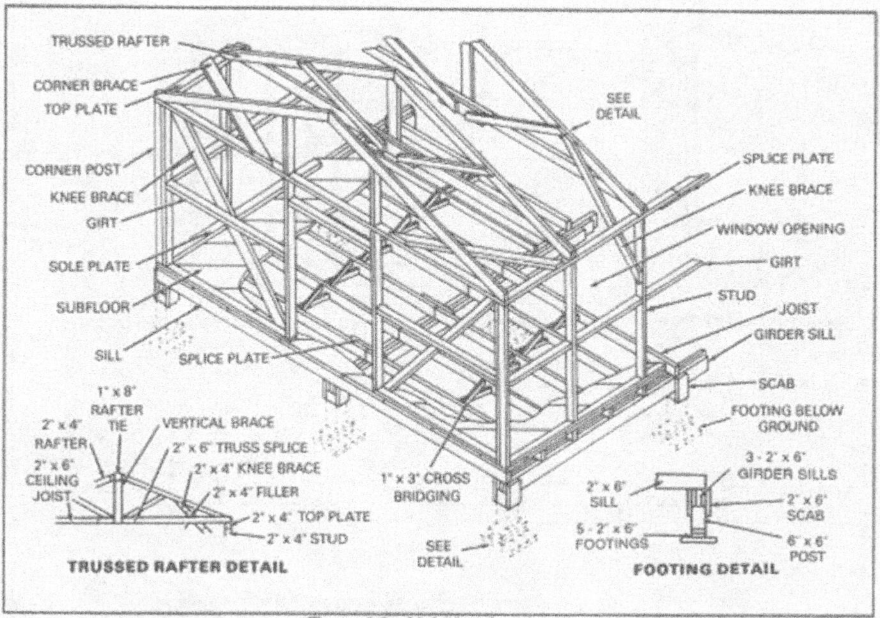

Figure 6-2. Light-framing details

Braced Frame

A braced frame (Figure 6-1, 3) is much more rigid than a balloon frame. Exterior studs extend only between floors and are topped by girts that form a sill for the joists of the succeeding floor. Girts are usually 4 x 6 inches. With the exception of studs, braced frame members are heavier than those in balloon framing. Sills and corner posts are customarily 4 x 6 inches. Unlike the studs, corner posts extend from sill to plate. Knee braces, usually 2 x 4 inches, are placed diagonally against each side of the corner posts. Interior studding for braced frames is the same as for balloon-frame construction.

HEAVY FRAMING

Heavy-frame buildings are more permanent, and are normally used for warehouses and shops. Heavy framing is seldom used in TO construction. Figure 6-3, page 6-4, shows the details of heavy framing. Heavy framing consists of framing members at least 6 inches in dimension (timber construction). Long, unsupported areas between walls are spanned by built-up roof trusses.

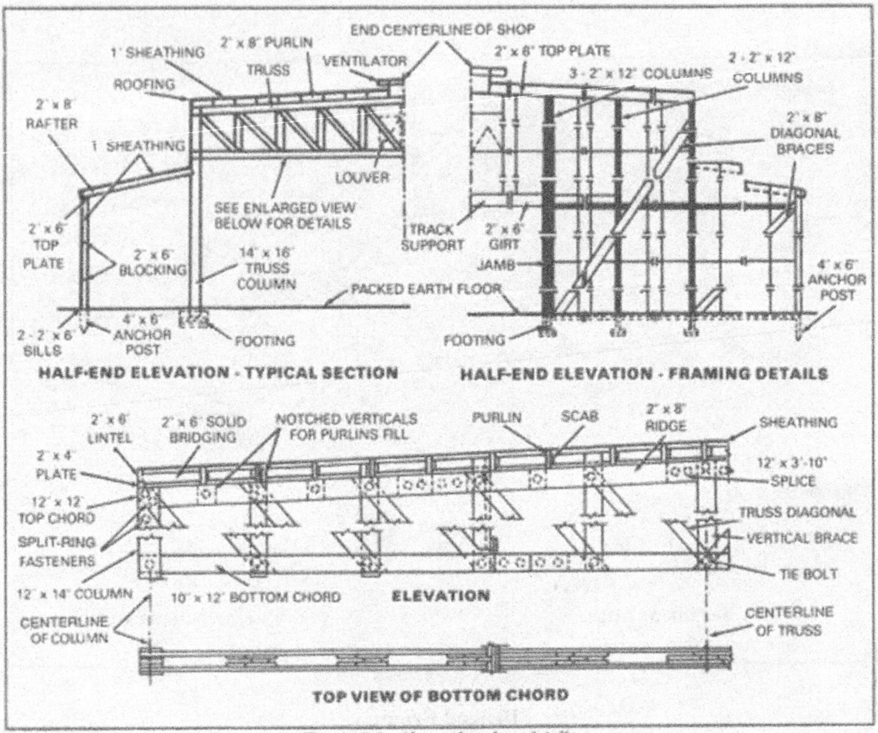

Figure 6-3. Heavy-framing details

EXPEDIENT FRAMING

Some field conditions require expedient framing techniques. For example—

- Light siding. Chicken wire and water-resistant bituminous paper can be sandwiched to provide adequate temporary framing in temperate climates.
- Salvaged framing. Salvaged sheet metal, such as corrugated material or gasoline cans, can be used as siding in the construction of emergency housing.
- Local timber. Poles trimmed from saplings or bamboo can be constructed into reasonably sound framing and may be secured with native vines if necessary.
- Wood-substitute framing. Adobe (soil, straw, and water—mixed until spreadable) can be used to form walls, floors, and foundations. A similar mixture may be used to form sun-dried bricks.
- Excavations. Proper excavation and simple log cribbing may also be covered with sod and carefully drained to give adequate shelter.

CONNECTIONS

Weak points in a structure usually occur at the connections (joints and splices) between pieces of lumber. However, these connections can be structurally sound if done correctly. Such weak points are usually a sign of poor workmanship.

JOINTS

Joints are connections between two pieces of timber that come together at an angle. The types of joints most commonly used in carpentry are butt joints and lap joints.

Butt Joints

A butt joint is formed by placing the end of one board against another board so that the boards are at an angle (usually a right angle), forming a corner. The types of butt joints are shown in Figure 6-4 and are described below.

Straight Butt Joint. This joint is formed by placing the square-cut end of one board against the square face of another. The butt end of one board should be square and the face of the other board smooth so that they fit perpendicular to each other. Select the right type of nails or screws to hold such a joint securely. For framing, butt joints are secured by 8d or 10d nails that are toenailed to strengthen the joint. The end grain is the weakest part of a piece of wood when used in joints. A butt joint is made at either one or two end-grain parts. It will be no stronger than the quality of those parts. A butt joint is, therefore, the weakest type of joint. This is especially true if the joint is made of two pieces of wood only.

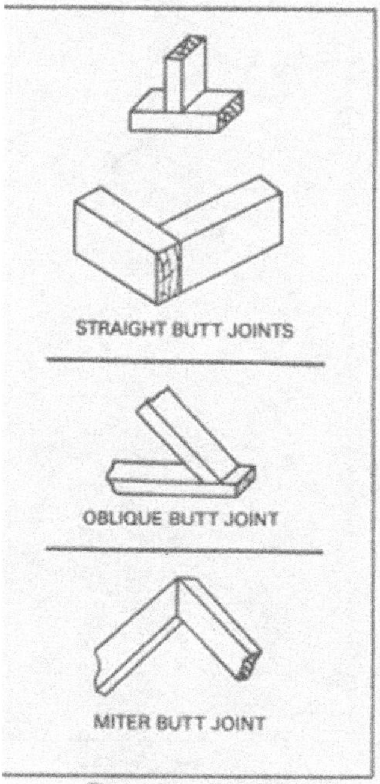

Figure 6-4. Butt joints

Oblique Butt Joint. This joint is formed by butting the mitered end of one board against the face of another board. Bracing is typically made with this joint. It should not be used where great strength is required. The strength of the oblique butt joint depends upon the nailing. The nail size

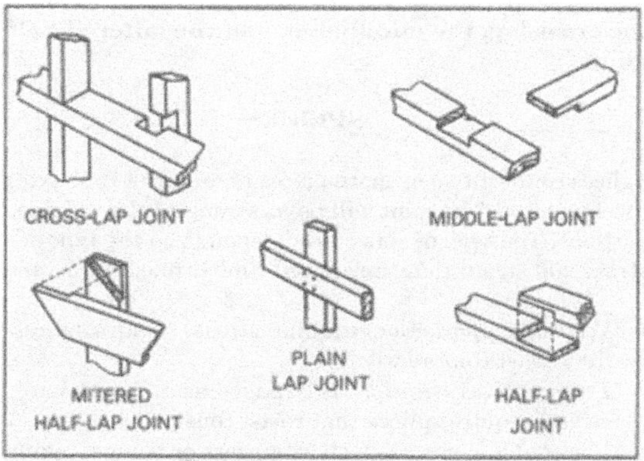

Figure 6-5. Lap joints

depends upon the timber size. Nails should be toenailed to strengthen the joint; not too many nails should be used.

Miter Butt Joint. This joint is formed by bringing the mitered ends of two boards together to form the desired angle. This joint is normally used at corners where a straight butt joint would not be satisfactory. To form a right-angle miter joint (the most commonly used), cut each piece at a 45° angle so that when the pieces are joined they will form a 90° angle. The miter joint is used mostly in framing. *However, it is a very weak joint and should not be used where strength is important.*

Lap Joints

The lap joint is the strongest joint. Lap joints (Figure 6-5) are formed in one of two ways: a plain lap joint or a half-lap splice joint.

Plain Lap Joint. This joint is formed by laying one board over another and fastening the two with screws or nails. This is the simplest and most often used method of joining. This joint is as strong as the fasteners and material used.

Half-Lap Splice Joint. This joint is formed by cutting away equal-length portions (usually half) from the thickness of two boards. The two are then joined so that they overlap and form a corner. Overlapping surfaces must fit snugly and smoothly. Overlaps should be sawed on the waste side of the gauge line, to avoid cutting laps oversize by the thickness of the cut. *This joint is relatively strong and easy to make.*

NOTE: Some useful variations of the half-lap joint are the cross-lap, the middle-lap, and the mitered half-lap joints.

SPLICES

Splices connect two or more pieces of material that extend in the same line. The joint will be as strong as the unjoined portions. The type of splice used depends on the type of stress and strain that the spliced timber must withstand.

- *Vertical supports* (longitudinal stress) require splices that resist compression.
- *Trusses, braces, and joists* (transverse and angular stress) require splices that resist tension.
- *Horizontal supports,* such as girders or beams, require splices that resist bending tension and compression.

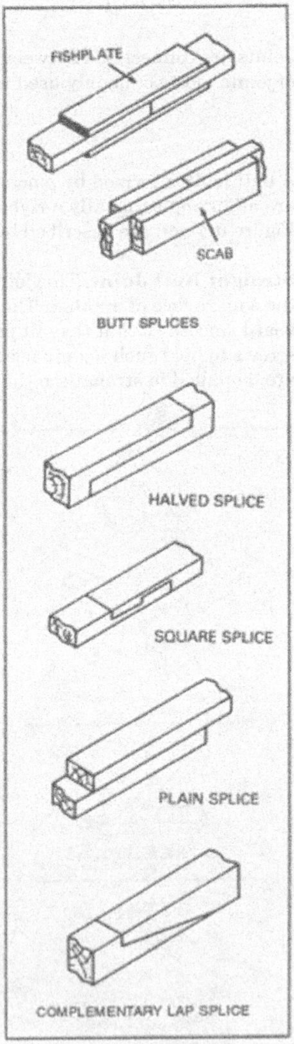

Figure 6-6. Splices

For example, splices for resisting compression are usually worthless for resisting tension or bending. Figure 6-6 shows splice types; Figure 6-7 shows splice stresses.

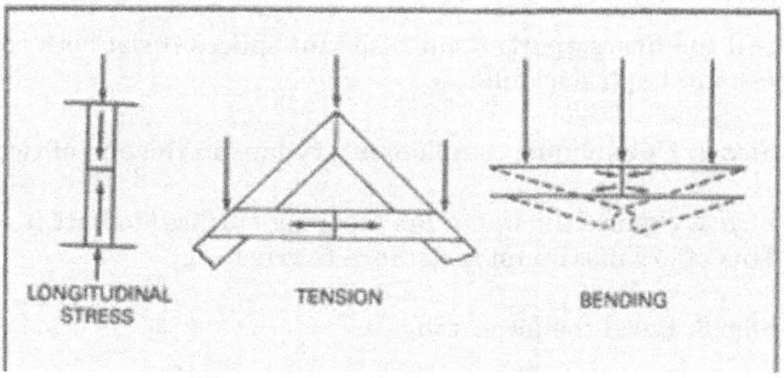

Figure 6-7. Splice stresses

Compression-Resistant Splices. Compression-resistant splices support weight or exert pressure and will resist compression stress only. The most common types of compression-resistant splices are the *butt splice* and the *halved splice.*

Butt Splice. This splice is constructed by butting the squared ends of two pieces of timber together and securing them in this position with two wood or metal pieces fastened on opposite sides of the timber. The two short supporting pieces keep the splice straight and prevent buckling. Metal plates used as supports in a butt splice are called *fishplates.* Wood plates are called *scabs* and are fastened in place with bolts or screws. Bolts, nails, or corrugated fasteners may be used to secure scabs. If nails are used with scabs, they are staggered and driven at an angle away from the splice. Too many nails, or nails that are too large, will weaken a splice.

Halved Splice. This splice is made by cutting away half the thickness of equal lengths from the ends of two pieces of timber, then fitting the tongues (laps) together. The laps should be long enough to provide adequate bearing surfaces. Nails or bolts may be used to fasten the halved splice. **Note: To give the halved splice resistance to tension as well as compression, fishplates or scabs may be used.**

Tension-Resistant Splices

In members such as trusses, braces, and joists, the joint undergoes stress in more than one direction; this creates tension, buckling the member in a predictable direction. Tension-resistant splices provide the greatest practical number of bearing surfaces and shoulders within the splice.

Square Splice. This splice is a modification of the compression halved splice. Notches are cut in the tongues or laps to provide an additional locking shoulder. The square splice may be fastened with nails or bolts. **Note: It may be greatly strengthened by using fishplates or scabs.**

Long, Plain Splice. This splice is a hasty substitute for the square splice. A long overlap of two pieces is desirable to provide adequate bearing surface and enough room for fasteners to make up for the lack of shoulder lock.

Bend-Resistant Splices

Horizontal timbers supporting weight undergo stress at a splice that results in compression of the upper part; this has a tendency to crush the fibers. Tension of the lower part also tends to pull the fibers apart. Bend-resistant splices resist both compression and tension. Make a bend-resistant splice as follows:

Step 1. Cut oblique, complementary laps in the end of two pieces of timber.

Step 2. Square the upper lap (bearing surface) to butt it against the square of the other lap. This offers maximum resistance to crushing.

Step 3. Bevel the lower tongue.

Step 4. Fasten a scab or fishplate along the bottom of the splice to prevent separation of the pieces.

NOTE: When this splice cannot be done, a butt joint, halved splice, or square splice secured by fishplates or scabs may be used.

SILLS

There are four types of wood sill construction: platform construction, balloon-framed construction, braced-framed construction, and the builtup sill. The sill is the foundation that supports a building and is the first part of a building to be set in place. It rests directly on the foundation posts or on the ground and is joined at the corners and spliced when necessary. Figure 6-8, page 6-8, shows the most common sills. The type of sill used depends on the type of construction used in the frame. To prevent air from entering into the building, spread a thin bed of mortar on top of the foundation wall. This also provides a solid base for the sill. Another technique is to use a sill sealer made of fiberglass. Place insulation material and a termite shield under the sill if desired.

PLATFORM CONSTRUCTION

Box sills are commonly used with platform framing, which is the most common type of framing. These may be used with or without the sill plate.

The sill or sill plate is anchored to the foundation wall for supporting and fastening joists with a header at the end of the joists resting on the foundation wall. In this type of sill, the sill is laid edgewise on the outside edge of the sill plate.

BALLOON-FRAMED CONSTRUCTION

"T" sills are usually used in balloon framing. There are two types of T-sills: one for dry, warm climates and one for colder climates. They are made the same, except that in the latter case the joists are nailed directly to the studs and sills and headers are used between the floor joists.

BRACED-FRAMING SILLS

Braced-framing sills (Figure 6-8) are usually used in braced-framing construction. The floor joists are notched and nailed directly to the sill and studs.

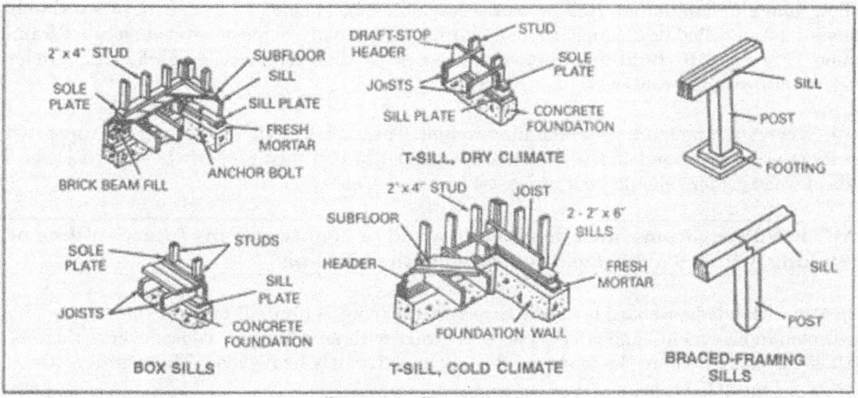

Figure 6-8. Types of sills

BUILT-UP SILLS

If posts are used in the foundation, use either sills made of heavy, single timbers or built-up sills. Built-up sills are made with two or more light timbers, such as 2 x 4s. A built-up sill is used when heavy, single timbers are not available and lighter lumber (such as a 2 x 4) alone would not support the building load. Figure 6-9 shows how to make a corner joint for a builtup sill.

Whether heavy timber or built-up sills are used, the joints should be over posts. The size of the sill depends on the load to be carried and the spacing of the posts. The sill plates are laid directly on the post or, in expedient framing, directly on graded earth. When earth floors are used, nail the studs directly to the sill.

GIRDERS

The distance between two outside walls is usually too great to be spanned by a single joist. A *girder is* used for intermediate support when two or more joists are needed to cover the span. A girder is a large beam that supports other smaller beams or joists. A girder may be made of timber, steel, reinforced concrete, or a combination of these materials.

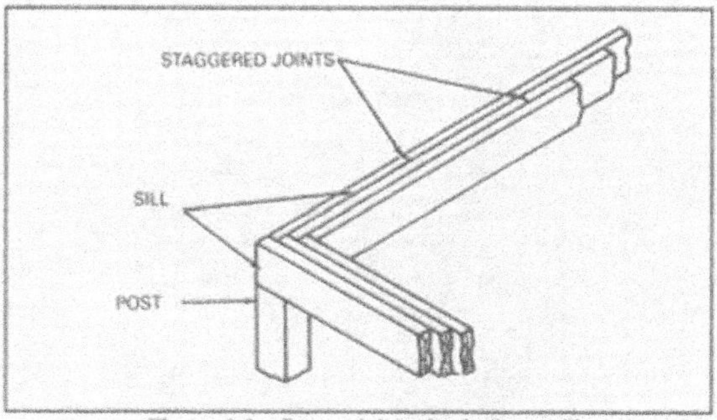

Figure 6-9. Corner joint of a built-up sill

Wooden girders are more common than steel in light-frame buildings. Built-up and solid girders should be of seasoned wood. Common types of wood girders include solid, built-up, hollow, and glue-laminated. Hollow beams resemble a box made of 2 x 4s, with plywood webs. They are often called box beams. Built-up girders are usually made of several pieces of framing lumber (Figure 6-10). Built-up girders warp less easily than solid wooden girders and are less likely to decay in the center.

Girders carry a large part of the building weight. They must be rigid and properly supported at the foundation walls and on the columns. They must be installed properly to support joists. The ends of wood girders should bear at least 4 inches on posts.

CAUTION Precautions must be taken to avoid or counteract any future settling or shrinking, which would cause distortion of the building.

A girder with a ledger board is used where vertical space is limited. This provides more headroom in basements and crawl spaces. A girder with joist hangers is used where there is little headroom or where the joists must carry an extremely heavy load. These girders are shown in Figure 6-11, page 6-10.

SIZE REQUIREMENTS

Carpenters should understand the effect of length, width, and depth on the strength of wood girders before attempting to determine their size.

Principles that govern the size of a girder are the—

- Distance between girder posts.
- Girder load area.
- Total floor load on the girder per square foot.
- Load on the girder per linear foot.
- Total load on the girder.
- Material to be used.
- Wood moisture content and types of wood used, since some woods are stronger than others.

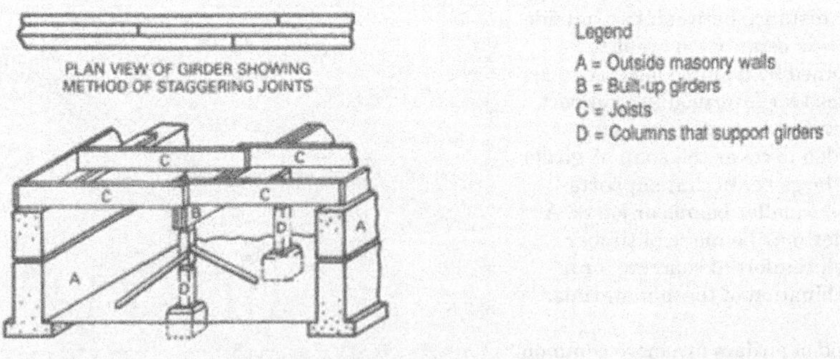

Figure 6-10. Built-up girder details

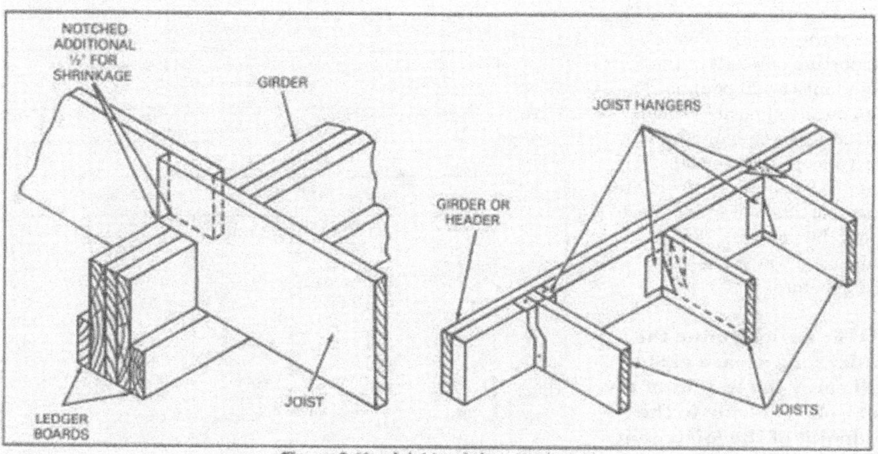

Figure 6-11. Joist-to-girder attachment

A girder should be just large enough to support an ordinary load. Any size larger than that wastes material. For greater carrying capacity, it is better to increase a girder's depth (within limits) than its width. When the depth of a girder is doubled (the width of lumber, such as 2 x 8 or 2 x 6), the safe load increases four times. For example, a girder 3 inches wide and 12 inches deep will carry four times as much weight as a girder 3 inches wide and 6 inches deep. Table 6-1 gives the sizes of built up wood girders for various loads and spans.

LOAD AREA

A building load is carried by foundation walls and the girder. Because the ends of each joist rest on the girder, there is more weight on the girder than on any of the walls. Before considering the load on the girder, it may be well to consider a single joist.

Example 1. Suppose a 10-foot plank weighing 5 pounds per foot is lifted by two men. If the men were at opposite ends of the plank, they would each support 25 pounds.

Now assume that one of these men lifts the end of another 10-foot plank of the same weight as the first one. A third man lifts the opposite end of that plank. The two men on the outside are each now supporting one-half of the weight of one plank (25 pounds apiece), but the man in the center is supporting one-half of each of the two planks (50 pounds).

The two men on the outside represent the foundation walls. The center man represents the girder. The girder carries one-half of the weight, and the other half is equally divided between the outside walls. However, the girder may not always be located halfway between the outer walls.

Example 2. Imagine the same three men lifting two planks that weigh 5 pounds per foot. One of the planks is 8 feet long and the other is 12 feet long. The total length of these two planks is the same as before. The weight per foot is the same, so the total weight in both cases is 100 pounds.

One of the outside men is supporting one-half of the 8-foot plank) or 20 pounds. The man on the opposite outside end is supporting one-half of the 12-foot plank, or 30 pounds. The man in the center is supporting one-half of each plank (50 pounds). This is the same total weight he was lifting before.

NOTE: To determine the girder load area: a girder will carry the weight of the floor on each side to the midpoint of the joists that rest upon it.

FLOOR LOAD

After the girder load area is known, the total floor load per square foot must be determined, for **safety purposes.** Both dead and live loads must be considered.

Dead Load

The dead load consists of all building structure weight. The dead load per square foot of floor area is carried directly or indirectly to the girder by bearing partitions. The dead load varies according to the construction method and building height. The structural parts in the dead load are—

Table 6-1. Sizes of built-up wood girders

(Based on Douglas fir 4-square guideline framing; deflection not over 1/360 of span; allowable fiber stress 1,600 lb/in²)

LOAD PER LINEAR FOOT OF GIRDER	LENGTH OF SPAN (FEET)				
	6	7	8	9	10
	NOMINAL SIZE OF GIRDER REQUIRED (INCHES)				
750	6 x 8	6 x 8	6 x 8	6 x 10	6 x 10
900	6 x 8	6 x 8	6 x 10	6 x 10	8 x 10
1050	6 x 8	6 x 10	8 x 10	8 x 10	8 x 12
1200	6 x 10	8 x 10	8 x 10	8 x 10	8 x 12
1350	6 x 10	8 x 10	8 x 10	8 x 12	10 x 12
1500	8 x 10	8 x 10	8 x 12	10 x 12	10 x 12
1650	8 x 10	8 x 12	10 x 12	10 x 12	10 x 14
1800	8 x 10	8 x 12	10 x 12	10 x 12	10 x 14
1950	8 x 12	10 x 12	10 x 12	10 x 14	12 x 14
2100	8 x 12	10 x 12	10 x 14	12 x 14	12 x 14
2250	10 x 12	10 x 12	10 x 14	12 x 14	12 x 14
2400	10 x 12	10 x 14	10 x 14	12 x 14	
2550	10 x 12	10 x 14	12 x 14		
2700	10 x 12	10 x 14	12 x 14		
2850	10 x 14	12 x 14	12 x 14		
3000	10 x 14	12 x 14			
3150	10 x 14	12 x 14			
3300	12 x 14	12 x 14			

NOTES:

1. **The 6" girder is figured as being made with three pieces 2" dressed to 1 3/4" thickness.**
2. **The 8" girder is figured as being made with four pieces 2" dressed to 1 3/4" thickness.**
3. **The 10" girder is figured as being made with five pieces 2" dressed to 1 3/4" thickness.**
4. **The 12" girder is figured as being made with six pieces 2" dressed to 1 3/4" thickness.**
5. **For solid girders, multiply the above loads by 1.130 when a 6" girder is used; 1.150 when an 8" girder is used; 1.170 when a 10" girder is used; and 1.180 when a 12" girder is used.**

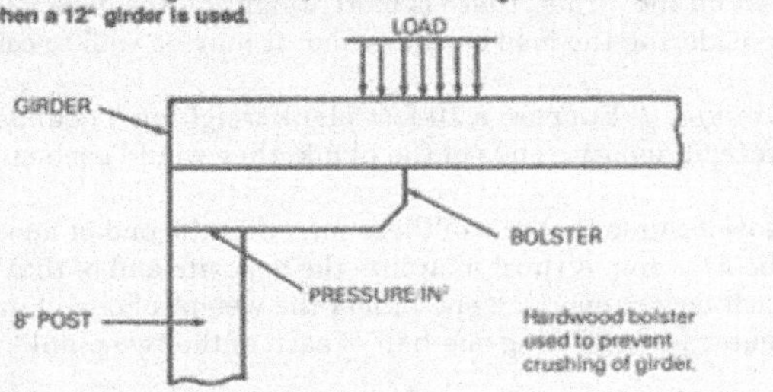

- Floor joists for all floor levels.
- Flooring materials, including the attic if it is floored.
- Bearing partitions.
- Attic partitions.
- Attic joists for the top floor.
- Ceiling laths and plaster, including the basement ceiling if it is plastered.

The total dead load for a light-frame building similar to an ordinary frame house is the dead-load allowance per square foot of all structural parts added together.

- The allowance for an average subfloor, finish floor, and joists without basement plaster should be 10 pounds per square foot.
- If the basement ceiling is plastered, allow an additional 10 pounds per square foot.
- If the attic is unfloored, make a load allowance of 20 pounds for ceiling plaster and joists when girders or bearing partitions support the first-floor partition.
- If the attic is floored and used for storage, allow an additional 10 pounds per square foot.

Live Load

The live load is the weight of furniture, persons, and other movable loads, not actually a part of the building but still carried by the girder. The live load per square foot will vary according to the building use and local weather conditions. Snow on the roof is also a part of the live load.

- Allowance for the live load on floors used for living purposes is 30 pounds per square foot.
- If the attic is floored and used for light storage, allow an additional 20 pounds per square foot.
- The allowance per square foot for live loads is usually governed by local building specifications and regulations.

The load per linear foot on the girder is easily figured when the total load per square foot of floor area is known.

Example. Assume that the girder load area of the building shown in Figure 6-12 is sliced into 1-foot lengths across the girder. Each slice represents the weight supported by 1 foot of the girder. If the slice is divided into 1-foot units, each unit will represent 1 square foot of the total floor area. To determine the load per linear foot of girder, multiply the number of units by the total load per square foot.

Note in Figure 6-12 that the girder is off-center. Remember that half of the load is supported by the girder and half by the foundation walls. Therefore, the joist length to be supported on one side of the girder is **7 feet** (one half of 14 feet) and the other side is 5 **feet** (one half of 10 feet), for a total distance of **12 feet** across the load area. Since each slice is 1 foot wide, it has a total floor area of 12 square feet.

Assume that the total floor load for each square foot is 70 pounds. Multiply the length times the width to get the total square feet supported by the girder (*7 feet x 12 feet = 84 square feet).*

84 square feet x 70 pounds per square feet (live and dead load) = 5,880 pounds total load on the girder

BUILT-UP GIRDERS

Figure 6-10, page 6-9, shows a *built-up girder.* Notice that the joists rest on top of the girder. This type of girder is commonly used in frame building construction. To make a built-up girder, select lumber that is as free from knots and other defects as possible.

Built-up girders are usually made of three pieces of framing lumber nailed together. (The pieces must be nailed securely to prevent individual buckling.) For proper arrangement of the pieces of lumber, the end grains should match the example in Figure 6-13. The nailing pattern should be square across the ends of the board (1 1/2 inches from each end) and then diagonal every 16 inches as shown in Figure 6-13. This pattern increases the strength of the girder. A typical two- or three-piece girder of 2-inch lumber should be nailed on both sides with 16d common nails.

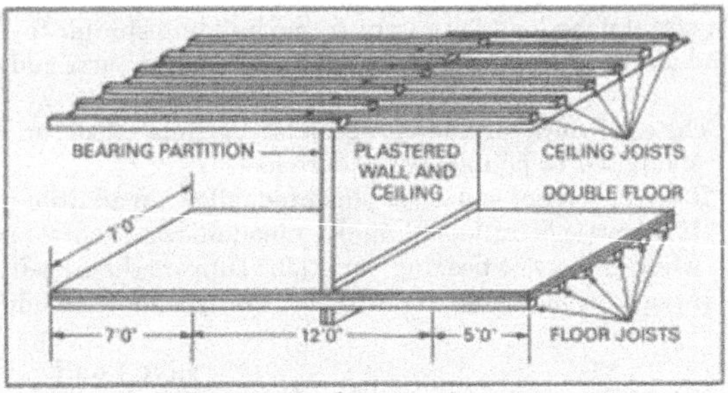

Figure 6-12. Girder load area

SPLICING

Methods for splicing girders differ according to whether the girder is built-up or solid-beam.

Built-Up Girders

The lumber for a built-up girder should be long enough so that no more than one joint will occur over the span between footings. The joints in the beam should be staggered, and the planks must be squared at each joint and butted tightly together.

Solid-Beam Girders

To splice solid beams, use half-lap joints or butt joints (Figure 6-14.) See Splices on page 6-6.

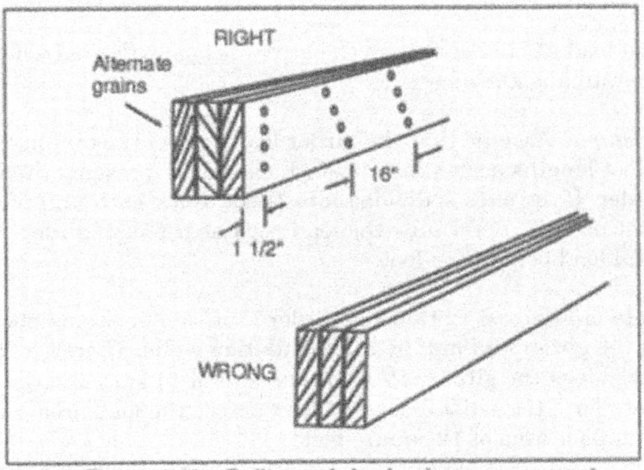

Figure 6-13. Built-up girder lumber arrangement

Half-Lap. Sometimes a half-lap joint is used to join solid beams (Figure 6-14). This is done by performing the following steps:

Step 1. Place the beam on one edge so that the annual rings run from top to bottom.

Step 2. Lay out the lines for the half-lap joint as shown in Figure 6-14.

Step 3. Make cuts along these lines, then check with a steel square to ensure a matching joint.

Step 4. Repeat the process on the other beam.

Step 5. Nail a temporary strap across the joint to hold it tightly together.

Step 6. Drill a hole through the joint with a drill bit about 1/16 inch larger than the bolt to be used, and fasten the joint with a bolt, a washer, and a nut.

Butt Joints. When a strapped *butt joint is* used to join solid beams (Figure 6-14, page 6-13), the ends of the beams should be cut square. The straps, which are generally 18 inches long, are bolted to each side of the beams.

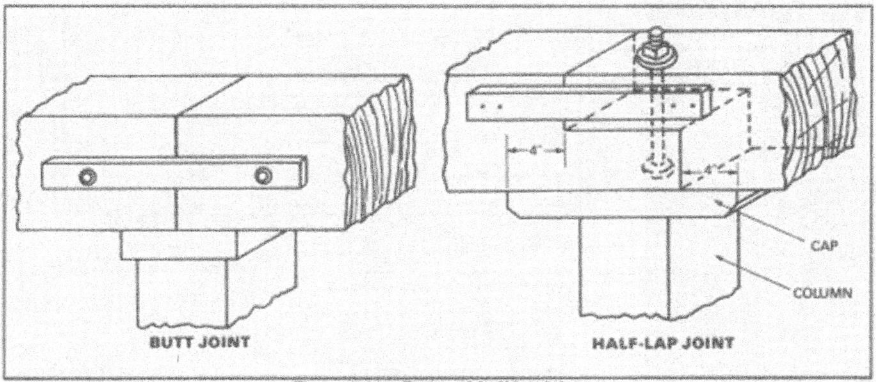

Figure 6-14. Butt and half-lap joints

GIRDER SUPPORTS

When building a small frame building, the carpenter should know how to determine the proper size of girder supports (called *columns* or *posts).*

A *column* or *post is* a vertical member that supports the live and dead loads placed upon it. It may be made of wood, metal, or masonry.

- *Wooden columns* may be solid timbers or several pieces of framing lumber nailed together with 16d or 20d common nails.
- *Metal columns* are made of heavy pipe, large steel angles, or I-beams.

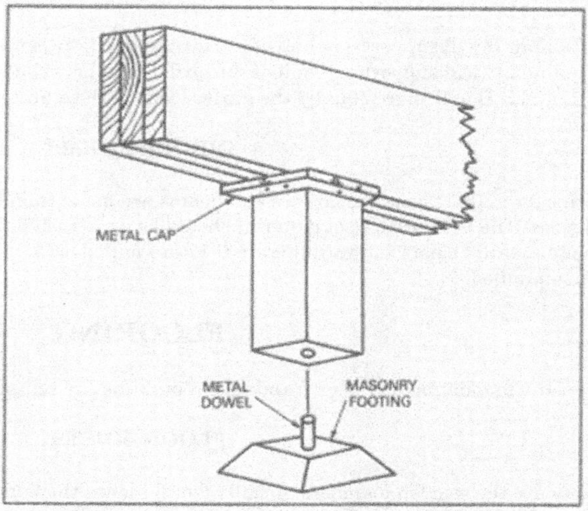

Figure 6-15. Use of metal cap and masonry footing

A column must have a bearing plate at the top and bottom which distributes the load evenly across the column. Basement posts that support girders should be set on masonry footings. Columns should be securely fastened at the top to the load-bearing member and at the bottom to the footing on which they rest.

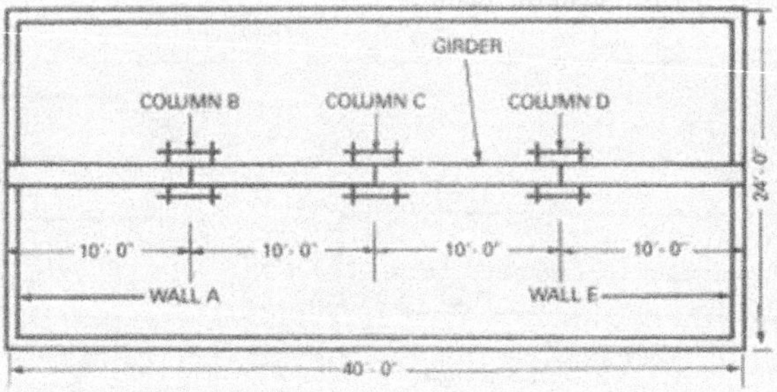

Figure 6-16. Column spacing

Figure 6-15 shows a solid wooden column with a metal bearing cap drilled to permit fastening it to the girder. The bottom of this type of column may be fastened to the masonry footing by a metal dowel. The dowel should be inserted in a hole drilled in the bottom of the column and in the masonry footing. The base is coated with asphalt at the drilling point to prevent rust or rot.

A good arrangement of a girder and supporting columns for a 24- x 40-foot building is shown in Figure 6-16.

- Column B will support one-half of the girder load between wall A and column C.
- Column C will support one-half of the girder load between columns B and D.
- Column D will share equally the girder loads with column C and wall E.

GIRDER FORMS

Forms for making concrete girders and beams are made from 2-inch-thick material dressed on all sides. The bottom piece of material should be constructed in one piece to avoid using cleats. The temporary cleats shown in Figure 6-17 are nailed on to prevent the form from collapsing when handled.

FLOORING

After the foundation and deck framing of a building are completed, the floor is built.

FLOOR JOISTS

Joists are the wooden members, usually 2 or 3 inches thick, that make up the body of the floor frame (Figure 6-18, page 6-16). The flooring or subflooring is nailed to the joists. Joists as small

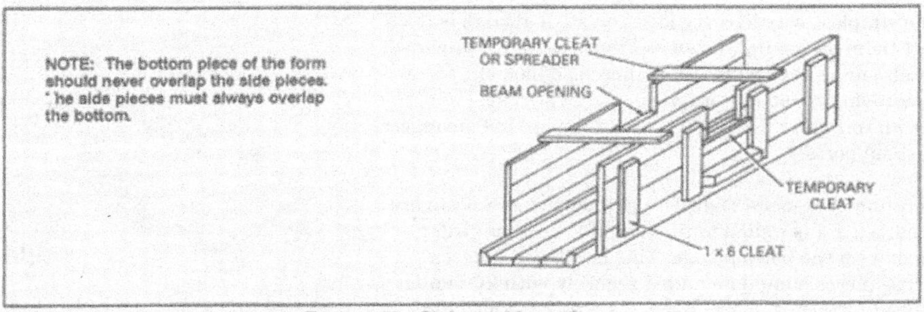

Figure 6-17. Girder and beam form

as 2 x 6 are sometimes used in light frame buildings. These are too small for floors with spans over 10 feet, but are frequently used for ceiling joists.

Joists usually carry a uniform load of materials and personnel; these are live loads. The weight of joists and floors is a dead load. The joists carry the flooring load directly on ends nearest the sills, girders, bearing partitions, or bearing walls. Joists are spaced 16 or 24 inches apart, center to center. Sometimes the spacing is 12 inches, but where such spacing is made necessary by the load, heavier joists should be used.

To support heavily concentrated loads or a partition wall, you may need to double the joist or place two joists together. Two typical reinforced joists are shown in Figure 6-19.

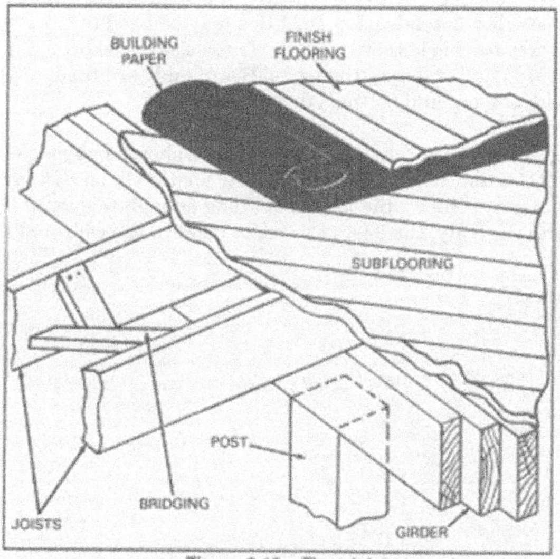

Figure 6-18. Floor joists

In joining joists to sills, be sure that the connection can hold the load that the joist will carry. The joist-connecting method in Figure 6-20, A, is used most often because it provides the strongest joint. The methods shown in Figure 6-20, B and C, are used when it is not desirable to use joists on top of the sill. The ledger plate should be securely fastened. If the joist must be notched, it should not be notched to the sill and girder over one-third of its depth to prevent splitting (Figure 6-20, D).

Joists must be level when framed to girders. If the girder is not the same height as the sill, the joist must be notched as shown in Figure 6-20, C. If the girder and sill are the same height, the joists must be framed to keep the joist level. The joist is always placed crown up. This counteracts the weight on the joists. In most cases there will be no sag below a straight line.

The simplest way to carry joists on steel girders is to rest them on top (as shown in Figure 6-20, E), provided headroom is not restricted. If there is a lack of headroom, use straps or hangers (iron stirrups) as shown in Figure 6-20, F. These art among the strongest joist supports.

In connecting joists to girders and sills where posts are used, a 2 x 4 is nailed to the face of the sill or girder, flush with the bottom edge. This is called a *ledger.* These pieces should be nailed securely with 20d nails spaced 12 inches apart. When 2 x 6 or 2 x 8 joists are used, it is better to use 2 x 4 ledgers. This prevents joists from splitting at the notch.

When joists are 10 inches or more deep, 2 x 4s may be used as ledgers without reducing the strength of the joists. If a notch is used, joist ties may be used to overcome this loss of strength. These ties are short 1 x 4 boards nailed across the joists. Board ends are flush with the top and bottom edges of the joists.

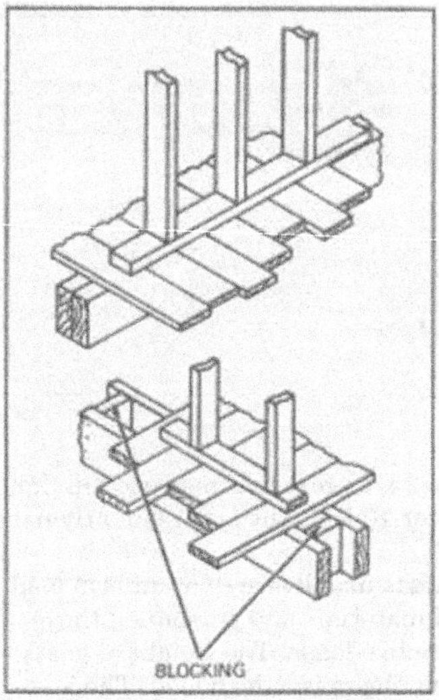

Figure 6-19. Reinforced joists

Overhead joists are joined to plates as shown in Figure 6-21, A and B. The inner end of the joist rests on the partition wall plates. If a joist is to rest on plates or girders, the joist is cut long enough to extend the full width of the plate or girder. Alternatively, the joists are cut to meet in the center of the plate or girder and connected with

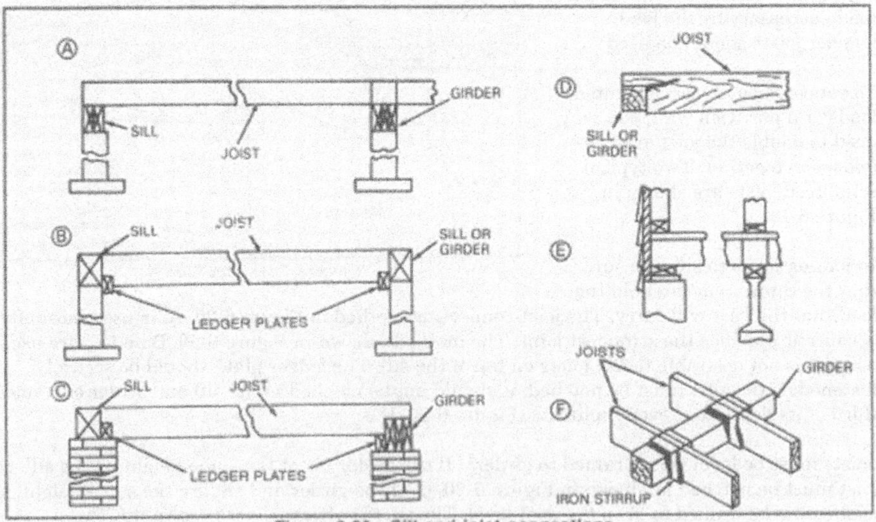

Figure 6-20. Sill and joist connections

a scab. When the ends of two joists lie side by side on the plate, they should be nailed together. Joists may also be joined to girders with ledgers (Figure 6-21, C and D).

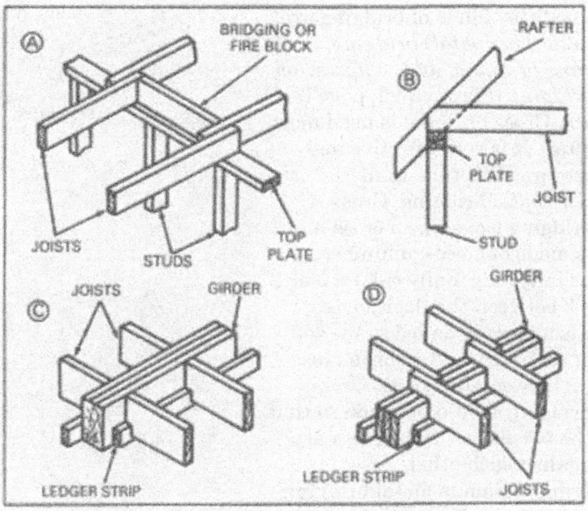

Figure 6-21. Top plates and ledgers

FLOOR JOISTS FOR PLATFORM CONSTRUCTION

Check the plans to determine the size and direction of the joists. If the sizes for joists are not specified on the plans, consult Tables 6-2 and 6-3 to determine the appropriate size.

FLOOR BRIDGING

Joists tend to twist from side to side, especially when used over a long span. Floor frames are bridged for stiffening and to prevent unequal deflection of the joists. This stiffening also enables an overloaded joist to receive some help from the joists on either side of it. A pattern for the bridging stock is obtained by placing a piece of material between the joists, then marking and sawing it. When sawed, the cut will form the correct angle.

Table 6-2. Allowable spans for floor joists using nonstress-graded lumber

Size of Floor Joists (Inches)	Spacing of Floor Joists (Inches)	Maximum Allowable Span (Feet and Inches)							
		Group I		Group II		Group III		Group IV	
		Plastered Ceiling Below	Without Plastered Ceiling Below	Plastered Ceiling Below	Without Plastered Ceiling Below	Plastered Ceiling Below	Without Plastered Ceiling Below	Plastered Ceiling Below	Without Plastered Ceiling Below
2 x 6	12	10-6	11-6	9-0	10-0	7-6	8-0	5-6	6-0
	16	9-6	10-0	8-0	8-6	6-6	7-0	5-0	5-0
	24	7-6	8-0	6-6	7-0	5-6	6-0	4-0	4-0
2 x 8	12	14-0	15-0	12-6	13-6	10-6	11-6	8-0	8-6
	16	12-6	13-6	11-0	11-6	9-0	10-0	7-0	7-6
	24	10-0	11-0	9-0	9-6	7-6	8-0	6-0	6-6
2 x 10	12	17-6	19-0	16-6	17-6	13-6	14-6	10-6	11-6
	16	15-6	16-6	14-6	15-6	12-0	13-0	9-6	10-0
	24	13-0	14-0	12-0	13-0	10-0	10-6	7-6	8-6
2 x 12	12	21-0	23-0	21-0	21-6	17-6	19-6	13-0	14-6
	16	18-0	20-0	18-0	19-6	15-6	16-6	12-0	13-0
	24	15-0	16-6	15-0	16-6	12-6	13-6	10-0	10-6

Note: The group classifications in this table refer to the species and minimum grades of nonstress-graded lumber. See Table 6-3.

The three kinds of bridging are: *solid (horizontal) bridging, cross bridging,* and *compression bridging* (Figure 6-22, page 6-20). Cross bridging is used most often. It is very effective and requires less time than horizontal bridging. Cross bridging looks like a cross and is made of pieces of lumber, usually diagonally cut 1 x 3 or 2 x 3 between the floor joists. Each piece is nailed to the top of each joist and forms a cross (x) between the joists. Cross bridging should be made so that the two pieces of the cross are against each other. Compression is metal bridging between joists.

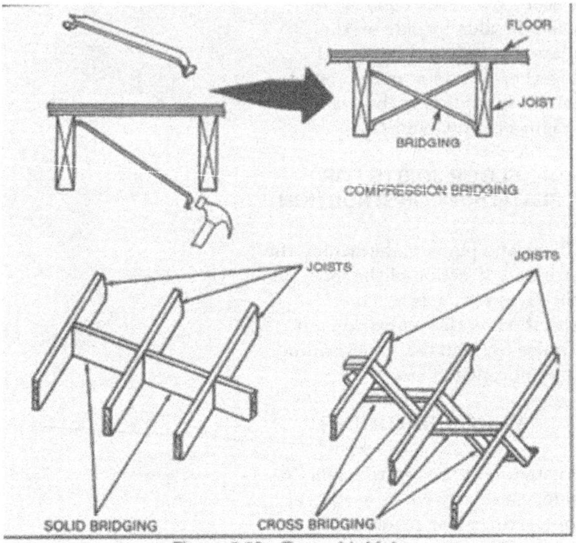

Figure 6-22. Type of bridging

Bridging should be nailed at the tops with 8d or 11 Id nails, and the bottoms should be left free until the subfloor is laid. This allows the joists to adjust to their final position and keeps the bridging from pushing up the joists and causing unevenness in the floor. The bottom ends of the bridging may then be nailed, forming a continuous truss across the floor. This prevents overloaded joists from sagging.

Cutting and fitting the bridging by hand is a slow process. A power saw should be used if it is available. One line of bridging should be placed on joists more than 8 feet long. On joists more than 16 feet long, two lines should be used.

Table 6 3. Group classification—nonstress-graded lumber

Species	Minimum Grade
Group I	
Douglas Fir and Larch	Construction
Group II	
Bald Cypress	No. 2
Douglas Fir (South)	Construction
Fir, White	Construction
Hemlock, Eastern	No. 1
Hemlock, West Coast and Western	Construction
Pine, Red (Norway Pine)	No. 1
Redwood, California	Select Heart
Spruce, Eastern	No. 1
Spruce, Sitka	Construction
Spruce, White and Western White	Construction

Group III	
Cedar, Western	Construction, West Coast Studs
Cedar, Western Red and Incense	Construction
Douglas Fir and Larch	Standard, West Coast Studs
Douglas Fir (South)	Standard
Fir, Balsam	No. 1, Standard
Fir, White	West Coast Studs
Hemlock, Eastern	No. 2, Standard
Hemlock, West Coast and Western	West Coast Studs
Pine, Ponderosa, Lodgepole, Sugar, and Idaho White	Construction
Redwood, California	Construction
Redwood, California (studs only)	Two Star
Spruce, Engelmann	Construction, Standard
Spruce, Sitka	West Coast Studs
Spruce, White and Western White	Standard
Group IV	
Cedar, Western	Utility
Cedar, Western Red and Incense	Utility
Douglas Fir and Larch	Utility
Douglas Fir (South)	Utility
Fir, White	Utility
Hemlock, West Coast and Western	Utility
Pine, Ponderosa, Lodgepole, Sugar, and Idaho White	Utility
Redwood, California	Merchantable
Redwood, California (studs only)	One Star
Spruce, Engelmann	Utility
Spruce, Sitka	Utility
Spruce, White and Western White	Utility

FLOOR OPENINGS

Floor openings for stairwells, ventilators, and chimneys are framed by a combination of *headers* and *trimmers.* Headers run at right angles to the direction of the joists and are doubled. Trimmers run parallel to the joists and are actually doubled joists. The joists are framed at right angles to the headers of the opening frame. These shorter joists, framed to headers, are called *tail beams, tail joists,* or *header joists.* The number of headers and trimmers needed at any opening depends upon—

- The shape of the opening—whether it is a simple rectangle or contains additional angles.
- The direction in which the opening runs, in relation to the joist direction.
- The position of the opening, in relation to partitions or walls.

Figure 6-23, page 6-20, shows examples of openings. One runs parallel to the joist and requires two headers and one trimmer. The other runs at right angles to the joists and requires one header and two trimmers. The openings shown in Figure 6-24, page 6-21, are constructed with corner angles supported in different ways. The cantilever method (shown on the right of Figure 6-24) requires that the angle be fairly close to a supporting partition with joists from an adjacent span that run to the header.

To frame openings of the type shown in Figure 6-25—

Step 1. Headers 1 and 2 are nailed to trimmers A and C with three 20d nails.

Step 2. Headers 1 and 2 are nailed to short joists X and Y with three 20d nails.

Step 3. Headers 3 and 4 are nailed to headers 1 and 2 with 16d nails spaced 6 inches apart.

Step 4. Trimmers A and C are nailed to headers 3 and 4 with three 20d nails.

Step 5. Trimmers B and D are nailed to trimmers A and C with 16d nails spaced 12 inches apart.

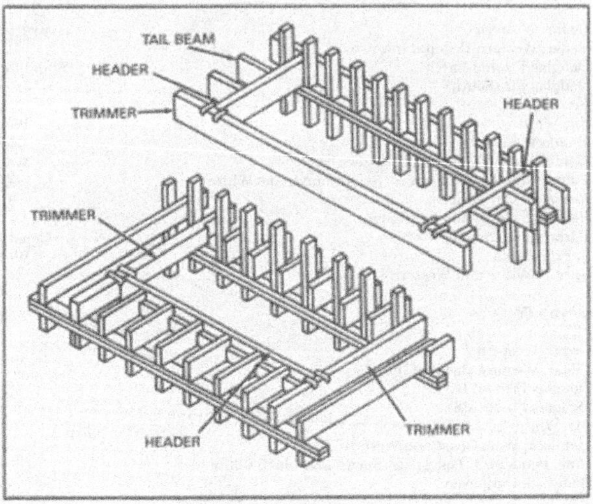

Figure 6-23. Floor openings

SUBFLOORS

The subfloor (Figure 6-26, page 6-22), if included in the plans, is laid diagonally on the joist framework and nailed with 8d to 10d nails. Subflooring boards 8 inches wide or more should have at least three nails per joist. Where the subfloor is more than 1 inch thick, larger nails should be used. The subfloor is normally laid before the walls are framed so that it can be walked on while walls are being framed.

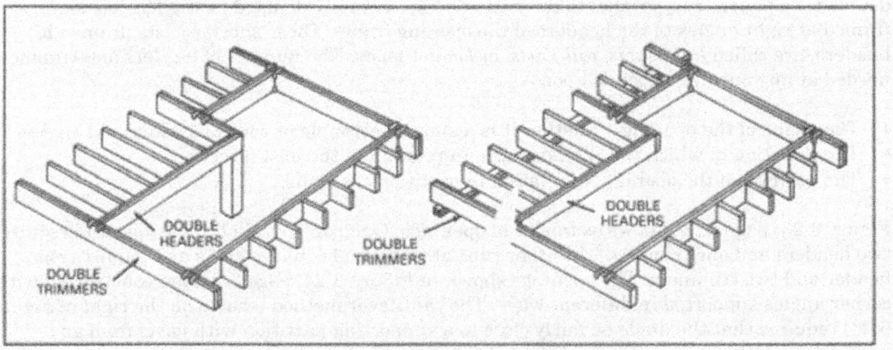

Figure 6-24. Double headers and trimmers

FINISH FLOORS

A finish floor in the TO is normally of 3/4-inch material, square-edged or tongue-and-groove (Figure 6-27, page 6-22). Finish flooring varies from 3 1/2 to 7 1/2 inches wide. It is laid directly on floor joists or on a subfloor and nailed with 8d common nails in every joist. When a subfloor exists, building paper is used between it and the finish floor to keep out dampness and insects.

In warehouses, where heavy loads are to be carried on the floor, 2-inch material should be used for the finish floor. Such flooring is also face-nailed with 16d or 20d nails. It is not tongue-and-groove, and it ranges in width from 4 to 12 inches. The joints are made on the center of the joist.

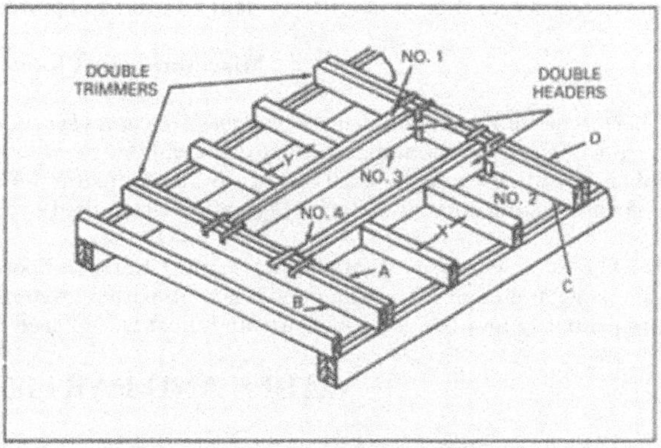

Figure 6-25. Floor-opening construction

Wood Floors

Wood floors must be strong enough to carry the load. The type of building and its intended use determine the arrangement of the floor system, the thickness of the sheathing, and the approximate spacing of the joists.

Concrete Floors

Concrete floors may be constructed for shops where earthen or wood floors are not suitable. These include aircraft repair and assembly shops, shops for heavy equipment, and certain kinds of warehouses.

After the earth has been graded and compacted, concrete is placed on the ground. The floor should be reinforced with steel or wire mesh. The foundation wall may be placed first and the concrete floor placed after the building is completed. This gives protection to the concrete floor while it sets.

Drainage is provided for the floor area around the footing and the area near the floor, to prevent flooding after heavy

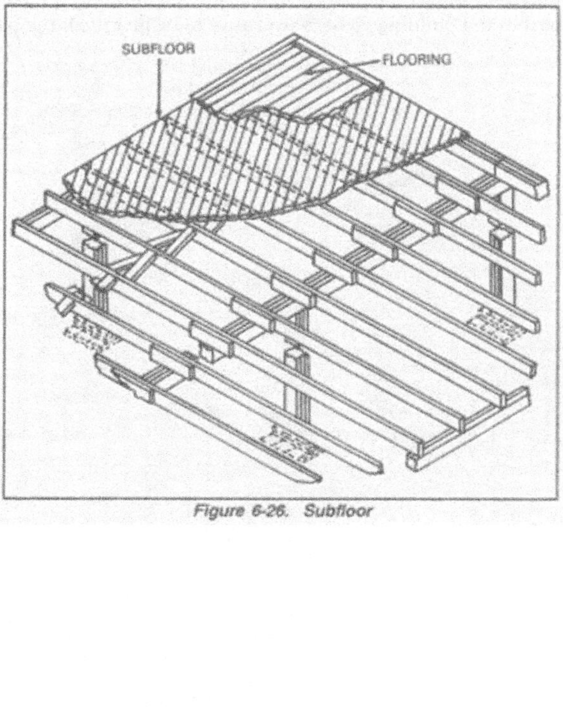

Figure 6-26. Subfloor

rains. A concrete floor is likely to be damp unless it is protected.

Miscellaneous Floors

These types of floors include *earth, adobe brick, duckboard, or rushes.* Miscellaneous flooring is used when conventional materials are unavailable or where there is a need to save time or labor. Such floors may be used if facilities are temporary or if required by the special nature of a structure. Selection of material is usually determined by availability.

Duckboard is widely used for shower flooring. Earthen floors are common; they conserve both materials and labor if the ground site is even without extensive grading. Rush or thatch floors are primarily an insulating measure and must be replaced frequently.

WALLS AND PARTITIONS

Once the floor is in place, it is used to support the wall frame. Wall framing (Figure 6-28) consists of studs, diagonal bracing, cripples, trimmers; headers, and fire blocks. It is supported by the floor sole plate. The vertical members of the wall framing are the studs. They support the top plates and the upper part of the building, or everything above the top plate line. Studs support the lath, plaster, and insulation on the inside and wall sheathing on the outside.

Walls and partitions, which are classed as framed constructions, are composed of studs. Studs are usually closely spaced, slender, 2 x 4 vertical members. They are arranged in a row with their ends bearing on a long horizontal member called a *bottom plate* or *sole plate,* and their tops are capped with another plate, called a *top plate.* Double top plates are used to tie walls and partitions together. The bearing strength of stud walls is determined by the strength of the studs. Figure 6-29, page 6-24, shows a typical wall construction.

Partition walls divide the inside space of a building. In most cases, these walls are framed as part of the building. Where floors are to be installed, the partition walls are left unframed.

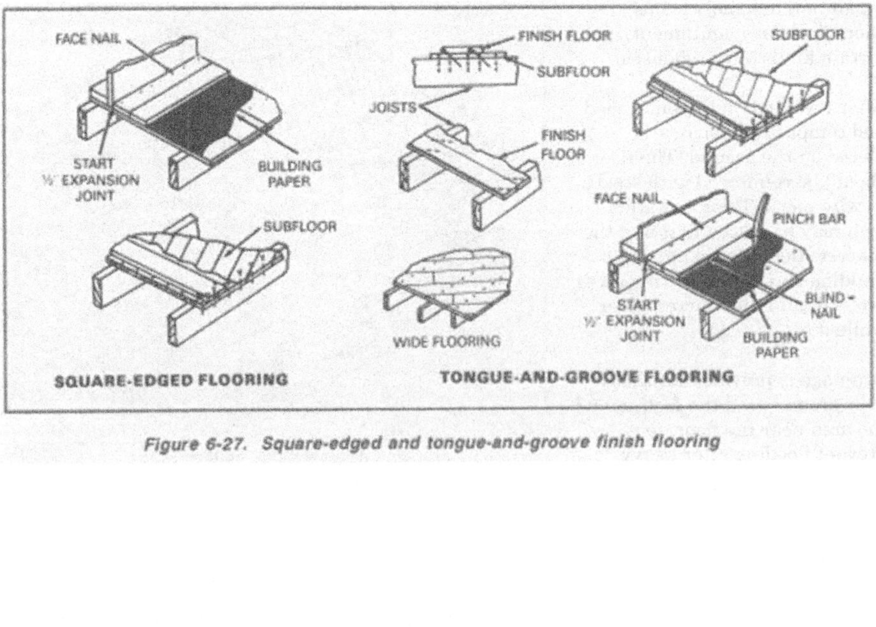

Figure 6-27. Square-edged and tongue-and-groove finish flooring

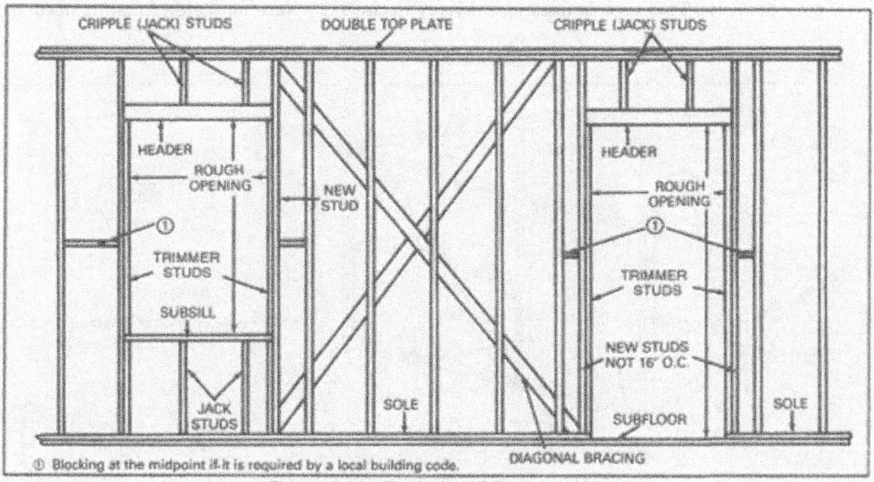

Figure 6-28. Typical wall-frame details

The two types of partition walls are bearing and nonbearing. The bearing type supports ceiling joists; the nonbearing type supports only itself, and may be put in at any time after the other framework is installed. Only one cap or plate is used. A sole plate should be used in every case, as it helps distribute the load over a larger area.

Partition walls are framed the same as outside walls; door openings are framed as outside openings. Where there are corners or where one partition wall joins another, corner posts or T-posts are used as in the outside walls. These posts provide nailing surfaces for the inside wall finish. Partition walls in a TO one-story building may or may not extend to the roof. The top of the studs has a plate when the wall does not extend to the roof. If the wall extends to the roof, the studs are joined to the rafters.

CORNER POSTS

A corner post forms an inside corner and an outside corner which provides a good nailing base for inside wall coverings. Figures 6-30, page 6-24, and 6-31, page 6-25, show two of the most common types of corner posts as they would appear constructed. The studs used at the corners of frame construction are usually built up from three or more ordinary studs to provide greater strength. These built-up assemblies are called *corner posts*. They are set up, plumbed, and temporarily braced. Corner posts may also be made in any of the following ways (Figure 6-32):

- A 4x 6 with a 2 x 4 nailed on the board side, flush with one edge (Figure 6-32, A). This type of corner is for a 4-inch wall. Where walls are thicker, heavier timber is used.
- A 4 x 4 with a 2 x 4 nailed to each of two adjoining sides (Figure 6-32, B).
- Two 2 x 4s nailed together with blocks between them and a 2 x 4 flush with one edge (Figure 6-32, C). This is the most common method.

- A 2 x 4 nailed to the edge of another 2 x 4, the edge of one flush with the side of the other (Figure 6-32, D). This type is used extensively in the TO where no inside finish is needed.

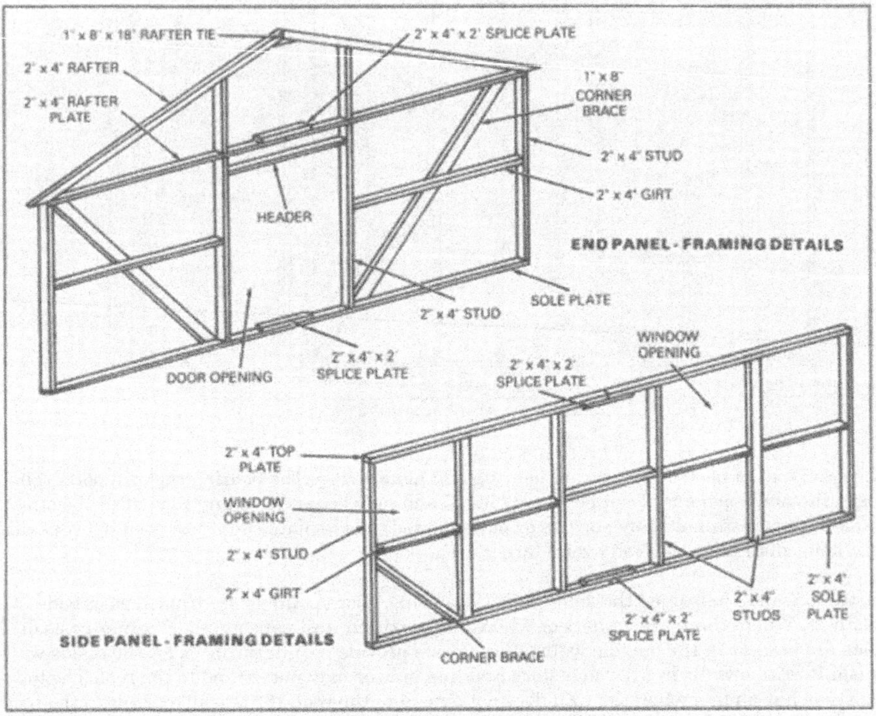

Figure 6-29. Typical wall construction

PARTITION POSTS

There are two types of partition posts—T-posts and double T-posts.

T-Posts

Whenever a partition meets another wall, a stud wide enough to extend beyond the partition on both sides is used. This provides a solid nailing base for the inside wall finish. This type of stud is called a T-

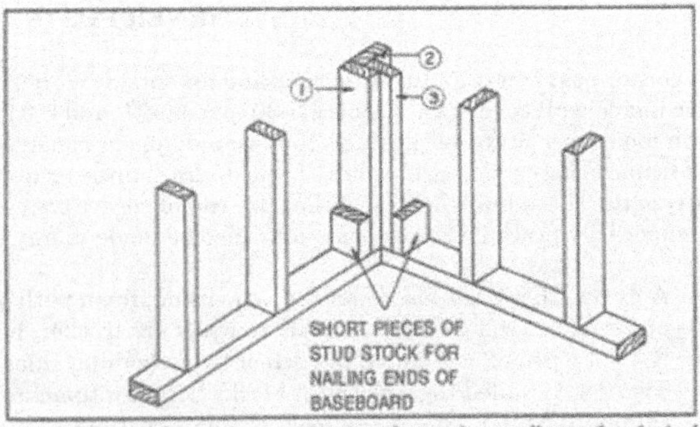

Figure 6-30. Corner-post construction using ordinary 2 x 4 studs

post and is made in any of the following ways (Figure 6-33):

- A 2 x 4 may be nailed and centered on the face side of a 4 x 6 (Figure 6-33, A).
- A 2 x 4 may be nailed and centered on two 4 x 4s nailed together (Figure 6-33, B).
- Two 2 x 4s may be nailed together with a block between them and a 2 x 4 centered on the wide side (Figure 6-33, C).
- A 2 x 4 may be nailed and centered on the face side of a 2 x 6, with a horizontal bridging nailed behind them to give support and stiffness (Figure 6-33, D).

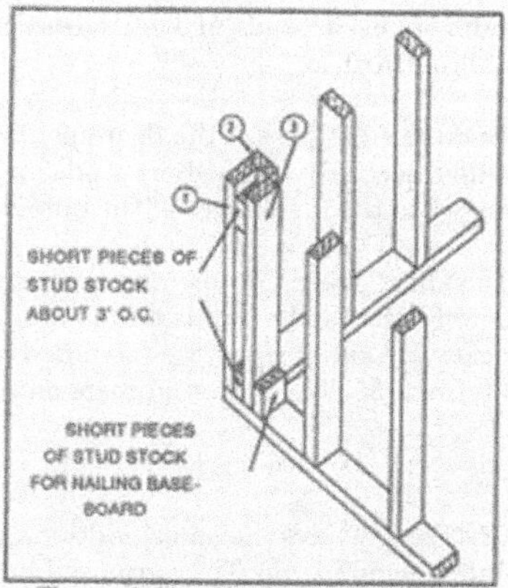

Figure 6-31. Alternate corner-post construction using ordinary 2 x 4 studs

Double T-Posts

When a partition is finished on one side only, the partition post used is a simple stud, set in the outside wall, in line with the side of the partition wall, and finished as shown in Figure 6-34, page 6-26. These posts are nailed in place along with the corner posts. The exact position of the partition walls must be determined before the posts are placed. When walls are more than 4 inches thick, wider timber is used.

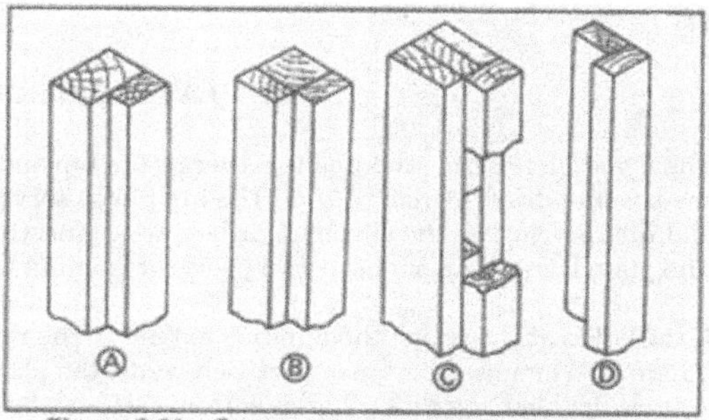

Figure 6-32. Corner-post construction using both 2-inch and 4-inch lumber

In special cases (for example, where partition walls cross), a double T-post is used. It is made as described above, and by nailing another 2 x 4 to the opposite wide side, as shown in Figure 6-34, A, B, and C (C is the most common).

STUDS

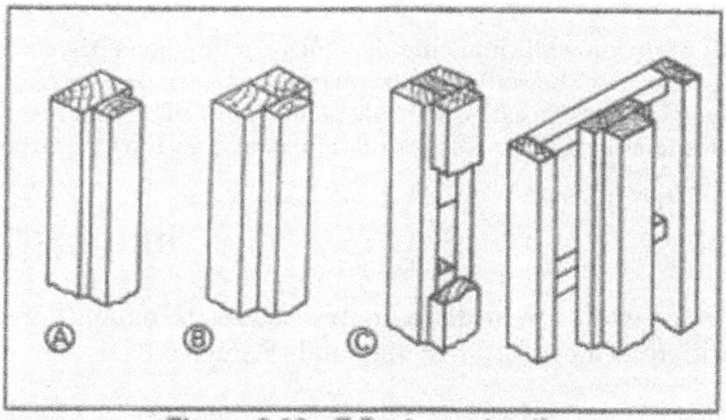

Figure 6-33. T-Post construction

After the sills, plates, and braces are in place and the window and door openings are laid out, the studs are placed and nailed with two 16d or 20d nails through the plates. The remaining studs are laid out on the sills or soles by measuring, from one corner, the distances the studs are to be set apart. Studs are normally spaced 12, 16, or 24 inches on center, depending upon the outside and inside finish material. If vertical siding is used, studs are set wider apart since the horizontal girts between them provide a nailing surface.

To double the post of the door opening, the outside studs are first placed into position and nailed securely. Then short studs (or *trimmers)* the size of the vertical opening are cut and nailed to the inside face of the outside studs as shown in Figure 6-35 on the sole plate.

The sill of a window opening must be framed. This sill is specified single or double. When it is double, the top header is nailed to the opening studs at the correct height and the trimmer next. The sill headers are toenailed to the trimmer. The door header is framed as shown in Figure 6-35. The jack stud rests on the sole plate.

GIRTS

Girts are always the same width as the studs and are flush with the face of the stud, both outside and inside. They are used in hasty construction when the outside walls are covered with vertical siding. Studs are placed from 2 to 10 feet apart, with girts spaced about 4 feet apart, running horizontally between them. The vertical siding acts in the same way as the studs and helps carry the weight of the roof. This type of construction is used extensively in the TO.

TOP PLATE AND SOLE PLATE

The *top plate* ties the studding together at the top and forms a finish for the walls. It supports the lower ends of the roof rafters. The top plates serve as connecting links between the wall and the roof, just as the sills and girders are connecting links between the floors and the walls. The plate is made up of one or two pieces of framing lumber the same size as the studs.

If the studs at the end of the building extend to the rafters, no plate is used at the end of the building. When used on top of partition walls, the plate is sometimes called the cap. Where the plate is doubled, the first plate or bottom section is nailed with 16d or 20d nails to the top of the corner posts and to the studs. The connection at the corner is made as shown in Figure 6-36. After the single plate is nailed securely and the corner braces are nailed into place, the top part of the plate is nailed to the bottom section with 10d nails. The plate may be nailed over each stud or spaced with two nails every 2 feet. Care must be taken to make sure all joints are staggered. The edges of the top section and the corner joints are lapped.

All partition walls and outside walls are finished either with a 2 x 4 or with a piece of lumber the same thickness as the wall. This lumber is laid horizontally on the floor or joists. It carries the bottom end of the studs, and is called the *sole* or *sole plate.* The sole should be nailed with two 16d or 20d nails at each joist it crosses. If it is laid lengthwise on top of a girder or joist, it should be nailed with two nails every 2 feet.

BRIDGING

Frame walls are bridged, in most cases, to make them more sturdy. There are two methods of bridging—horizontal or diagonal (Figure 6-37).

Horizontal

Horizontal bridging is nailed between the studs horizontally and halfway between the sole and top plates. This bridging is cut to lengths that correspond to the distance between the studs at the bottom. Such bridging not only stiffens the wall but also helps straighten studs.

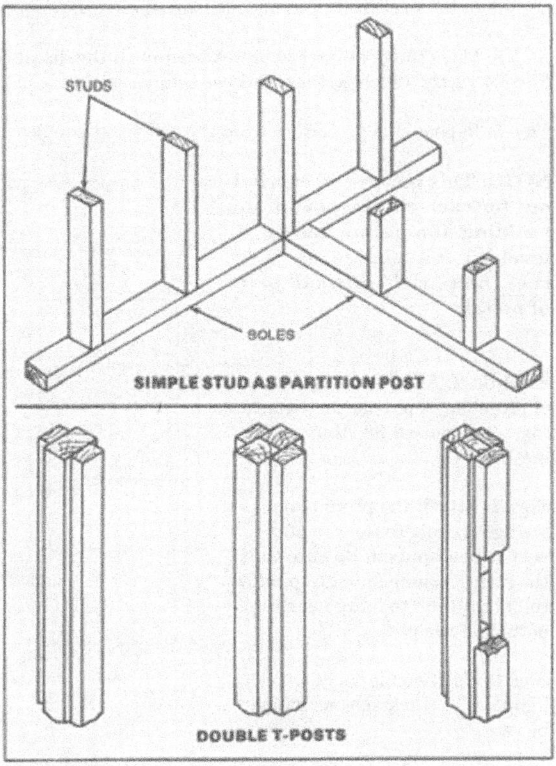

Figure 6-34. Partition posts

Diagonal

Diagonal bridging is nailed between the studs at an angle. It is more effective than the horizontal type since it forms a continuous truss and keeps the walls from sagging. Whenever possible, interior partitions and exterior walls should be bridged alike.

PLUMBED POSTS AND STRAIGHTENED WALLS

After the corner post, T-post, and intermediate wall studs have been nailed to the plates or girts, the walls must be plumbed and straightened so that permanent braces and rafters may be installed. This is done by using a level or plumb bob and a chalk line.

Plumbing Posts

There are two methods for plumbing posts.

Method 1. To plumb a corner with a plumb bob—

Step 1. Attach a string to the bob. The string should be long enough to extend to or below the bottom of the post.

Step 2. Lay a rule on top of the post so that 2 inches of the rule extend over the post on the side to be plumbed.

Step 3. Hang the bob line over the rule so that the line is 2 inches from the post and extends to the bottom of it, as shown in Figure 6-38.

Step 4. With another rule, measure the distance from the post to the center of the line at the bottom of the post. If it does not measure 2 inches, the post is not plumb.

Step 5. Move the post inward or outward until the distance from the post to the center of the line is exactly 2 inches, then nail the temporary brace in place.

Step 6. Repeat this procedure from the other outside face of the post. The post is then plumb.

NOTE: This process is carried out for each corner post of the building. If a plumb bob or level is not available, use a rock, half-brick, or small piece of metal.

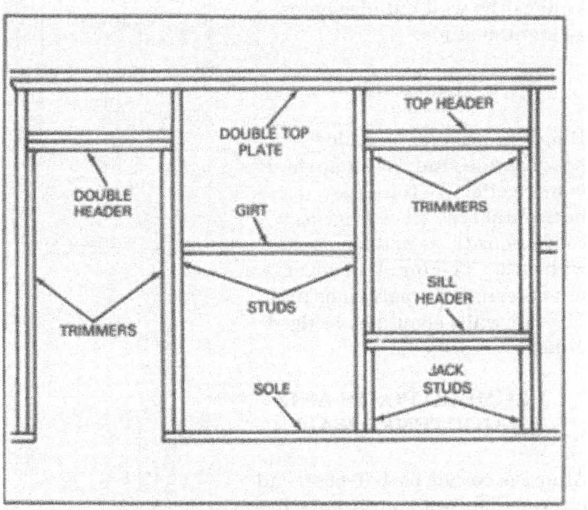

Figure 6-35. Door and window framing

Method 2. An alternative method of plumbing a post is shown in the inset in Figure 6-38. 1b use this method—

Step 1. Attach the plumb-bob string securely to the top of the post to be plumbed. Be sure that the string is long enough to allow the plumb bob to hang near the bottom of the post.

Step 2. Use two blocks of wood, identical in thickness, as gauge blocks.

Step 3. Tack one block near the top of the post between the plumb-bob string and the post

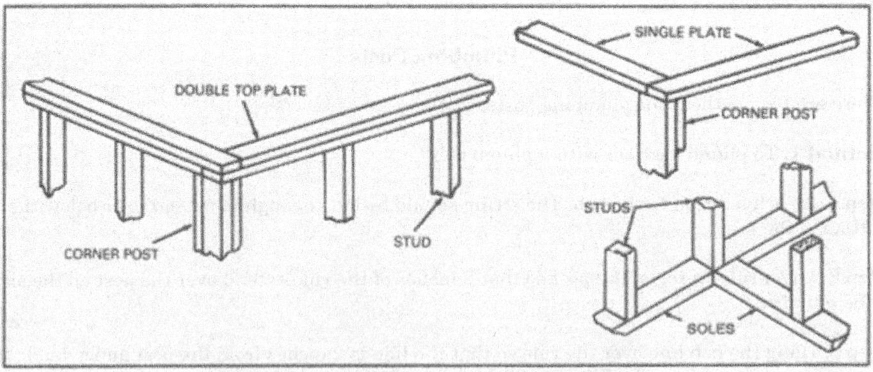

Figure 6-36. Top-plate and sole-plate construction

(gauge block 1).

Step 4. Insert the second block between the plumb-bob string and the bottom of the post (gauge block 2).

Step 5. If the entire face of the second block makes contact with the string, the post is plumb.

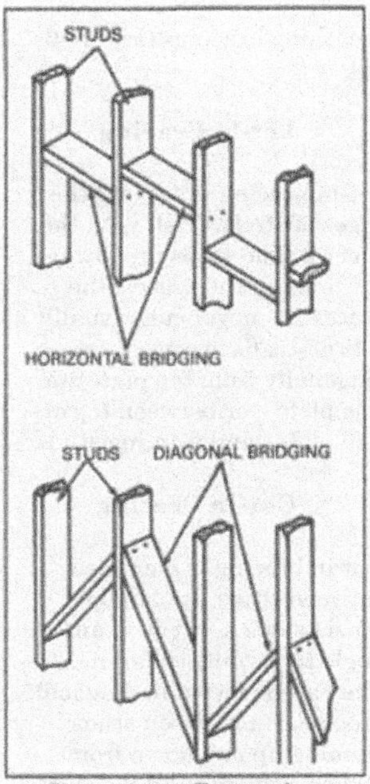

Figure 6-37. Types of wall bridging

Straightening Walls

The following procedure is carried out for the entire perimeter of the building. Inside partition walls should be straightened the same way (Figure 6-39).

Step 1. Plumb one corner post with a level or a plumb bob. Nail temporary braces to hold the post in place. Repeat this procedure for all corner posts.

Step 2. Fasten a chalk line to the outside of one corner post at the top and stretch the line to the corner post at the opposite end of the building. Then fasten the line to this post.

Step 3. Place a 3/4-inch block under each end of the line for clearance.

Step 4. Place temporary braces at intervals small enough to hold the wall straight.

Step 5. Nail the brace when the wall is far enough away from the line to permit a 3/4-inch block to slide between the line and the plate.

BRACING

Bracing is used to stiffen framed construction and make it rigid. Bracing may be used to resist winds, storms, twists, or strains. Good bracing keeps corners square and plumb. Bracing prevents warping, sagging, and shifting that could otherwise distort the frame and cause badly fitting doors and windows and cracked plaster. The three methods commonly used to brace frame structures are let-in, cut-in, and diagonal-

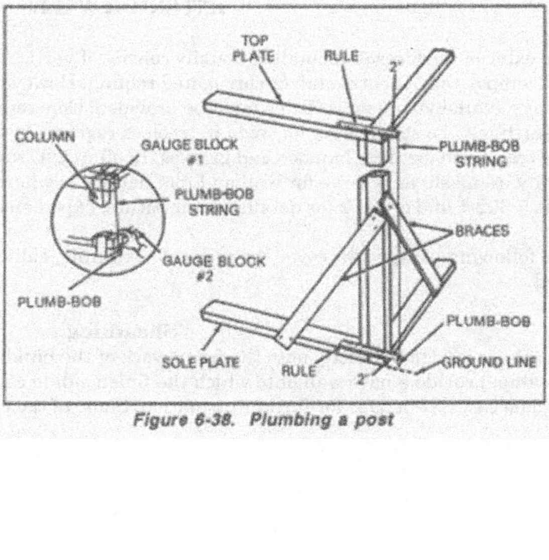

Figure 6-38. Plumbing a post

sheathing bracings (Figure 6-40).

Let-In Bracing

Let-in bracing is set into the edges of studs, flush with the surface. The studs are always cut to let in the braces; the braces are never cut. Usually 1 x 4s or 1 x 6s are used, set diagonally from top plates to sole plates, or between top or sole plates and framing studs.

Cut-In Bracing

Cut-in bracing is toenailed between studs. It usually consists of 2 x 4s cut at an angle to permit toenailing. They are inserted in diagonal progression between studs running up and down from corner posts to the sill or plates.

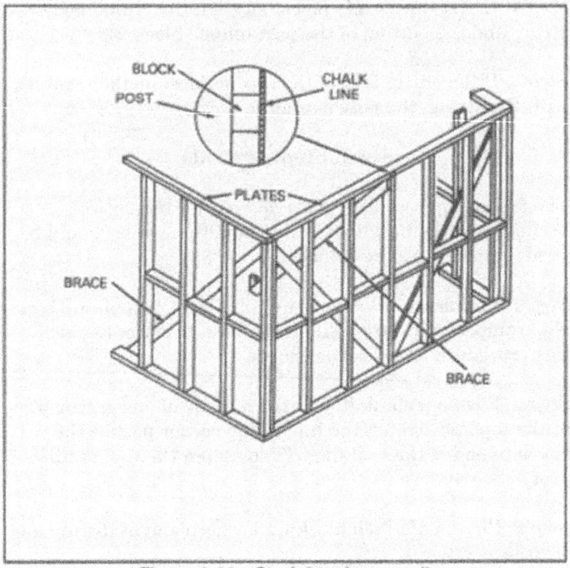

Figure 6-39. Straightening a wall

Diagonal-Sheathing Bracing

The strongest type of bracing is diagonal sheathing. Each board braces the wall. If plywood sheathing 5/8 inch thick or more is used, other methods of bracing may be omitted.

EXTERIOR WALLS

The exterior surfaces of a building usually consist of vertical, horizontal, or diagonal sheathing and composition, sheet metal, or corrugated roofing. However, in TOs, those materials are not always available and substitutes must be provided. Concrete blocks, brick, stone rubble, metal, or earth may be substituted for wood in treeless regions. In the tropics, improvised siding and roofs can be made from bamboo and grasses. Roofing felt, sandwiched between two layers of light wire mesh, may serve for wall and roof materials where the climate is suitable. Refer to TMs 5-302-1 and 5-302-2 for details on substitute, expedient, and improvised construction.

The following paragraphs cover the types of sheathing, siding, and building paper that may be used.

Sheathing

Sheathing is nailed directly onto the framework of the building. It is used to strengthen the building; provide a base wall onto which the finish siding can be nailed; act as insulation; and, in some cases, be a base for further insulation. Some of the common types of sheathing are—

- Wood, 1 inch thick by 6, 8, 10, or 12 inches wide of No. 1 common square or matched-edge material.
- Gypsum board, 1/2 inch thick by 4 feet wide and 8 feet long.
- Fiberboard, 25/32 inch thick by 2 or 4 feet wide and 8, 9, 10, or 12 feet long.
- Plywood, 1/4, 3/8, 1/2, or 5/8 inches thick by 4 feet wide and 8, 9, 10, or 12 feet long.

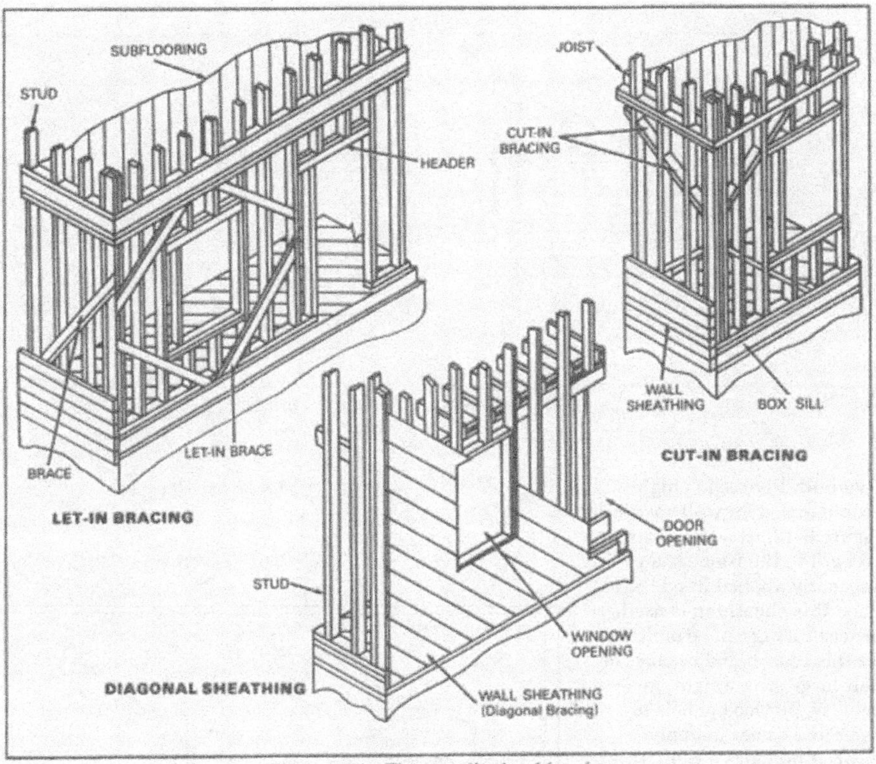

Figure 6-40. Three methods of bracing

Wood. Wood wall sheathing comes in almost all widths, lengths, and grades. However, it is normally 6 to 12 inches wide and 3/4 inch thick. Lengths are selected for economical use and the sheathing is either square- or matched-edge. Sheathing 6 or 8 inches wide should be nailed with two 8d nails at each stud crossing. Wider boards should be nailed with three 8d nails. It is laid on tight, with all joints made over the studs. It may be nailed on horizontally or diagonally (Figure 6-41); however, diagonal application adds much greater strength to the structure. If the sheathing is to be put on horizontally, start at the foundation and work toward the top. If it is to be put on diagonally, start at a bottom corner of the building and work toward the opposite wall.

Gypsum Board. This type of sheathing is made by casting a gypsum core into a heavy, water-resistant, fibrous envelope. The long edges of the 4 x 8 boards are tongue and grooved. Each board is 1/2 inch thick. Gypsum board is generally used with wood siding. Gypsum board should be nailed with 13/4- or 2-inch galvanized roofing nails spaced 7 inches on center. Gypsum board can be nailed (together with the wood siding) directly to the studs (Figure 6-42). Gypsum sheathing is fireproof, water resistant, and windproof. It does not warp or absorb water and does not require the use of building paper.

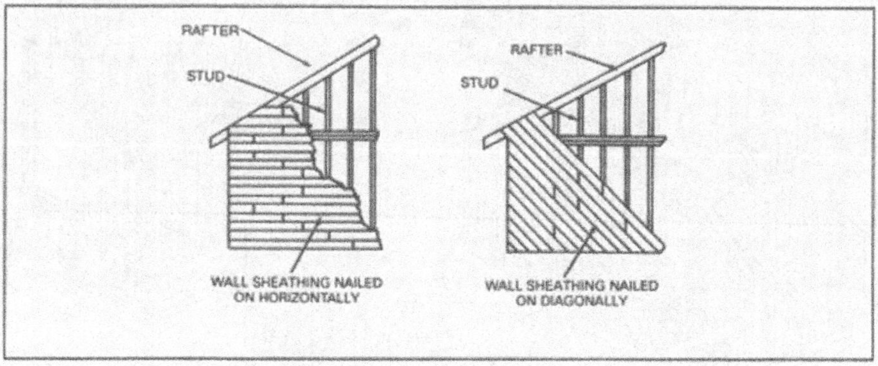

Figure 6-41. Horizontal and diagonal sheathing

Plywood. Plywood is highly recommended for wall sheathing (Figure 6-42). It adds a lot more strength to the frame than using diagonally applied wood boards. When this sheathing is used, corner bracing can be omitted. For this reason and because of their large size, weight, and stability, plywood panels are faster and easier to apply. Plywood provides a tight, draft-free installation of high insulation value.

The minimum thickness of plywood wall sheathing should be 1/4 inch for 16-inch stud spacing, and 3/8 inch for 24-inch stud spacing. The panels should be installed with the face grain **parallel** to the studs. A little more stiffness can be gained by

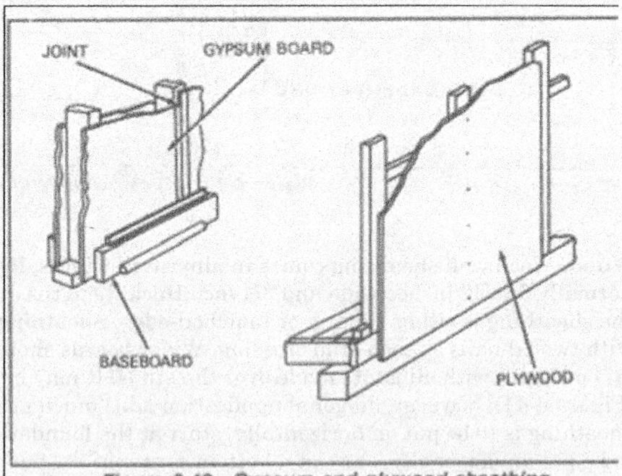

Figure 6-42. Gypsum and plywood sheathing

installing them across the studs, but this requires more cutting and fitting. Use 6d common nails for 1/4-, 3/8-, and 1/2-inch panels. At the edges of the panels, space the nails not more than 6 inches on center; elsewhere, not more than 12 inches on center.

Siding

The siding for exterior walls should be decay-resistant, hold tight at the joints, and take and hold paint well.

Wood Siding. Wood siding should be decay-resistant, well-seasoned lumber. It should hold tight at the joints and take and hold paint well. It ranges from 1/2 to 3/4 inch thick by 12 inches wide.

Vertical Wood Siding. Vertical wood siding (Figure 6-43) is nailed securely to girts with 8d or 10d nails. The cracks are covered with wood strips called battens. To make this type of wall more weatherproof, some type of tar paper or light roll roofing may be applied between the siding and the sheathing.

Horizontal Wood Siding Horizontal wood siding (Figure 6-43) is cut to various patterns and sizes to be used as the finished outside surface of a structure. There are two types bev*eled siding* and *drop siding* (Figure 6-43).

Beveled. Beveled siding is made with beveled boards, thin at the top edge and thick at the butt. It is the most common form of wood siding. It comes in 1 inch for narrow widths, and 2 inches and over for wide types. Beveled siding is usually nailed at the butt edge, through the top edge of the board below. Very narrow siding is quite often nailed near its thin edge, like shingles. It is nailed to solid sheathing over which building paper has been attached. Window and door casings are first framed. The siding butts are put against the edges of these frames. Corners may be mitered, or the corner boards may first be nailed to the sheathing. Siding is then fitted against the edges.

Drop. Drop siding is used as a combination of sheathing and siding or with separate sheathing. It comes in a wide variety of face profiles and is either shiplapped or tongue and grooved. If used as a combined sheathing and siding material, tongue-and-groove lumber is nailed directly to the studs with the tongue up. When sheathing is not used, the door and window casings are set after the siding is up. If sheathing is first used and then building paper is added, drop siding is applied with beveled siding, **after** the window and door casings are in place.

Corrugated-Metal Siding. Corrugated metal is often used as wall cover since it requires little framing, time, and labor. It is applied vertically and nailed to girts. Nails are placed in the ridges. Sheathing can be used behind the iron with or without building paper. Since tar paper used behind metal will cause the metal to rust, a resin-sized paper should be used.

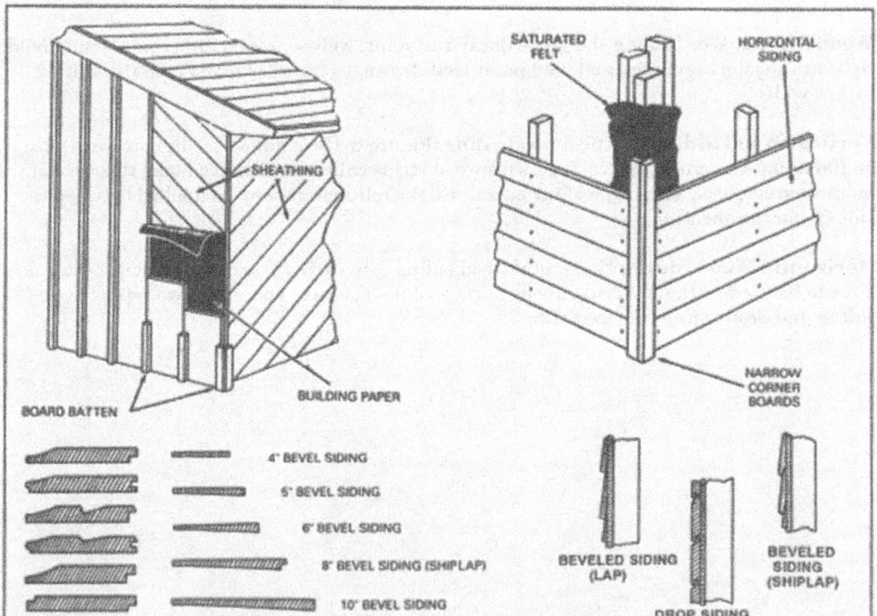

Figure 6-43. Vertical and horizontal wood siding

Vinyl and Aluminum Siding. Vinyl and aluminum sidings are popular, low-maintenance, low-cost wall covering. They can be backed with polystyrene or other board reinforcement, both to give the siding a strong base and an insulating R factor value.

Figure 6-44, page 6-34, shows the most common vinyl and aluminum sidings and installation accessories. Some variations exist between manufacturers, but the basic rules for installation are universal. They are—

- Nail in the center of slots.
- Do not nail tightly, leave loose for contracting and expanding.
- Leave at least 1/4-inch clearance at all stops.
- Do not stretch tight.
- Strap and shim all uneven walls.

Step 1. Place outside and inside corners with the bottom of the trim even with the area to weatherproof. Use a level or transit to maintain a constant horizontal line.

Step 2. Using a level, transit, or chalk line, place the bottom of the starter strip level with the bottom of the corner trim. The starter strip will butt the edge of the trim.

Step 3. Snap the bottom of the siding onto the starter strip and slide it tight into the corner trim, then out 1/4 inch to allow for expansion and contraction of materials in changing temperatures. Nail with roofing nails, 16 inches on center, in the center of the slot, without driving the nails home (leaving approximately 1/16 of an inch between the nailhead and the wall sheathing).

Step 4. Attach additional pieces of siding in a like manner, except the additional pieces will be placed on top of earlier placed pieces (as top end and bottom ends, like male and female ends in tongue-and-groove materials). Lap tight, then pull away 1/4 inch.

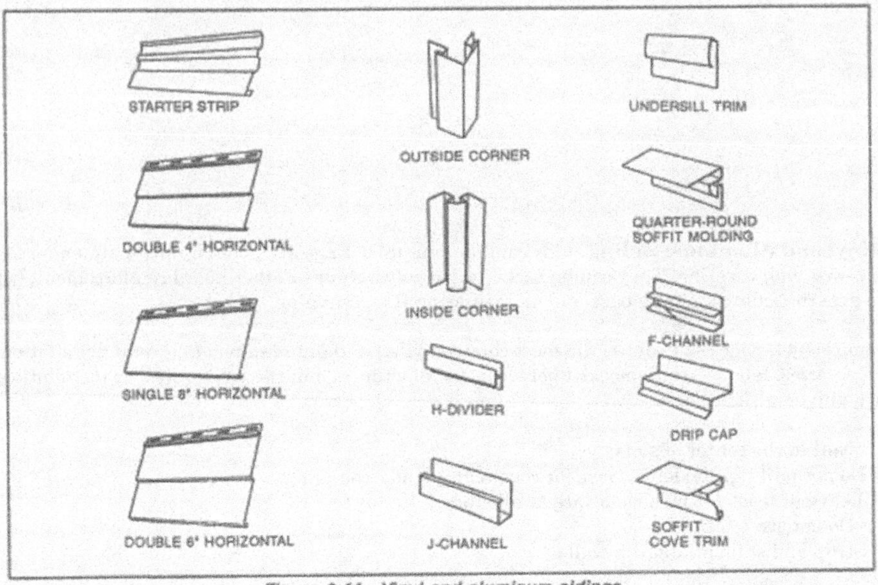

Figure 6-44. Vinyl and aluminum sidings

NOTE: When ending a "run" into a corner or a J-channel, cut so that installation is 1/4 inch from butt to trim.

Step 5. Install J-channels on surfaces where the siding run breaks (as in a window or door). The J-channel provides a weatherproof surface.

On vertical breaks, butt the siding as described in the previous paragraph. On horizontal breaks, install the undersill trim inside the J-channel. The undersill trim is a fastening device. On surfaces such as the top of the door or window trim, this channel will hold pieces of siding that were cut (removing the part of siding that "snaps" onto the top of previous pieces) tight, keeping them from flapping in the breeze. On surfaces such as where the siding butts into the soffit or below windows, a slotting tool is used to pierce the part of the siding that slides into the undersill trim. This pierce pushes out part of the siding and forms a catch. This is used when siding is pushed into the undersill trim, providing fastening for the top of a section of siding where nailing is not possible.

Building Paper

Building paper comes in several types. The most common type is resin-sized. It is generally red or buff in color (sometimes black). It comes in rolls, usually 36 inches wide; each roll containing 500 square feet and weighing 18 to 50 pounds. Normally, this building paper is not waterproof.

Another type is heavy paper saturated with a coal-tar product, sometimes called sheathing paper. This type is waterproof and insulates against heat and cold. In wood-frame buildings to be covered with siding, shingles, or iron, building paper is used to protect against heat, cold, or dampness. Building paper is applied horizontally from the bottom of the wall, and nailed with roofing nails at the laps. Overlapping the paper helps water runoff. Care must be taken not to tear the paper. The waterproof paper is also used in the built-up roof when the roof is nearly flat. Several layers are used, with tar between the layers.

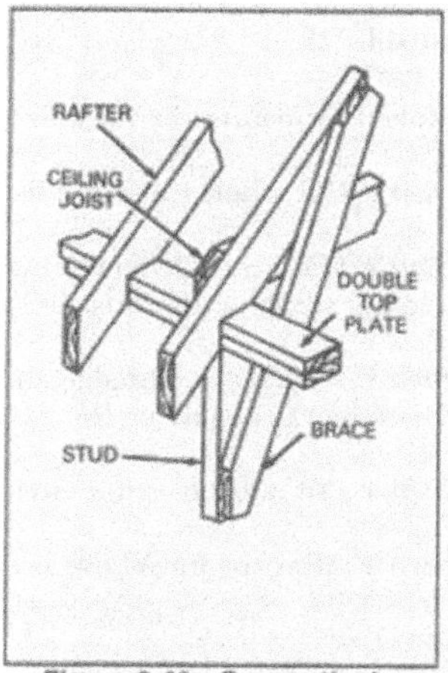

Figure 6-46. One method of bracing attic plate

CEILINGS

Ceiling joists form the framework of the ceiling of the room. They are usually lighter than floor joists but must be large enough and strong enough to resist bending and buckling.

Ceiling joists are usually installed 16 inches or 24 inches on center, starting at one side of the building and continuing across, parallel to the rafters (Figures 6-45 and 6-46). Extra joists, if needed, may be placed without affecting the spacing of the prime joists.

Selecting and installing the ceiling joists are much the same as for floor joists. Ceiling joists are nailed to both the plates and the rafters, if possible, and lapped and nailed over bearing partitions. Joists that lie beside rafters on a plate are cut at the same pitch as the rafter, flush with the top of the rafter. Joists are installed crown or camber up.

WALL OPENINGS

In addition to doors and windows (Chapter 8), other wall openings are needed.

STOVEPIPES

Stovepipes may extend outside a building through a side wall to eliminate the need for flashing and waterproofing around the pipe (Figures 6-47 and 6-48). The opening should be cut in an area selected to avoid cutting studs, braces, plates, or other structural members. Sheathing must be cut back in a radius 6 inches greater than that of the pipe. Safety thimbles or other insulation must be used on the inside and outside of the

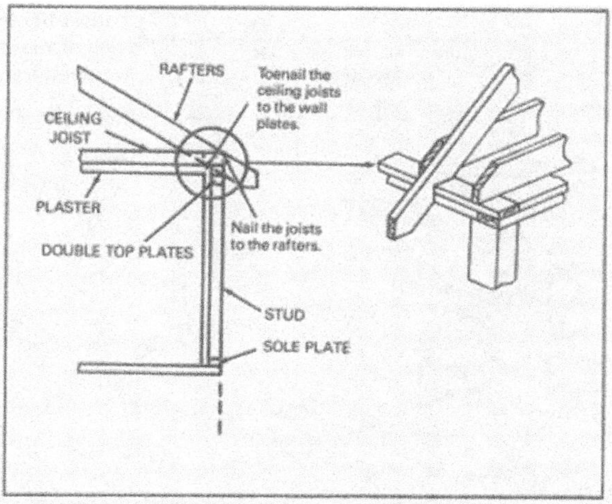

Figure 6-45. Built-up plate supporting joist and roof rafters

sheathing. Sheet metal insulation may be constructed and used as a single insulator on the outside.

Make the opening for the stovepipe as follows:

Step 1. Cut a hole through the sheet metal where the stovepipe will penetrate.

Step 2. Mark a circle on the metal 1/2 inch larger in diameter than the pipe. Then make another circle within this circle, with a diameter 2 inches less than the diameter of the first.

Step 3. With a straightedge, draw lines through the center of the circle from the circumference. These marks should be from 1/2 to 3/4 inch apart along the outer circumference.

Step 4. Cut out the center circle, then cut to the outside of the circle along the lines drawn.

Step 5. After the lines have been cut, bend the metal strips outward at a 45° angle.

Step 6. Force the pipe through the hole to the desired position.

NOTE: Very little water will leak around this joint.

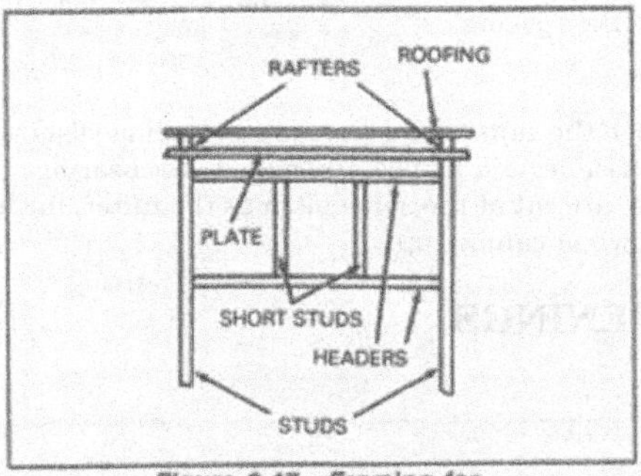

Figure 6-47. Framing for stovepipe

VENTILATORS

Ventilation is necessary to prevent condensation in buildings. Condensation may occur in the walls, in the crawl space under the structure, in the basement, on windows, and in many other places. Condensation is most likely to occur during the first six to eight months after a building is built and in extreme cold weather when interior humidity

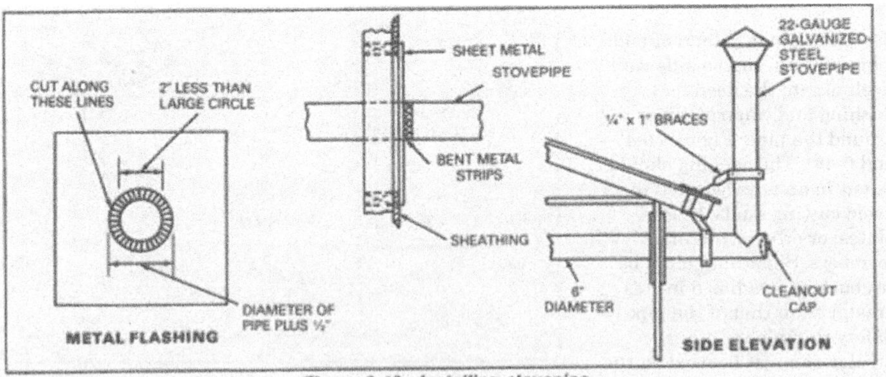

Figure 6-48. Installing stovepipe

is high. Proper ventilation under the roof allows moisture-laden air to escape during the winter heating season and allows the hot, dry air of the summer season to escape. The upper areas of a structure are usually ventilated by louvers or ventilators. (Types of ventilators are shown in Figure 6-49.)

Upper Structure

One of the most common methods of ventilating is to use wood or metal louver frames. There are many types, sizes, and shapes of louvers.

Determine the size and number of ventilators by the size of the area to be ventilated. One square foot of vent should be placed for each 150 square feet of floor space **without** soffit vents and for each 300 square feet of floor space **with** soffit vents. The minimum net open area should be 1/4 square inch per square foot of ceiling area.

Louver frames are usually 5 inches wide. The back edge of the frame should be rabbeted out for a screen, a door, or both. Louvers have ¾-inch slats, which are spaced about 1 3/4 inches apart. The slats should have sufficient slant or slope to prevent rain from driving in. For the best results, upper-structure louvers should be placed as near to the top of the gable as possible.

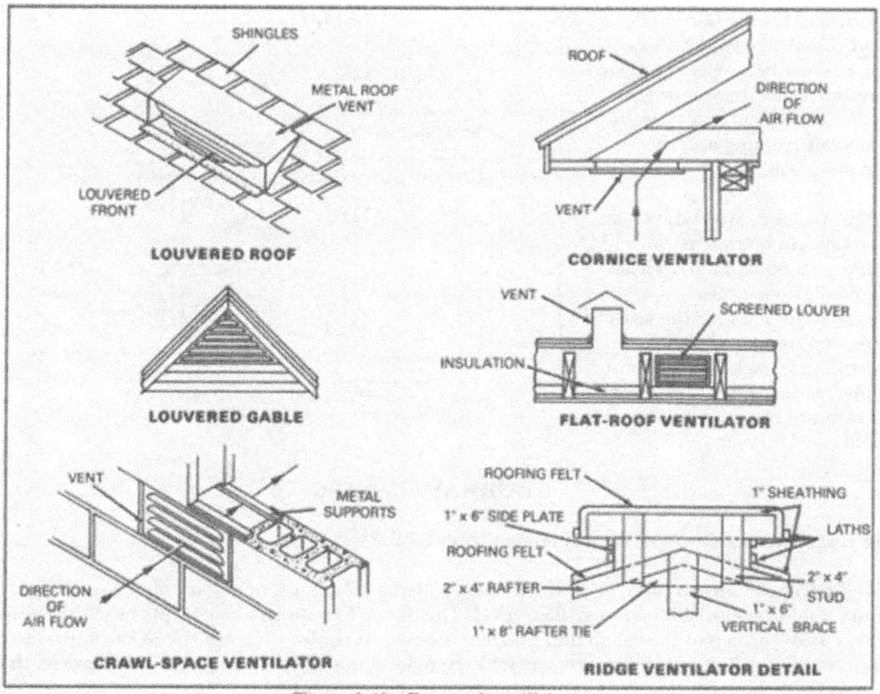

Figure 6-49. Types of ventilators

Crawl Spaces

Crawl spaces under foundations (of structures without basements) should be well ventilated. Air circulation under the flooring prevents excessive condensation which causes warping, swelling, twisting, and rotting of the lumber. The crawl space ventilators are usually called *foundation louvers.* They are set into the foundation as it is built. A good foundation vent should be equipped with a copper or bronze screen and adjustable shutters for opening and closing the louver. Louver sizes should be figured on the same basis as upper-structure louvers 1/4 square inch for each square foot of under-floor space.

STAIRWAYS

Stair work is made up of the framing on the sides, known as stringers (or carriages), and the steps known as treads. Sometimes risers are framed into the stairs at the back of the treads. The stringers (or carriages) are 2 to 3 inches thick and 8 or more inches wide. They are cut to form the step of the stairs.

There are usually three stringers to a stair if stairs are more than 36 inches wide—one at each of the two outer edges and one at the center. Floor joists must be properly framed around the stairwell (or well hole) to have enough space for the stair framing and the finished trim.

Step stringers are fastened at the top and bottom as shown in Figures 6-50 and 6-51. These figures also show the foundation and give the sizes of the step treads, as well as installation methods and post construction. The types of steps shown are the most common in field construction.

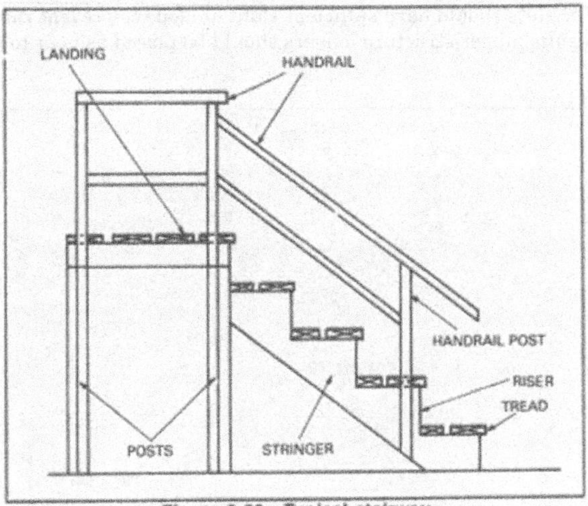

Figure 6-50. Typical stairway

STAIRWAY FRAMING

To frame simple, straight, string stairs (Figure 6-52,)—

Step 1. Take a narrow piece of straight stock, called a story pole, and mark on it the distance from the lower-floor to the upper-floor level. This is the lower-room height, plus the thickness of the floor joists and the rough and finished flooring. It is also the total rise of the stairs. Keep in mind that a flight of stairs forms a right triangle. The rise is the height of the triangle, the run is the base, and the length of the stringers is the hypotenuse.

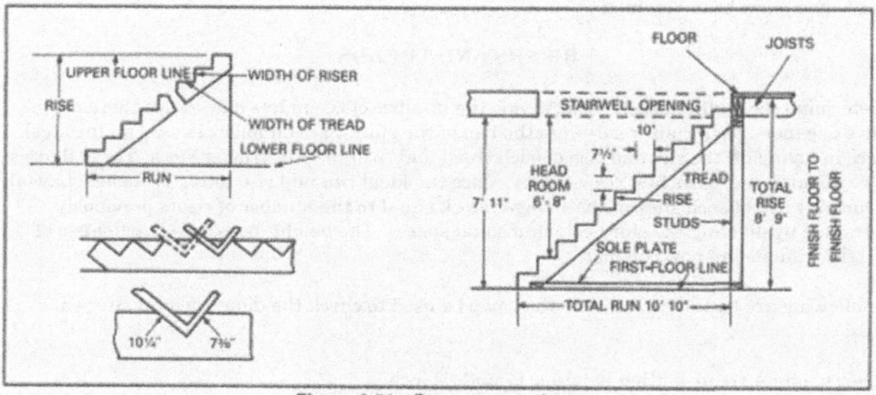

Figure 6-51. Step construction

Step 2. Set dividers at 7 inches, the average distance from one step to another.

Step 3. Step off this distance on the story pole.

Step 4. Adjust the divider span slightly if this distance will not divide evenly into the length of the story pole. Step off this distance again.

Step 5. Continue this adjusting and stepping off until the story pole is marked off evenly. 1he span of the dividers must be near 7 inches. This represents the rise of each step.

Step 6. Count the number of spaces stepped off evenly by the dividers on the story pole. This will be the total number of risers on the stairs.

Step 7. Measure the length of the stairwell opening for the length of the run of the stairs. This length may also be obtained from the plans. The stairwell-opening length forms the base of a right triangle. The height and base of the

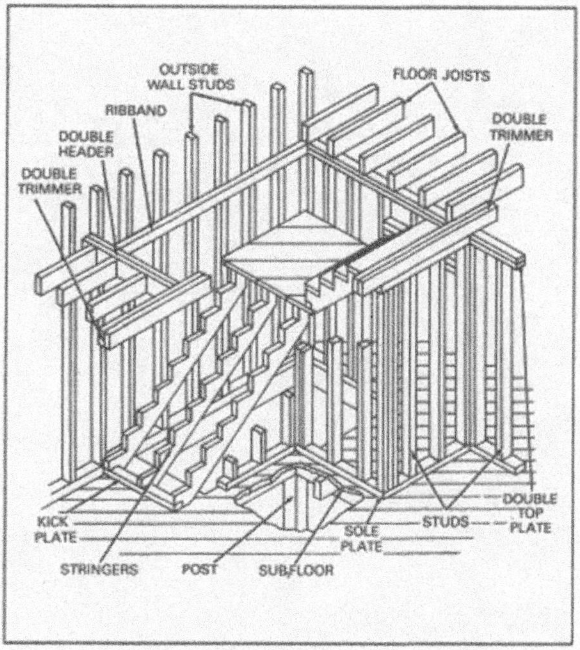

Figure 6-52. Complete stairway detail

triangle have now been obtained.

RISERS AND TREADS

To determine the width of each tread, divide the number of risers less one—since there is always one more riser than tread—into the run of the stairs. This number is used on the steel square in laying off the run and rise of each tread and riser on the stringer stock. These figures will be about 7 and 10 inches, respectively, since the ideal run and rise totals 17 inches. Lay off the run and rise of each step on the stringer stock equal to the number of risers previously determined by dividing the story pole into equal spaces. The height, base, and hypotenuse of the right triangle are now complete.

The following are two rules of thumb that may be used to check the dimensions of risers and treads:

- Riser height + tread width = between 17 and 19 inches.
- Riser height x tread width = between 70 and 75 inches.

If the sum of the height of the riser and the width of the tread falls between 17 and 19 inches and the product of the height of the riser and the width of the tread equals between 70 and 75 inches, the design is satisfactory

ROOF SYSTEMS AND COVERINGS CHAPTER 7

The roof's main purpose is to keep out the rain, cold, or heat. It must be strong enough to with stand high winds; sloped to shed water; and, in areas of heavy snow, it must be constructed more rigidly to bear the extra weight. This chapter will familiarize carpenters with the most common types of roof construction and materials. This chapter also covers reroofing.

ROOF FRAMING

Roofs for TOs are chosen to suit the building; the climate; the estimated length of time the building will be used; and the material, time, and skill required for construction. TO constraints dictate simple design as shown in Figure 7-1.

GABLE ROOF. This roof has two roof slopes that meet at the center (ridge), forming a gable. It is the most common roof because it is simple, economical, and may be used on any type of structure.	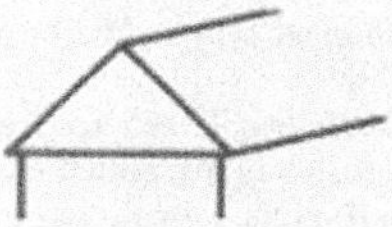
LEAN-TO OR SHED ROOF. This roof is used where hasty or temporary construction is needed and where sheds or additions to buildings are erected. The pitch of this roof is in one direction only. The roof is held up by the walls or posts on four sides. One wall, or the posts on one side, is higher than those on the opposite side.	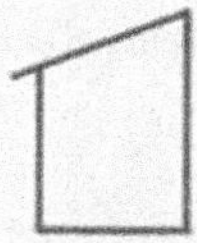
HIP ROOF. This roof has four sides or slopes running upward toward the center of the building to create a ridge (or peak). Rafters at the corners run diagonally from the bottom edge to meet at the center (ridge). Other rafters are then framed into these.	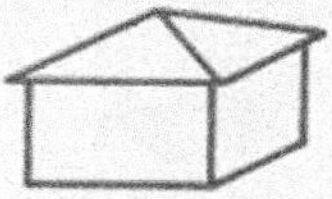
VALLEY ROOF. This roof is formed of two intersecting hip or gable roofs. The two roofs meet at a valley. Each roof slants in a different direction. This roof is seldom used, since it is complicated and requires much time and labor.	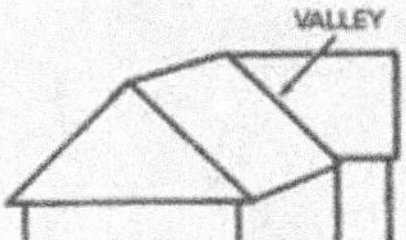

Figure 7-1. Types of roofs

ROOFING TERMS

When framing a roof (Figure 7-2), carpenters must be familiar with commonly used roofing terms (Figures 7-3 and 7-4).

RAFTERS

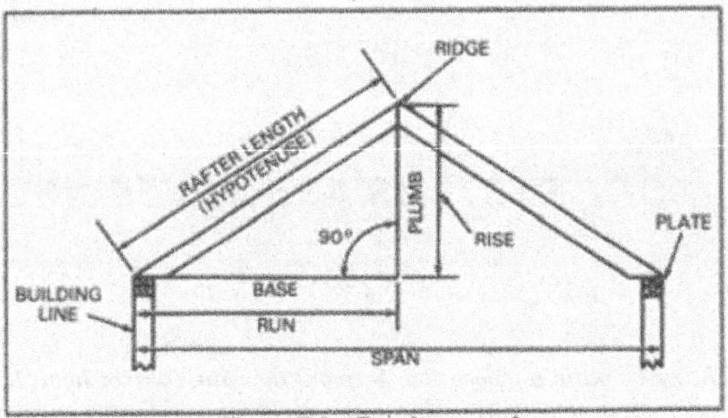

Figure 7-2. Framing a roof

Rafters make up the main framework of all roofs. They are inclined members spaced from 16 to 48 inches apart. They vary in size, depending on length and spacing. The tops of inclined rafters are fastened to the ridge or another rafter, depending on the type of roof. Rafters rest on the top wall plate.

Rafters are nailed to the plate, not framed into it. Some are cut to fit the plate, while in hasty construction they are merely laid on top of the plate and nailed in place. They may extend a short distance beyond the wall to form the eaves and protect the sides of the building.

Types of Rafters

Examples of most types of rafters are shown in Figure 7-3. The four types are—

Common Rafters. These are framing members that extend at right angles from the plate line to the roof ridge. They are called common rafters because they are common to all types of roofs and are used as the basis for laying out other types of rafters.

Hip Rafters. These are roof members that extend diagonally from the corner of the plate to the ridge.

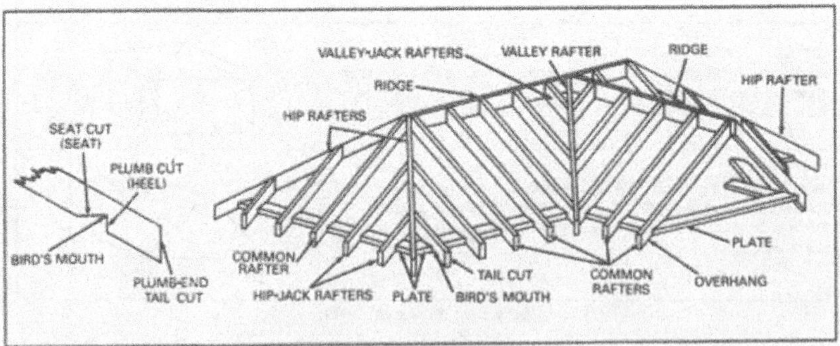

Figure 7-3. Common roof-framing terms

Basic Triangle. The basic triangle is the most elementary tool used in roof framing. When framing a roof, the basic right triangle is formed by the horizontal lines (or run), the rise (or altitude), and the length of the rafter (the hypotenuse). Any part of the triangle can be computed if the other two parts are known. Use the following equation:

The square of the hypotenuse of a right triangle is equal to the sum of the squares of the two sides. In roofing terms—

$$\text{Rafter length}^2 = \text{run}^2 + \text{rise}^2$$

Rise. The rise of a rafter is the vertical (or plumb) distance that a rafter extends upward from the plate.

Span of a Roof. The span of any roof is the shortest distance between the two opposite rafters' seats.

Cut of a Roof. The cut of a roof is the rise over the run (such as 4/12 roof) or the pitch of the roof.

Horizontal Line. A horizontal line is one level with the building foundation.

Line Length. In roof framing, line length is the hypotenuse of a triangle whose base is the run and whose altitude is the total rise.

Total Rise. The total rise is the vertical distance from the wall plate to the top of the ridge.

Run. Run always refers to the level distance any rafter covers—normally, one-half the span.

Unit of Run (or unit of measurement). Unit of measurement, 1 foot (or 12 inches), is the same for the roof as for any other part of the building. Using this common unit of measurement, the framing square is used in laying out large roofs (Figure 7-5, page 7-4).

Pitch. Pitch signifies the amount that a roof slants and the ratio of rise (in inches) to run (inches). Using this method, 4, 6, or 8 inches rise per foot of run would give a pitch of 4-12, 6-12, or 8-12. (Figure 7-6, page 7-4, shows how to determine roof pitch.)

Plumb Line. The line formed by the cord on which the plumb bob is hung.

Bird's Mouth. A cutout, near the bottom of a rafter, that fits over the top plate. The cut that fits the top of the plate is called the seat; the cut for the side of the plate is called the heel.

Overhang. The part of a rafter that extends past the outside edge of the walls of a building. When laying out a rafter, this portion is in addition to the length of a rafter and is figured separately. The overhang is often referred to as the tailpiece.

Plate. The wall-framing member that rests on the top of the wall studs.

Ridge. The highest horizontal roof member. It ties the rafters together at the upper end.

Figure 7-4. Roofing terms

Valley Rafters. These rafters extend from the plate to the ridge along the lines where two roofs intersect.

Jack Rafters. These are a common rafter. The three kinds of jack rafters are the—

- Hip jack, which extends from the plate to the hip rafter.
- Valley jack, which extends from the ridge of the valley rafter.
- Cripple jack, which is placed between a hip rafter and a valley rafter. The cripple jack rafter is also part of a common rafter, but it touches neither the ridge of the roof nor the rafter plate.

Collar Tie and Beam

A collar tie or beam (Figure 7-7) is a piece of stock (usually 1 x 4, 1 x 6, l x 8, or 2 x 4) fastened in a horizontal position to a pair of rafters between the plate and the ridge of the roof. This type of beam keeps the building from spreading. Most codes and specifications require them to be 5 feet apart or every third rafter, whichever is less. Collar ties are nailed to common rafters with four 8d nails to each end of a 1-inch tie. If 2-inch material is used for the tie, they are nailed with three 16d nails at each end. This type of bracing is used on small roofs where no ceiling joists are used

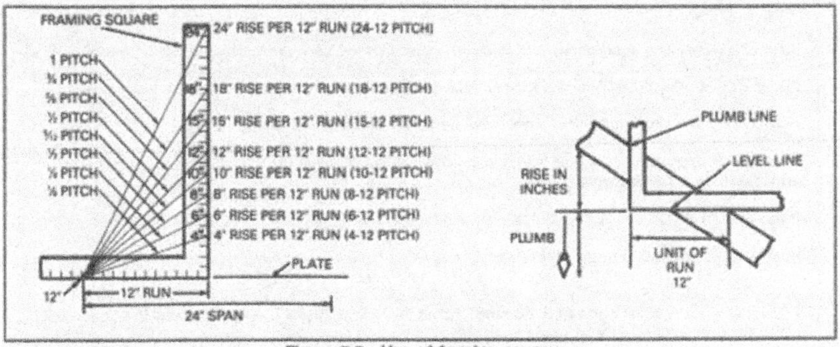

Figure 7-5. Use of framing square

and the building is not wide enough to require a truss.

In small roofs that cover only narrow buildings in which the rafters are short, there is no need for interior support or bracing. In long spans, the roof would sag in the middle if it were not strengthened in some way. To support long rafters, braces or other types of supports must be installed.

Rafter Layout

Rafters must be laid out and cut with slope, length, and overhang exactly right so that they will fit when placed in the roof.

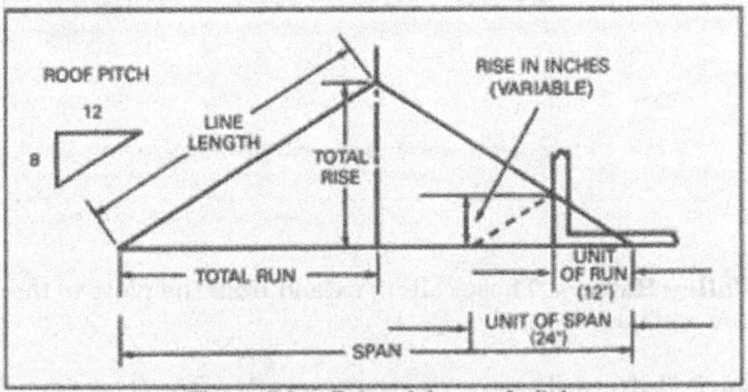

Figure 7-6. Determining roof pitch

Scale or Measurement Method. The carpenter should first determine the length of the rafter and the length of the lumber from which the rafter may be cut. If he is working from a roof plan, he learns the rafter lengths and the width of the building from the plan. If no plans are available, the width of the building must be measured.

Step 1. To determine the rafter length, first find one-half of the horizontal distance (total run) of the rafter. The amount of rise per foot will not be considered yet. (For example, if the building is 20 feet wide, half of the span will be 10 feet. See the example below.)

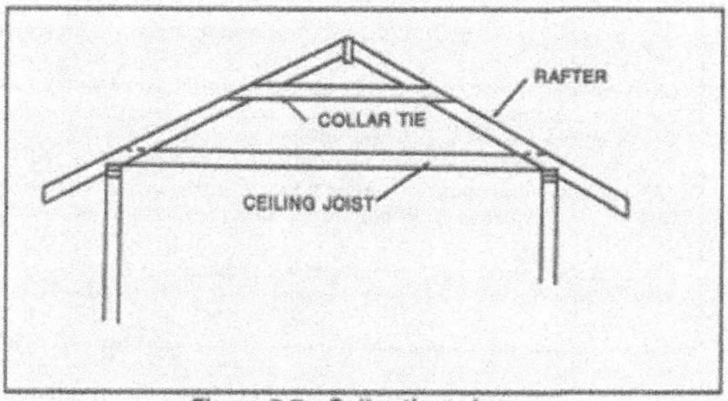

Figure 7-7. Collar tie or beam

Step 2. After the length has been determined, lay the timber on sawhorses (saw benches), with the crown or bow (if it has any) as the top side of the rafter. If possible, select a straight piece for the pattern rafter. If a straight piece is not available, have the crown away from the person laying out the rafter.

Step 3. Hold the square with the tongue in your left hand, the blade in your right, and the heel toward your body. Place the square as near the upper end of the rafter as possible. (For example, in Figure 7-8, page 7-6 (step 1) the figures 8 on the tongue and 12 on the blade are placed along the timber edge that is to be the top edge of the rafter.)

Step 4. Mark along the outside tongue edge of the square, which will be the plumb cut at the ridge.

Step 5. Since the length of the rafter is known to be 12 feet and 1/6 inch, measure the distance from the top of the plumb cut and mark it on the timber. Hold the square in the same manner with the 8 mark on the tongue directly over the 12-foot 1/6-inch mark. Mark along the tongue of the square to give the plumb cut for the seat (Figure 7-8, step 2).

Step 6. Measure off, perpendicular to this mark, the length of overhang along the timber. Make a plumb-cut mark in the same way, keeping the square on the same edge of the timber (Figure 7-8, step 3). This will be the tail cut of the rafter. Often, the tail cut is made square across the timber.

Step 7. The level cut or width of the seat is the width of the plate measured perpendicular to the plumb cut, as shown in Figure 7-8, step 4. Using a try square, square the lines down on the sides from all level and plumb-cut lines. Now the rafter is ready to be cut (Figure 7-8, step 5).

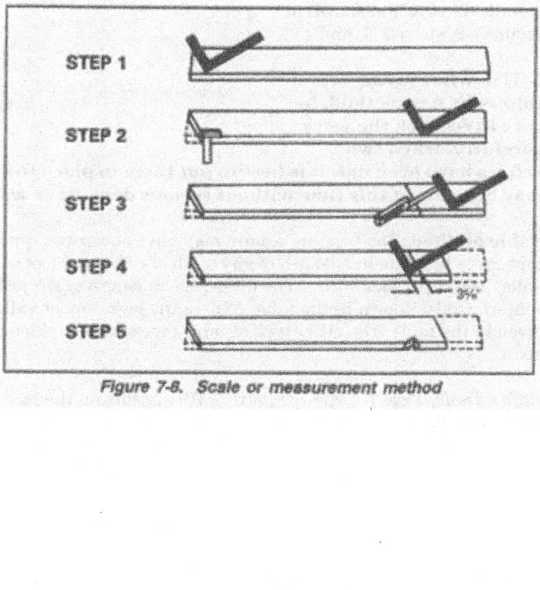

Figure 7-8. Scale or measurement method

EXAMPLE - Rafter layout, step 1: determine rafter length

As an example, use a rise per foot of 8 inches. To determine the approximate overall length of the rafter, measure on a steel carpenter square the distance between 8 on the tongue and 12 on the blade, because 8 is the rise and 12 is the unit of run. This distance is 14 5/12 inches. This represents the line length of a rafter with a total run of 1 foot and a rise of 8 inches. Since the run of the rafter is 10 feet, multiply 10 by the line length for 1 foot. The answer is 144 2/12 inches, or 12 feet and 1/6 inch. The amount of overhang (normally 1 foot) must be added if an overhang is to be used. This makes a total of 13 feet, which is an odd length for timber. Use a 14-foot timber.

Rafter length = total run x line length + overhang
10 x 14 5/12 inches = 144 2/12 inches = 12 feet + 1 foot = 13 feet

Step-Off Method. The rafter length of any building may be determined by "stepping it off by successive steps with the square, as follows:

Step 1. Step off the same number of steps as there are feet in the run. For example, if a building is 20 feet 8 inches wide, the run of the rafter would be 4 inches over 10 feet.

Step 2. This 4 inches is taken care of in the same manner as the full-foot run; that is, with the square at the last step position, make a mark on the rafters at the 4-inch mark (Figure 7-9, step 1).

Step 3. With the square held for the same cut as before, make a mark along the tongue. This is the line length of the rafter. The seat cut and hangover are made as described above and shown in Figure 7-9, steps 2, 3, and 4.

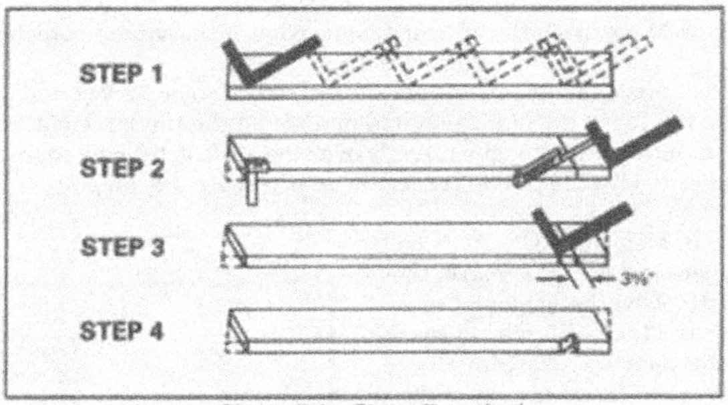

Figure 7-9. Step-off method

NOTE: When laying off rafters by any method, be sure to recheck the work carefully. When two rafters have been cut, it is best to put them in place to see if they fit. Minor adjustments may be made at this time without serious damage or waste of material.

Table Method. The framing square may have one or two types of rafter tables on the blade. One type gives both the line length of any pitch of rafter per foot of run and the line length of any hip or valley rafter per foot of run. The difference in length of the jack rafter, spaced 16 or 24 inches (on center), is also shown in the table. Where the jack, hip, or valley rafter needs side cuts, the cut is given in the table. The other type of table gives the actual length of a rafter for a given pitch and span.

Rafter Table, Type 1. Type 1 (Figure 7-10) appears on the face of the blade. This type is used to determine the length of the common, valley, hip, and jack rafters and the angles at which they must be cut to fit at the ridge and plate.

To use the table, the carpenter must first know what each figure represents.

- The row of figures in the first line represents the length of common rafters per foot of run, as the title at the left-hand end of the blade indicates.
- Each set of figures under each inch division mark represents the length of rafter per foot of run, with a rise corresponding to the number of inches over the number. (For example, under the 16-inch mark appears the number 20.00 inches. This number equals the length of a rafter with a run of 12 inches and a rise of 16 inches. Under the 13-inch mark appears the number 17.69 inches, which is the rafter length for a 12-inch run and a 13-inch rise.) See the Type 1 layout example below.

NOTE: The other five lines of figures in the table will not be discussed, as they are seldom used in the TO.

The remaining procedure for laying out the rafters after the length has been determined is as described previously.

EXAMPLE - Rafter layout, table method, Type 1

NOTE: To use the table for laying out rafters, the width of the building must first be known.

Suppose the building is 20 feet 8 inches wide and the rise of the rafters is to be 13 inches per foot of run. The total run of the rafter will be 10 feet 4 inches (or 10 1/3 feet). Look at the first line of Figure 7-10. Under the 13-inch mark appears the number 17.69, which is the length (in inches) of a rafter with a run of 1 foot and a rise of 13 inches. To find the length of this rafter, use the following formula:

Rafter length = total run x length of common rafter per ft of run ÷ 12

10 1/3 x 17.69 inches = 182.79 inches
182.79 inches ÷ 12 = 15 3/12 feet

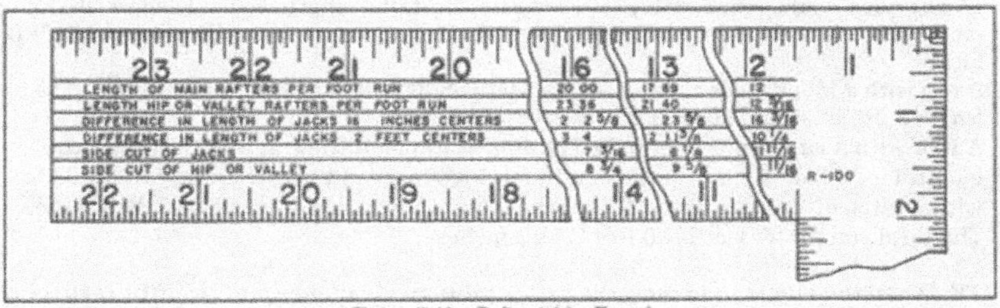

Figure 7-10. Rafter table, Type I

Rafter Table, Type 2. Type 2 (Figure 7-11, page 7-8) appears on the back of the blade of some squares. This shows the run, rise, and pitch of rafters of the seven most common pitches of roof.

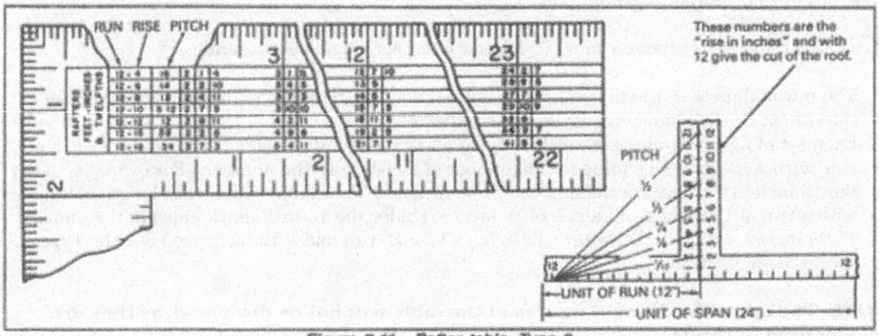

Figure 7-11. Rafter table, Type 2

The figures are based on the horizontal measurement of the building from the center to the outside.

The rafter table and the outside edge of the back of the square, both the body and tongue, are in twelfths. (The inch marks may represent twelfths of an inch or twelfths of a foot.) This table is used in connection with the marks and figures on the **outside** edge of the square. At the left end of the table are figures representing the run, the rise, and the pitch:

- In the first column, the figures are all 12 (12 inches or 12 feet). They represent the run of 12.
- The second column of figures represents various rises.
- The third column of figures (in fractions) represents the various pitches.

These three columns of figures show that a rafter with a run of 12 and a rise of 4 has a 1/6 pitch, 12 and 6 has a 1/4 pitch, and 12 and 12 has a 1/2 pitch. For example, use this scale for—

- A roof with a 1/6 pitch (or the rise of 1/6 the width of the building) and a run of 12 feet. Find 1/6 in the table, then follow the same line of figures to the right until directly beneath the figure 12. Here appear the numbers 12, 7, 1O, which is the rafter length required and which represents 12 feet 7 inches, and 10/12 of an inch. They are written as follows: 12 feet 7 10/12 inches.
- A roof with a 1/2 pitch (or a rise of 1/2 the width of the building) and a run of 12 feet. The rafter length is 16, 11, 6, or 16 feet 11 6/12 inches.
- A roof with a run of more than 23 feet. For example, if the run is 27 feet, find the length for 23 feet, then find the length for 4 feet and add the two. The run for 23 feet with a pitch of 1/4 is 25 feet 8 5/12 inches. For 4 feet, the run is 4 feet 5 8/12 inches. The total run for 27 feet is 30 feet 2 1/12 inches.

NOTE: When the run is in inches, the rafter table reads inches and twelfths instead of feet and inches.

See the Type 2 rafter table layout example at the top of the next page.

After the length of the rafter has been found, the rafter is laid out as explained previously.

EXAMPLE - Rafter layout, table method, Type 2

If the pitch is 1/2 and the run is 12 feet 4 inches, add the rafter length of a 12-foot run to that of a rafter length of a 4-inch run, as follows:

- *For a run of 12 feet and 1/2 pitch, the length is 16 feet 11 6/12 inches.*
- *For a run of 4 inches and 1/2 pitch, the length is 5, 7, 11. In this case, the 5 is inches, the 7 is twelfths, and the 11 is 11/12 of 1/12, which is nearly 1/12.*
- *Add it to the 7 to make it 8, making a total of 5 8/12 inches, then add the two lengths together.*
- *This sum is 17 feet 5 2/12 inches.*

Rafter lengths are given in the table. The overhang must be added. ***NOTE: When the roof has an overhang, the rafter is usually cut square to save time. When the roof has no overhang, the rafter cut is plumb, but no notch is cut in the rafter for a seat. The level cut is made long enough to extend across the plate and the wall sheathing. This type of rafter saves material, although little protection is given to the side wall.***

TRUSSES

A truss is a framed or jointed structure composed of straight members connected only at their intersections in such a way that if loads are applied at these intersections, the stress in each member is in the direction of its length. Straight, sound timber should be used in trusses. The types of trusses used in building construction are shown in Figure 7-12. (The Howe and Fink trusses are most commonly used.) Truss terms are listed in Figure 7-13, page 7-10.

Trusses are used for large spans to give wide, unobstructed floor space for such large buildings as shops and hangars. Sometimes small buildings are trussed to save material. These small trusses act as rafters and give the roof rigidity.

The web members of a truss divide it into triangles. The members indicated by heavy lines normally carry tensile stresses for vertical loads. Sometimes the top chords of these trusses slope slightly in one or two directions for roof drainage, but this does not change the type of truss. The necessary number of subdivisions, or panels, depends upon the length of the span and the type of construction.

Truss Supports

Trusses are supported by bearing walls, posts, or other trusses. To brace a truss to a wall or post, knee braces are used as shown in Figure 7-14. These braces tend to make a truss of the entire building by tying the wall to the roof.

Purlins

Purlins are used in roof construction to support corrugated sheet metal if it is used, or to support the sheathing of roofs framed with trusses.

- In small roofs, purlins are inserted between the rafters and nailed through the rafters.
- In large buildings where heavy trusses are used, the purlins are continuous members that rest

on the trusses and support the sheathing.

- In small buildings, such as barracks, mess halls, and small warehouses, 2 x 4s are used for

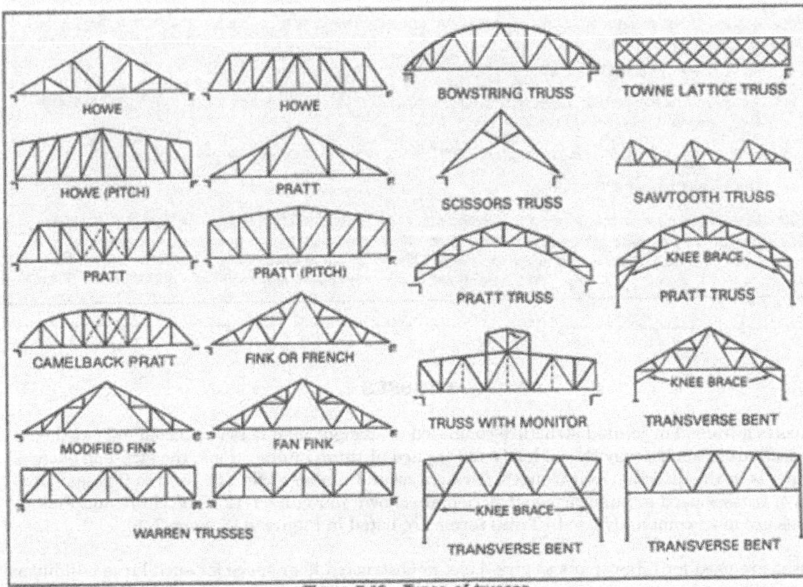

Figure 7-12. Types of trusses

purling, with the narrow side up.

Truss Layout

To lay out a truss, use Figure 7-15 and the following steps:

Step 1. Get the material to a level spot of ground where work benches will be almost level.

Step 2. Obtain from the blueprints the measurement of all pieces to be used in the truss.

Step 3. Lay out the length on the different sizes of timber and cut them accurately.

Step 4. After all lengths are cut, lay them in their correct position to form a truss.

Step 5. Nail them together temporarily.

Step 6. Lay out the location of all holes to be bored. Recheck the measurements for accuracy.

Step 7. Bore holes to the size called for on the print. Use a brace and bit or the woodborer that accompanies the air compressor. Bore holes perpendicular to the face of the timber.

Bottom Chord. A member that forms the lower boundary of the truss.

Top Chord. A member that forms the upper boundary of the truss.

Chord Member. A member that forms part of either the top or bottom chord.

Member. The component that lies between any adjacent joints of a truss; it can be of one or more pieces of structural material.

Web Member. A member that lies between the top and bottom chords.

Joint. Any point in a truss where two or more members meet; sometimes called a *panel point.*

Panel Length. The distance between any two consecutive joint centers in either the top or bottom chords.

Pitch. The ratio of the height of truss to the span length.

Height of Truss. The vertical distance at midspan from the joint center at the ridge of a pitched truss, or from the centerline of the top chord of a flat truss, to the centerline of the bottom chord.

Span Length. The horizontal distance between the centers of the two joints located at the extreme ends of the truss.

Figure 7-13. Truss terms

Step 8. After the holes have been bored, dismantle the truss and withdraw the nails.

Truss Assembly

Assembling a truss after it has been cut and bored is simple (Figure 7-16). In most cases, timber connectors are used where different members of the truss join. Straight, sound timber should be used in trusses.

- Assemble the truss with the timber connectors in place.
- Place the bolts in the holes and tighten them.
- Place washers at the head and nut ends of each bolt.

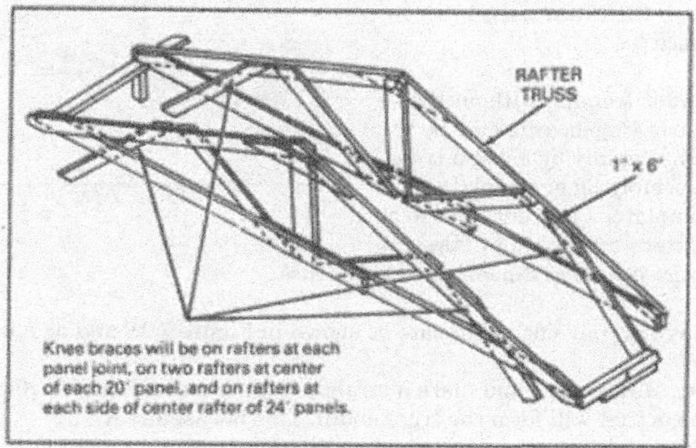

Figure 7-14. Knee braces

Rafters are usually made into trusses, as shown in Figure 7-17. Two rafters are connected at the top, using a collar tie well nailed into both rafters. Before any ties or chords are nailed, the rafters should be spread at the lower end to equal the width of the building. This is done by using a template or by measuring the distance between the seat cuts with a tape.

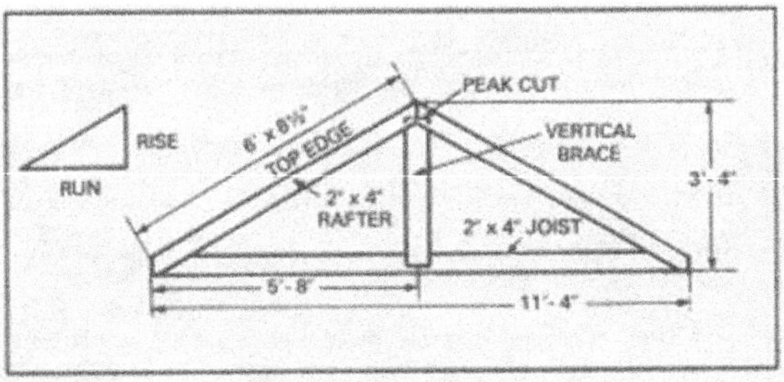

Figure 7-15. Truss layout

A 1 x 6 or 2 x 4 chord is nailed across the rafters at the seat cut to tie them together. This chord forms a truss with the two rafters. A hanger or vertical member of 1 x 6 is nailed to the rafter joint and extends to the chord at midpoint, tying the rafter to the chord.

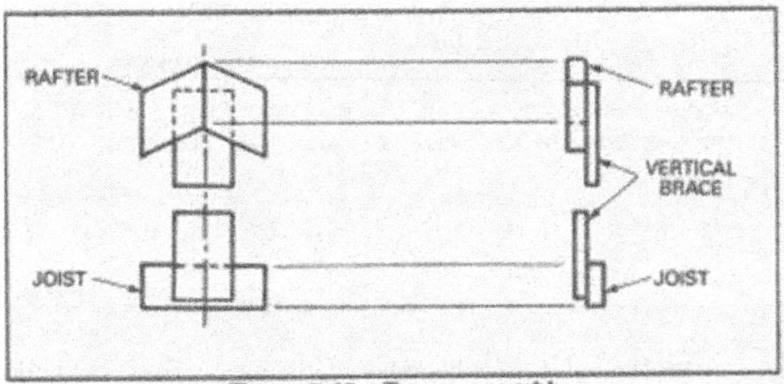

Figure 7-16. Truss assembly

In wide buildings where the joists or chords must be spliced and there is no support underneath, the rafter and joists support one another as shown in Figure 7-18.

If no additional bracing is needed, the truss is set in place on the plates. If additional bracing is needed, a knee brace is nailed to the chord. The knee brace forms a 45° angle with the wall stud. For easier erection, the knee brace may be omitted until the rafter truss is set in place.

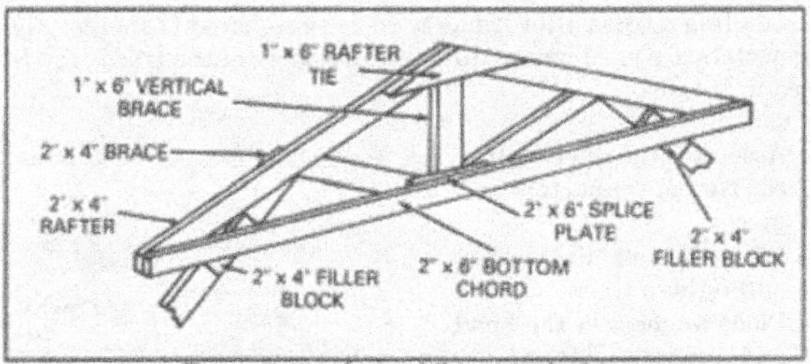

Figure 7-17. Assembling a typical truss

Rafter framing without the use of ridgeboards may be done rapidly by using a truss assembly jig or template. The template is laid out to form a pattern conforming to the exact **exterior** dimensions of the truss.

Layout. Lay out a template as shown in Figure 7-19 and as follows:

Step 1. Measure and mark a straight line on any selected surface. Mark the exact length of the joists that will form the truss chord. This is baseline A.

Step 2. From the center of the baseline and at right angles to it, lay out a centerline (C) to form the leg of a right triangle, the base of which is half the length of the baseline (A), and the hypotenuse of which (B) is the length of the rafter measured as indicated.

Step 3. Nail 2 x 4 x 8 blocks flush with the ends of baseline A and centerline C as shown in Figure 7-19. Mark the centerline on the center jig blocks.

Assembly. Assemble with a template as shown in Figure 7-19 and as follows:

Step 1. Start the assembly by setting a rafter in the jig with the plate cut fitted over the jig block at one end of the baseline. The peak is flush with the centerline on the peak jig block. Nail a holding block outside the rafter at point D.

Step 2. Lay one 2 x 4 joist or chord in place across the base blocks.

Step 3. Lay two 2 x 4 rafters in place over the joist.

Step 4. Center one end of a 1 x 6 hanger under the rafter peak. Center the rafters against the peak block.

Step 5. Nail through the rafters into the hanger using six 8d nails.

Step 6. Line up one end of the chord.

Step 7. Nail through the rafter with 16d nails.

Step 8. Line up the other end of the chord.

Step 9. Nail as above.

Step 10. Center the bottom of the hangers on top of the chord and nail with 8d nails.

Installation of Trusses

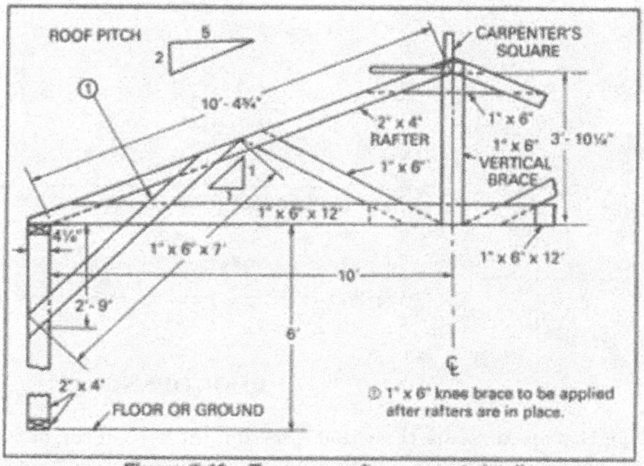

Figure 7-18. Truss- or rafter-support detail

After the rafters are assembled into trusses, they must be placed on the building (Figure 7-20, page 7-14). Assemble the first set of rafters either in the end section of the building or at the center. Raise rafter trusses into position (by hand) and nail them into place with 16d nails. (Temporary workbenches may be built for the workers to stand on while erecting trusses.) These trusses must be temporarily braced at the end section of the building until the sheathing is applied. Knee braces are not used on every rafter truss unless needed. Install trusses as follows:

Step 1. Mark the proper positions of all truss assemblies on the top plate. The marks must show the exact position on the face of all rafters (such as south or north).

Step 2. Rest one end of a truss assembly, peak down, on an appropriate mark on the top plate on one side of the structure.

Step 3. Rest the other end of the truss on the corresponding mark of the top plate on the other side of the structure.

Step 4. Rotate the assembly into position using a pole or rope.

Step 5. Line up and secure the rafter faces flush against the marks.

Step 6. Raise, align, and nail the three assemblies into position. Nail temporary 1 x 6 braces across these three assemblies. Repeat with the other assemblies as they are brought into position. Check the rafter spacing at the peaks as the braces are nailed on.

Step 7. Braces may be used as a platform when raising those trusses for which there is not enough room for rotation (Figure 7-21).

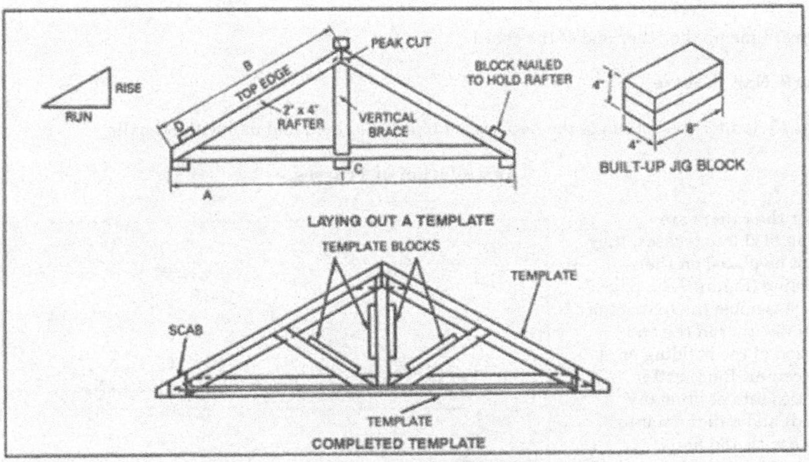

Figure 7-19. Laying out and assembling with a template

ROOF OPENINGS

Major roof openings are those that interrupt the normal run of rafters or other roof framing. Such openings may be for ventilators, chimneys, trap-door passages, or skylight or dormer windows. Figure 7-22 shows roof-opening construction.

Roof openings are framed by headers and trimmers. Double headers are used at right angles to the rafters, which are set into the headers in the same way as joists in floor-opening construction. Trimmers are actually double rafter construction in roof openings. Nailing strips may be added if needed.

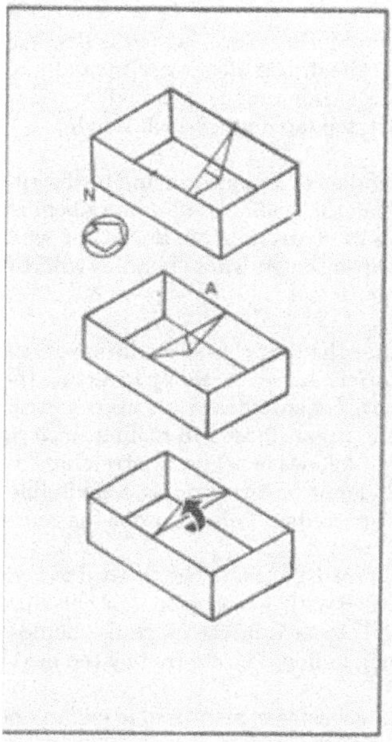

Figure 7-20. Erecting rafter trusses

ROOF DECKING

Procedures for installing plywood sheathing is similar to installing wall sheathing except it is laid perpendicular to the rafters and trusses.

ROOF COVERINGS

Asphalt and asbestos-cement roof coverings are most frequently used on pitched-roof structures. Built-up roofing is used mainly on flat or nearly flat roofs.

ASPHALT AND ASBESTOS-CEMENT ROOFING

Asphalt roofing comes in rolls (usually 3 feet wide) called *rolled roofing,* in rolled strips (usually 15 inches wide and 3 feet long), and as individual shingles. The type most commonly used is the flat strip, often called a strip shingle.

A 1 x 3 square-butt shingle is shown in Figure 7-23. This shingle should be laid 5 inches to *the weather,* meaning that 7 inches of each course should be overlapped by the next higher course. The lower, exposed end of a shingle is called the *butt.* The shingle shown in Figure 7-23 has a square butt divided into three tabs. Various other butt shapes are manufactured. Asbestos-cement roofing usually consists of individual shingles.

Laying Asphalt Roofing

The first step in covering a roof is to erect a scaffold to a height which will bring the eaves about waist-high to a man standing on the scaffold.

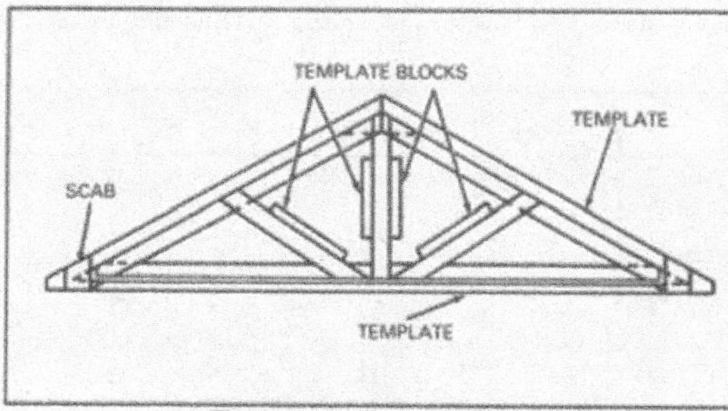

Figure 7-21. Temporary bracing

Before any roof covering is applied, the roof sheathing must be swept clean and carefully inspected for irregularities, cracks, holes, or other defects. No roofing should be applied unless the sheathing boards are absolutely dry.

An underlay of roofing felt is first applied to the sheathing. Roofing felt usually comes in 3-foot-wide rolls and should be laid with a 2-inch top lap and a 4-inch side lap.

Bundles of shingles should be distributed along the scaffold before work begins. There are 27 strips in a bundle of 1 x 3 asphalt strip shingles. Three bundles will cover 100 square feet.

After the first course at the eaves (the *starter course) is* laid by inverting the first course of shingles or the starter strip of mineral-surfaced roll roofing, each course that follows is begun by stretching a guideline or by snapping a chalk line from edge to edge. This positions the course.

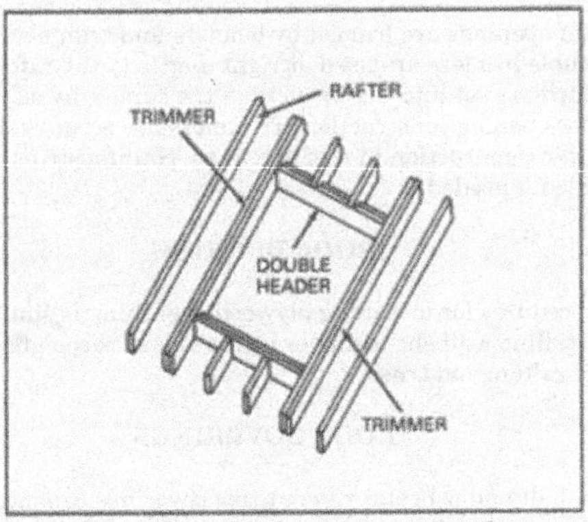

Figure 7-22. Roof-opening construction

Figure 7-24 shows the method of laying a 1 x 3 asphalt strip-shingle roof. Strip shingles should be nailed with 1-inch copper or hot-dipped, galvanized roofing nails, two to each tab; this means six nails to each full strip. Nails should be placed about 6 1/2 inches from the butt edges to ensure that each nail will be covered by the next course *(blind-nailing)* and driven through two courses.

An asbestos-cement roof is laid in about the same way as the asphalt strip shingles.

Applying Shingles at Hips and Valleys

One side of a hip or valley shingle must be cut at an angle to obtain an edge line that will match the line of the hip or valley rafter. One way to cut these shingles is to use a pattern made as follows:

Select a piece of 1 x 6 material about 3 feet long. Determine the unit length of a common rafter in the roof. Set the framing square back up on the piece to the unit run of a common rafter on the tongue and the unit length of a common rafter on the blade, as shown in Figure 7-25, A. Draw a line along the tongue. Saw the piece along this line and use it as a pattern to cut the shingles as

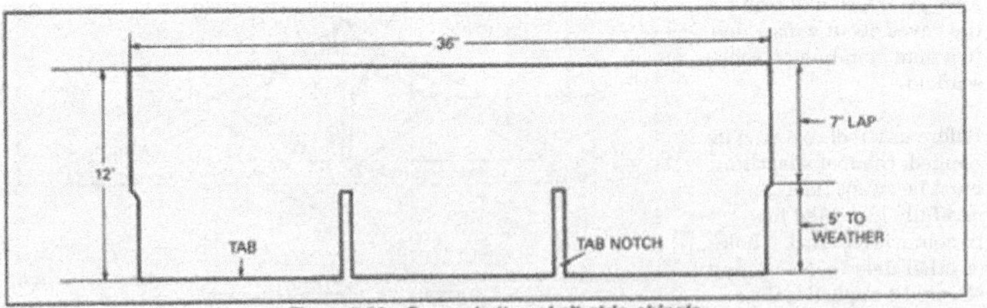

Figure 7-23. Square-butt asphalt strip shingle

shown in Figure 7-25, B.

Installing Flashing

Places especially susceptible to leakage in roofs and outside walls are made watertight by the installation of *flashing.* Flashing is sheets or strips of a watertight, rustproof material (such as galvanized sheet or sheet copper alloy for valleys and felt for hips). Flashing deflects water from places that are susceptible to leakage. The places in a roof most subject to leakage are the lines along which adjoining roof surfaces intersect (such as the lines followed by ridges, hips, and valleys) and the lines of intersection between roof surfaces and the walls of dormers, chimneys, skylights, and the like.

Ridge lines and hip lines naturally tend to shed water; therefore, they are only moderately subject to leakage. A strip of felt paper usually makes a satisfactory ridge or hip flashing. The ridge or hip is then finished as shown in Figure 7-26. Squares are made by cutting shingles into thirds. The squares are then blind-nailed to the ridge or hip as shown in

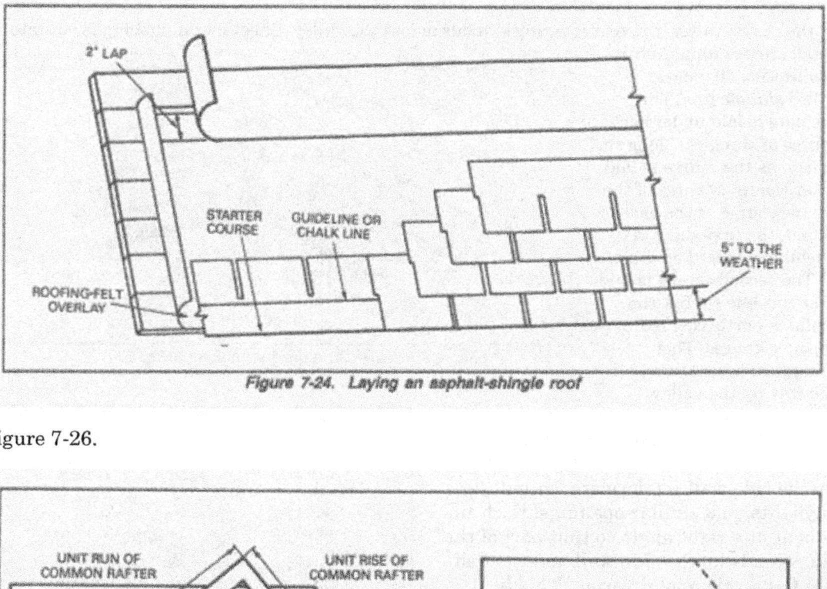

Figure 7-24. Laying an asphalt-shingle roof

Figure 7-26.

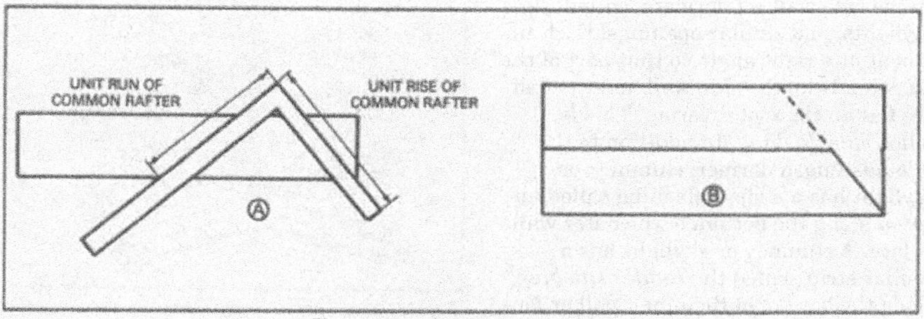

Figure 7-25. Laying out a pattern

Since water gathers in the roof valleys, they are more subject to leakage. Valley flashing varies with the manner in which the valley is to be finished. There are two common types of valley finish: the *open valley* and the *closed valley.*

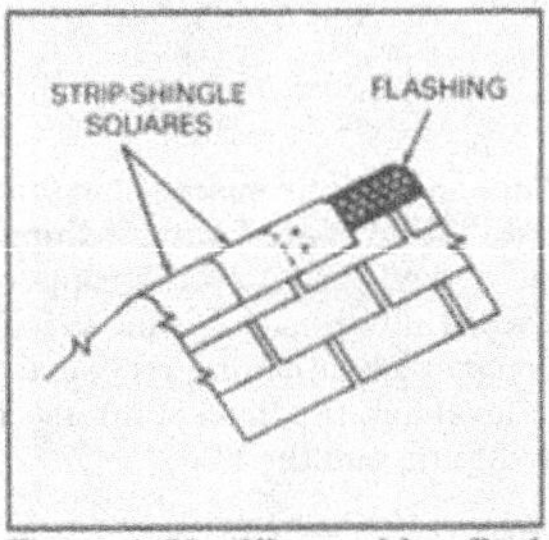

Figure 7-26. Hip or ridge flashing

Figure 7-27 shows part of an open valley. The roof covering does not extend across the valley. The flashing consists of a prefabricated piece of galvanized iron, copper, zinc, or similar metal, with a splash rib or ridge down the center and a small crimp at the edges. The flashing is nailed down to the valley, with nails driven in the edges (outside the crimps), as shown in Figure 7-27. Care must be taken not to drive nails through the flashing inside the crimps, to avoid leakage. Figure 7-28, page 7-18, shows an open valley using rolled roofing.

In the closed valley, the roof covering extends across the valley. Sheet metal flashing is cut into small **sheets measuring about** 18 x 10 inches, called *shingle tins.* This flashing is laid under each course of shingles, along the valley, as the course is laid. After the first course of the double course at the eaves is laid, the first sheet of flashing is placed on top of it. The second course is laid over this one so that the metal is partly covered by the next course. This procedure is continued all the way up the valley.

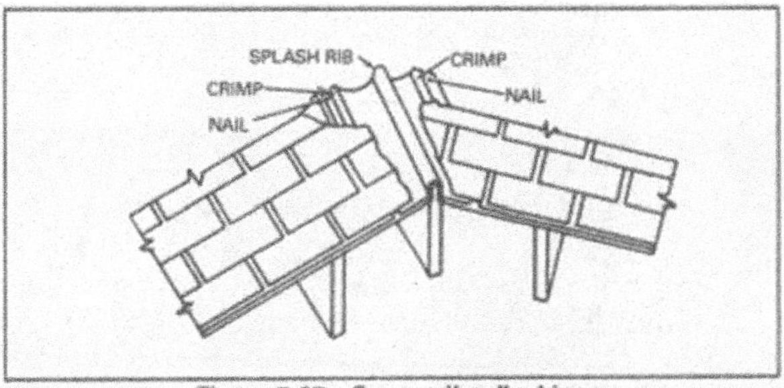

Figure 7-27. Open-valley flashing

Shingle tins measuring about 5 x 7 inches may also be used to lay flashing up the side walls of dormers, chimneys, skylights, and similar openings. Each tin is bent at a right angle so that part of the tin extends up the side wall and the rest lies flat on the roof covering. This is called *side flashing.* In addition to the side flashing, a dormer, chimney, or skylight has a strip of flashing called an *apron* along the bottom of the outer wall or face. A chimney or skylight has a similar strip, called the *saddle 1?ashing,* along the bottom of the inner wall or face. Figure 7-29 shows vertical wall flashing.

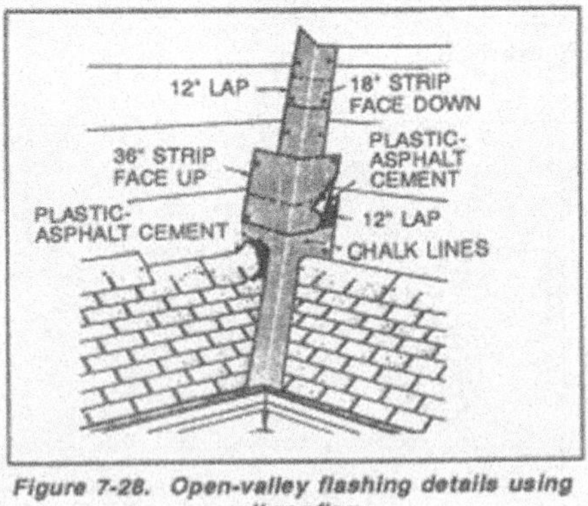

Figure 7-28. Open-valley flashing details using roll roofing

BUILT-UP ROOFING MATERIAL

The following building papers are used on a built-up roof. Their purpose is to prevent the seepage of bitumen through roof sheathing on which a built-up roof has been applied.

- *Rosin paper is* a felt paper, usually pale red, filled with rosin compound.
- *Kraft paper is* a light brown paper that is usually glazed.
- *Sisal kraft* consists of two layers of glazed kraft paper with a center section of sisal embedded in a black bituminous compound and laminated by heat and pressure.
- *Roofing felt is* a felt paper that has been saturated with a bituminous compound (heavy pitch or asphalt oils). The basic ingredients are usually either asbestos or rag felts. The roll may vary from 32 to 36 inches wide. Weights for built-up roofing vary from 15 to 65 pounds per square. The 15-pound felt is most commonly used because of its light weight.

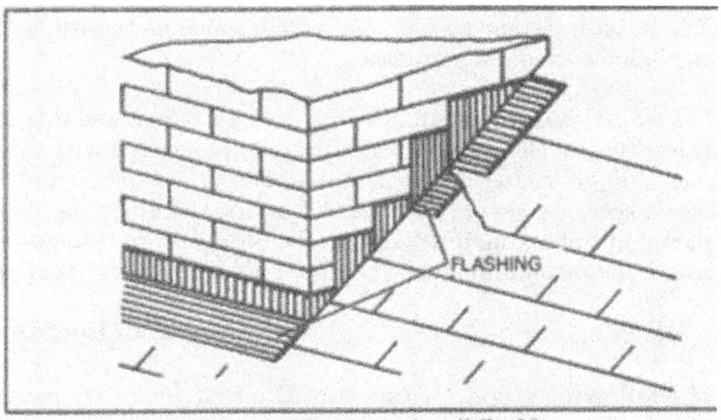

Figure 7-29. Vertical wall flashing

A binder is used to bond the roofing felt together and form a watertight seal. Asphalt and coal tar are the two main types of bituminous binders used. Drying out of the binder causes deterioration of built-up roofs. If this did not happen, a built-up roof would last indefinitely. Asphalt is the preferred binder. It is used on roofs sloping up to 6 inches per foot (1/4 pitch). Asphalt has a melting point of 350' to 41 0°F. A roof covered with asphalt should be protected with a covering of slag, gravel, or other protective material. Tar has a lower melting point (300° to 350°F) than asphalt, so it will move more easily; therefore, it is not recommended for roofs having a slope of more than 3 inches per foot (1/8 pitch).

Aggregate, crushed stone, or gravel from 1/4 to 5/8 inch in diameter is embedded in a coat of asphalt or tar to hold the roof covering down. It also prevents the binding from disintegrating because of the weather.

Gravel stops on slag or gravel-surfaced roofs, and metal-edge strips on smooth-surfaced built-up roofs are used to finish all exposed edges and eaves to prevent wind from getting under the edges and causing blowoffs. The gravel stop also prevents the loss of gravel or slag off the edge of the roof. The *flashing flange* of the gravel stop or edge strip is placed over the last ply of felt. It should be nailed securely to the roof deck and double felt stripped. Then the finished coat of bitumen and surfacing or cap sheet should be applied. The lip of the gravel stop should extend a minimum of 3/4 inch above the roof deck. The lip of the edge strip should be a maximum of 1/2 inch above the deck. Both should be securely fastened to the fascia board.

RE ROOFING

This section provides information on reroofing for the following types of roofs: asphalt-shingle roofs, asphalt-prepared roll roofings, built-up roofs, slate roofs, tile roofs, asbestos-cement roofs, metal roofs, and wood-shingle roofs.

ASPHALT-SHINGLE ROOFS

The following types of asphalt strip shingles are used for reroofing hospitals and mobilization-type buildings with pitched roofs. These shingles are applied directly over the existing roll roofings.

Standard-weight shingles should be four-tab, 10 x 36 inches, and intended for a 4-inch maximum exposure. Weight per square (100 square feet) applied should be approximately 210 pounds. Shingles are fastened with 1 1/4- or 1 1/2-inch nails with heads having a minimum diameter of 3/8 inch. Zinc-coated nails are best.

Thick-butt shingles should be three-tab, 1 x 3 feet, and intended for a 5-inch maximum exposure. The entire surface of the shingles should be covered with mineral granules. The bottom part of each shingle, including the part intended to be exposed and a section at least 1 inch above the cutout sections, should be thicker than the remainder of the shingle. Weight per square applied should be approximately 210 pounds. Shingles are fastened with 1 1/2- or 1 3/4-inch nails with heads having a minimum diameter of 3/8 inch. Zinc-coated nails are best.

Preparation of Roof Decks

The following steps assume that the roof decks are covered with smooth or mineral-surfaced, asphalt-prepared roofing and that the shingles will be applied directly over the existing roofing.

Step 1. Drive in all loose and protruding nails flush with the existing roll roofing.

Step 2. Cut out all vertical and horizontal buckles and wrinkles in the existing roofing. Nail down the edges of the cuts with 3/4-inch or 1-inch roofing nails so that the entire roof deck is smooth.

Step 3. If shingles are applied over smooth-surfaced roofing or over mineral-surfaced roofing which does not match the shingles, apply an 18-inch starting strip of mineral-surfaced roll roofing at the eaves. Use roofing surfaced with granules of the same type and color as the shingles.

Step 4. Before the strips are applied, unroll them carefully and lay them on a smooth, flat surface until they lie perfectly flat.

Step 5. Nail starter strips at the top at about 18-inch intervals. The lower edge, bent down and nailed to the edge of the sheathing board, should extend about 3/4 inch beyond the edge of the board to form a drip edge. Space the nails in the edge of the sheathing board 6 inches apart.

A starter strip need not be used if the shingles are the same color as the existing roofing and the existing roofing is not buckled. Figure 7-30, page 7-20, shows an example of roof replacement.

Application of Shingles

Shingles are attached in different ways, depending on the type.

Standard-Weight, Four-Tab, 10- x 36-Inch Shingles. Start the first course with a full shingle placed so that one edge, cut off flush with the tab, is also flush with the side of the roof. The bottoms of the tabs are placed flush with the eaves. Place nails about 3/4 inch above each cutout

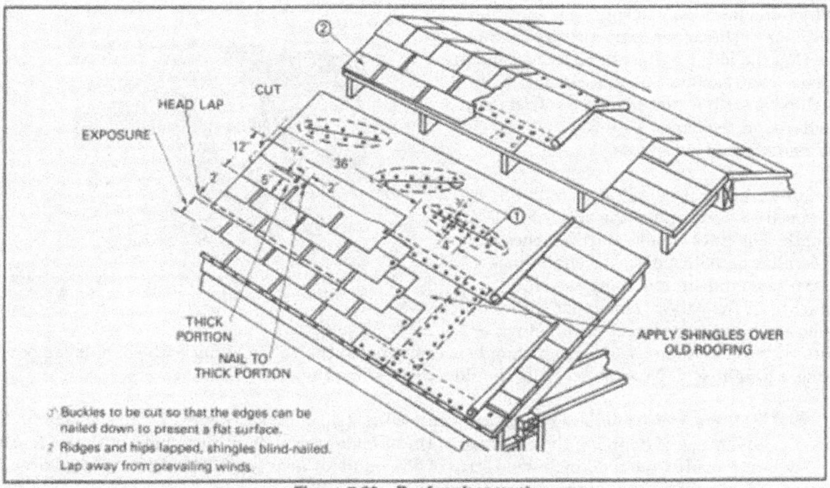

Figure 7-30. Roof replacement

section and in the same relative position at each end of the shingle. Use two nails at every cutout. Nail at the center first, then above the cutout sections nearest the center, and finally at the ends. Nailing may start at one end and proceed regularly to the other. Complete the first course with full-width shingles applied so that the ends barely touch each other.

Start the second course with a shingle from which half a tab has been cut. Place it so that the bottoms of the tabs are flush with the tops of the cutout sections of the shingle in the first course. Complete this course with full-width shingles.

Start the third course with a shingle from which one tab has been cut; the fourth with one from which one and one-half tabs have been cut; and so on, until eventually a full shingle is used again.

Thick-Butt, Three-Tab Shingles. Follow the same method described for standard shingles. Always nail these shingles through the thick part, about 3/4 inch above the cutout sections. The importance of nailing through the thick part of asphalt shingles cannot be emphasized too strongly. Practically all difficulties experienced with asphalt shingles on Army buildings have resulted from nailing the shingles too high.

Hips and Ridges. Finish hips and ridges with individual shingles furnished especially by the manufacturer or with shingles cut from strip shingles. Hips and ridges may also be finished with a strip of mineral-surfaced roofing 9 inches wide, bent equally on each side and nailed on 2-inch centers 3/4 inch from the edges.

Open Valleys. Open valleys may be flashed with 90-pound, 18-inch-wide mineral-surfaced asphalt roll roofing (Figure 7-31) placed over the valley underlayment. It is centered in the valley with the surfaced side down and the lower edge cut flush with the eaves flashing. When it is necessary to splice the material, the ends of the upper segments are laid to overlap the lower segments 12 inches and are secured with asphalt plastic cement. A 36-inch-wide strip is placed over the first strip, centered in the valley with the surfaced side up and secured with nails.

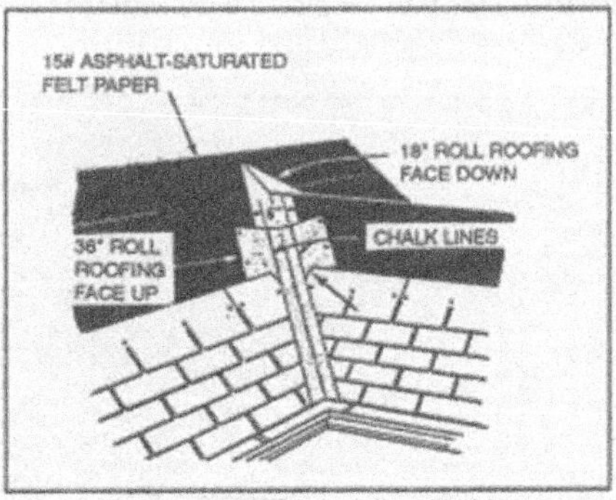

Figure 7-31. Open-valley flashing

Before shingles are applied, a chalk line is snapped on each side of the valley for its full length. The lines should start 3 inches from the valley on both sides. The chalk lines serve as a guide in trimming the shingle units to fit the valley. The upper corner of each end shingle is clipped to direct water into the valley (Figure 7-31). Each shingle is cemented to the valley lining with asphalt cement to ensure a tight seal. No exposed nails should appear along the valley flashing.

Closed Valleys. Closed valleys can be used only with strip shingles (Figure 7-32). This method has the advantage of doubling the coverage of the shingles throughout the length of the valley. A valley lining made from a 36-inch-wide strip of 55-pound (or heavier) roll roofing is placed over the valley underlayment and centered in the valley (Figure 7-32).

Valley shingles are laid over the lining by either of two methods:

- Applying both roof surfaces at the same time, with each course in turn woven over the valley.
- Covering each surface to the point approximately 36 inches from the center of the valley and the valley shingles woven in place later.

Either way, the first course at the valley is laid along the eaves of one surface over the valley lining and extended along the adjoining roof surface for at least 12 inches. The first course of the adjoining roof surface is then carried over the valley on top of the previously applied shingle. Each course thereafter is laid alternately, weaving the valley shingles over each other.

The shingles are pressed tightly into the valley and nailed in the usual manner except that no nail should be located closer than 6 inches to the valley centerline and two nails are used at the end of each terminal strip (Figure 7-32).

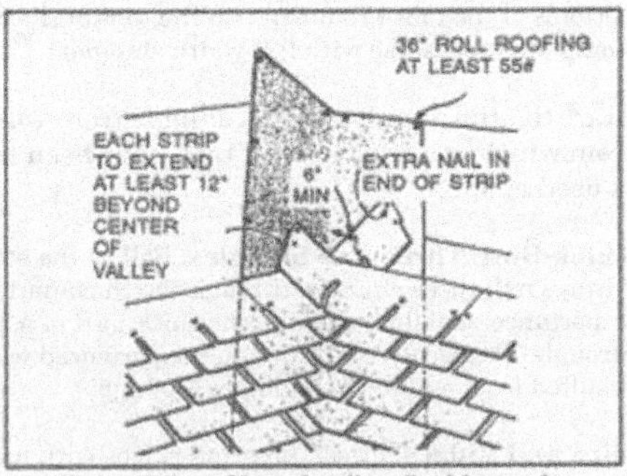

Figure 7-32. Closed-valley flashing using woven strip shingles

ASPHALT-PREPARED ROLL ROOFINGS

There are two types of asphalt-prepared roll roofing: mineral-surfaced and smooth-surfaced.

Mineral-Surfaced Roll Roofing

Mineral-surfaced, asphalt-prepared, two-ply roofing should consist of a layer of 15-pound asphalt-saturated felt and two plies of roll roofing, cemented together with hot asphalt. Cut roll-roofing material into lengths of 18 or 20 feet, stacked free from wrinkles and buckles in protected piles. Maintain the roofing material at a temperature of at least 50°F for 24 hours before laying.

First, cover the roof areas with a layer of 15-pound asphalt-saturated felt, with all joints lapped 2 inches. Nail as required to prevent blowing off during the application of roofing. Next, lay either plain, dry unsurfaced roofing or dry mineral-surfaced roofing as a starter sheet. Lay this upside down parallel to and flush with the eaves. Nail through tin or fiber disks on 12-inch staggered centers; that is, with one row of nails on 12-inch centers placed not more than 2 inches from the lower edge, and with a second row on 12-inch centers staggered with respect to the first and about 8 inches above the first.

Over the lower half of this sheet, apply a uniform coating of hot asphalt at the rate of 30 pounds per 100 square feet. Place the first sheet of roll roofing in the asphalt. Cover the entire roof area, lapping each successive sheet, to obtain a two-ply roofing with a 2-inch head lap. Cement the lower or mineral-surfaced portion of each sheet with hot asphalt to the preceding sheet. Nail the edge through tin or fiber disks on 12-inch staggered centers. Use two rows of nails. Place the first row on 12-inch centers not more than 2 inches above the mineral surfacing and the second row on 12-inch centers staggered with respect to the first and about 8 inches above the first.

Perform the work in such a way that no fastenings or asphalt will show on the finished surface. Apply the asphalt immediately before unrolling the sheet of roofing. Do not apply the asphalt more than 3 feet ahead of the roll. Step the edge of each sheet into the asphalt so that all laps are securely sealed. Place the end laps 6 inches in width with the underlying edges nailed on 6-inch centers, asphalt-cement the overlying edges, and step down firmly. Place one ply of roofing at eaves and edges, turn down neatly, and secure it with a wood member nailed on 8-inch centers.

Smooth-Surfaced Roll Roofing

Before laying the roll-roofing material, cut it into 18-or 20-foot lengths. Stack them free of wrinkles and buckles in protected piles, and maintain them at a temperature of at least 50°F for 24 hours.

For TO construction, apply single-ply roll roofing horizontally with at least 4-inch side laps and 6-inch end laps. Nail the underlying edges of laps through tin or fiber disks on 6-inch centers. Cement overlying laps with hot asphalt or an approved cold-applied sealing compound. Step down firmly on the edges to provide proper adhesion. Double the roofing over the ridge with at least 4-inch laps. Turn roofing down neatly at eaves and edges. Nail the roofing in place on 6-inch centers. Figure 7-33 shows an example.

BUILT-UP ROOFS

Buildings with roofs of relatively low pitch (less than 2 inches per foot), originally roofed with asphalt-prepared roll roofings, should be reroofed with smooth-surfaced asphalt built-up roofing or with coal-tar-pitch built-up roofing.

Use smooth-surfaced asphalt built-up roofing on buildings with original smooth-surfaced roll roofing.

Use asphalt built-up roofing or coal-tar-pitch built-up roofing on mobilization-type buildings with roofs of relatively low pitch (usually 1/2 inch per foot), originally roofed with wide-selvage, mineral-surfaced roll roofing. If the roof is nearly flat so water collects and stands, the latter type of roofing is best. Asphalt roofs may be smooth- or mineral-surfaced. Coal-tar-pitch roofs must be mineral-surfaced.

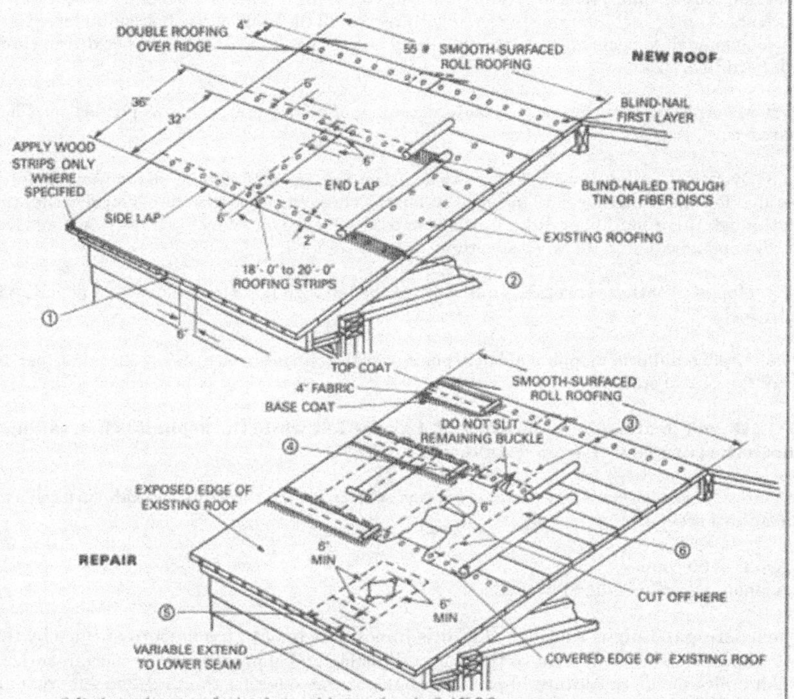

① At edges and eaves, roofing is to be turned down neatly and nailed 6" O.C.
② Blind-nailed joints should be cemented with hot asphalt or cold-applied sealing compound.
③ Repairs to buckles over seams; slit buckles for 6" only and repair as described in text.
④ Cover seams and nails with base coat of asphalt roof coating, then a 4" strip of asphalt-saturated fabric and top coating (fibrous adhesive compound of brushing consistency).
⑤ Repairs to small damaged areas; open the lower seam and insert the strip as shown.
⑥ Repairs to large damaged areas; remove the damaged section and replace it using full-width sheet, lapped, cemented, and nailed as described in text.

NOTES:
1. Material for steep-surfaced TO buildings will be smooth-surfaced, asphalt-prepared roll roofing conforming to federal specifications.
2. Buckles will be cut out and the roof shower tacked before reroofing.
3. Ridges and hips will have their roofing doubled. The first layer is to be blind-nailed 6" O.C. on both slopes. The top layer is to be cemented in place.

Figure 7-33. Smooth-surfaced roll roofing

Prepare the roof deck by driving in all loose and protruding nails and cutting out all buckles and wrinkles. Then apply a three-ply, smooth-surfaced asphalt built-up roof as follows:

Step 1. Lay one layer of 15-pound asphalt-saturated felt over the entire surface. Lap each sheet 3 inches horizontally and vertically and nail the laps on 12-inch centers. Also nail through the center of each sheet on 12-inch centers staggered with respect to the nails at the horizontal laps. Use nails long enough to penetrate into the sheathing at least 3/4 inch. They should be driven through tin or hard fiber disks.

Step 2. Mop the entire surface with a uniform coating of hot asphalt, using 25 pounds per 100 square feet.

Step 3. Over this coating of asphalt, lay two additional layers of 15-pound, 36-inch, asphalt-saturated felt. Lap each sheet 19 inches, and lap the sheet ends not less than 6 inches. Nail these felts through tin or hard fiber disks 1 inch from the back edge on 12-inch centers. Use nails long enough to penetrate into the wood sheathing at least 3/4 inch.

Step 4. Mop each of these sheets the full width of the lap with hot asphalt, using 25 pounds per 100 square feet.

Step 5. Apply a uniform mopping of hot asphalt over the entire surface, using 30 pounds per 100 square feet of roof surface.

NOTE: Do not heat asphalt above 400°F. Lay the felt while the asphalt is hot, taking care to keep the surface free from wrinkles or buckles.

The materials needed per 100 square feet of roof surface for the three-ply, smooth-surfaced asphalt built-up roof are—

- Asphalt: 80 pounds.
- Asphalt-saturated felt: 45 pounds.

If the existing roofing is so rough that it is impossible to obtain a smooth surface by the method described above, remove the original roofing and apply a three-ply, smooth-surfaced, asphalt built-up roof. Substitute 30-pound asphalt-saturated felt for the 15-pound felt originally specified.

If a slag or gravel-surfaced roof is desired for mobilization-type buildings, at step 5 above, apply 45 pounds of hot asphalt instead of 30 pounds per 100 square feet. Into this hot coating, place 300 pounds of roofing slag or 400 pounds of roofing gravel per 100 square feet of roof surface.

Coal-Tar-Pitch Built-Up Roofs

Prepare the roof surface as previously described, then apply a three-ply, coal-tar-pitch built-up roof as follows:

Step 1. Apply one layer of 15-pound coal-tar-saturated felt over the entire roof surface. Prepare it as described in step 1 of "Asphalt Built-Up Roofs."

Step 2. Mop the entire surface with a uniform coating of hot coal-tar pitch, using 30 pounds per 100 square feet.

Step 3. Over this coating of coal-tar pitch, lay two additional layers of 15-pound coal-tar-saturated felt 36 inches wide. Lap each sheet 19 inches over the preceding sheet. If 32-inch felt is used, lap each sheet 17 inches. Nail the felt 1 inch from the back edge on 12-inch centers through tin or hard fiber disks. Use nails long enough to penetrate into the wood sheathing at least 3/4 inch. Lap the ends of the sheets at least 6 inches.

Step 4. Mop each sheet the full width of the lap with hot coal-tar pitch, using 25 pounds per 100 square feet.

Step 5. Apply a uniform pouring of hot coal-tar pitch over the entire surface. Use 55 pounds per 100 square feet. While the pitch is hot, place over it 300 pounds of roofing slag or 400 pounds of roofing gravel per 100 square feet.

The materials required per 100 square feet of roof surface are—

- Coal-tar pitch: 110 pounds.
- Coal-tar-saturated felt: 45 pounds.
- Roofing slag: 300 pounds.

or

- Roofing gravel: 400 pounds.

NOTE: Do not heat the coal-tar pitch above 375°F. Lay the felt while the coal-tar pitch is still hot, taking care to keep the surface free from wrinkles or buckles.

SLATE ROOFS

Very old slate roofs sometimes suffer failure because of the nails used to fasten the slates. In such cases, remove and replace the entire roof, including the felt underlay materials. Remove or drive in any protruding nails. Make every effort to obtain a smooth, even deck similar to the original one. Apply 30-pound asphalt-saturated felt horizontally over the entire roof deck. Lap the sheets not less than 3 inches. Turn them up 6 inches or more on vertical surfaces and **over** 12 inches or more on ridges and hips. Secure the sheets along laps and exposed edges with large-head roofing nails spaced about 6 inches apart.

Re-lay all original slates that are in good condition. Replace defective slates with new slates of the same size, matching the original color and texture as nearly as possible.

Recommended slate sizes for large new buildings are 20 or 22 inches long; for small new buildings, 16 or 18 inches long. Use slates of uniform length, in random widths, and punched for a head lap of not less than 3 inches.

Lay roof slates with a 3-inch head lap. Fasten each slate with two large-head slating nails and drive them so that their heads just touch the slate. Do not drive the nails "home." The opposite is true of wood shingles; therefore, workmen accustomed to laying wood shingles must nail slate carefully.

Bed all slates in an approved elastic cement on each side of hips and ridges within 1 foot of the top and along gable rakes within 1 foot of the edge. Match slate courses on dormer roofs with those on the main roof. Lay slate with open valleys.

TILE ROOFS

Before reroofing with tiles, restore the roof deck as nearly as possible to its original condition. Replace defective boards and apply asphalt-saturated felt (30-pound type) or prepared roofing. Lap the sheets not less than 3 inches. Turn them up on vertical surfaces for at least 6 inches and **over** ridges and hips for at least 12 inches. Secure the sheets along laps and exposed edges with large-head roofing nails spaced about 6 inches apart.

Tiles must be free from fire cracks or other defects that will impair the durability, appearance, or weather tightness of the finished roof. Special shapes are provided for eaves starters, hips, ridges, top fixtures, gable rakes, and finials. Special shapes for field tile at hips and valleys may be factory-molded or may be job cut from whole tile and rubbed down to clean, sharp lines. Roof tiles for use on Army buildings are generally furnished in one or more of the following types:

- *Mission tiles* are straight-barrel-type, molded to a true arc and machine-punched for one nail and a 3-inch head lap. Use regular cover tile for ridges and hips. Finish with plain mission finials. Eaves closures and hip starters are available. Approved sizes are generally 8 inches wide by 14 to 18 inches long.
- *Spanish tiles* are S-shaped and machine-punched for two nails and a 3-inch head lap. Eaves closures and hip starters are available. Use mission-type cover tiles for hips and ridges. Approved sizes are generally 9 1/2 to 12 inches wide by 12 to 18 inches long.
- *Slab-shingle tiles* are the flat, noninterlocking type, punched for two nails and a 2-inch head lap. Approved sizes are 6 to 10 inches wide, 15 inches long, and 1/2 inch thick.

Mission and Spanish Tiles

Before starting to lay tiles, mop the wood nailing strips with hot asphalt. Fill the spaces in back of the cant strips with asphalt cement. Lay tiles with open valleys. Set eaves closures back 3 inches from the lower edge of eaves tiles. Lay pan tiles with uniform exposure to the weather. Lay cover tiles in a uniform pattern, except where otherwise necessary to match existing roofs. Give all tiles a minimum lap of 3 inches and extend pan tiles 1 inch over the rear edge of the gutter.

Cut the tiles so that they meet projections with finished joints and point them up with roofer's cement. Waterproof the spaces between field tiles and wood nailing strips at ridges and hips with a fill of roofer's cement. Fit all tiles properly and then secure them with nails long enough to penetrate at least 1 inch into the wood base.

Fill spaces between pan and cover tiles in the first row at the eaves with solid cement mortar made of one part Portland cement, three parts fine sand, and enough clean water to form a plastic mix. Wet all tiles before applying mortar, then press them firmly into the mortar bed. Match the tile courses on dormer roofs with those on the main roof. Cut surplus mortar off neatly. Point up all open joints. Remove loose mortar from exposed surfaces.

Where hurricane winds can be expected, consider reinforcing tile roofs by laying all field tiles in Portland cement mortar. To do this, fill the ends of tiles at eaves, hips, ridges, rakes, and spaces beneath ridges solid with cement mortar. Fill the full width of laps between the tiles, both parallel and perpendicular to the eaves, with cement mortar.

Slab-Shingle Tiles

Lay slab-shingle tiles with a 2-inch head lap. Secure each tile with two large-head roofing nails. Double the tiles at eaves and project them 1 inch over the rear edge of gutters. Lay all tiles within 1 foot of hips, ridges, and abutting vertical surfaces in roofer's cement. Lay 10- or 12-inch tiles with 1-inch head lap on the sides of dormers. Match the tile courses on dormer roofs with those on the main roof. Lay tile roofs with open valleys.

ASBESTOS-CEMENT ROOFS

Before reroofing with asbestos-cement shingles, restore the roof deck as nearly as possible to its original condition. Replace defective boards, and apply new 30-pound asphalt-saturated felt or prepared roofing in horizontal courses. Lap the sheets not less than 3 inches. Turn them up at least 6 inches on vertical surfaces and **over** at least 12 inches on ridges and hips. Secure the sheets along laps and exposed edges with large-head roofing nails spaced about 6 inches apart.

Re lay all asbestos-cement shingles that are in good condition. Replace defective shingles with new shingles of the same size, matching the originals as nearly as possible in color and texture.

Lay each shingle with a 2-inch head lap and secure it with two large-head slating nails. Drive the nails so that their heads just touch the shingles. Do not drive the nails "home" as in laying wood shingles. Bed all shingles on each side of hips and ridges within 1 foot of the edge in an approved elastic slater's cement. Project the shingles 1 inch over the rear edges of gutters. Lay shingles with a 1-inch head lap on sides of dormers. Match the shingle courses on dormer roofs with those on the main roof. Lay shingles with open valleys.

METAL ROOFS

To conserve critical materials, replace metal roofs with nonmetallic roofing materials.

WOOD-SHINGLE ROOFS

When old roofing is removed—

- Restore the roof deck as nearly as possible to its original condition. Replace rotted boards and pull out or drive down all protruding nails.
- Install flashings and apply new shingles.

If the existing shingle roofs can be made smooth and can be nailed properly, apply new wood shingles directly over weathered wood-shingle roofs. Reroof over existing wood shingles as follows:

Step 1. Nail down or cut off curled and warped shingles. Nail loose shingles securely, and remove or drive down protruding nails.

Step 2. Cut off the old first-course shingles at the eaves just below the butts of the second course. Replace them with a 1 x 3 or 1 x 4 strip nailed flush with the eaves line.

Step 3. Cut back the shingles at the gable ends about 3 inches. Replace them with a 1 x 2, 1 x 3, or 1 x 4 strip nailed flush with the gable end.

Step 4. Remove weathered shingles at the ridge. Replace them with a strip of beveled siding, thin edge down, to provide a solid base for nailing the ridge shingles. Treat the hip the same as the ridges.

Step 5. Fill open valleys with wooden strips level with the old shingle surface or with a narrow strip placed across the 'V" of the valley to act as a support for new flashings.

Step 6. Inspect flashings carefully. Replace terne and galvanized flashings. Reuse old flashings if they are in good condition.

Step 7. Use the following nails in applying shingles over an existing roof:

- 5d box or special overroofing nails, 14-gauge, and 1 3/4 inches long for 16- and 18-inch shingles.
- 6d nails, 13-gauge, and 2 inches long for 24-inch shingles.

One square of roofing will need about 3 1/2 pounds of nails.

Step 8. Apply new shingles as recommended by their manufacturer.

DOORS AND WINDOWS — CHAPTER 8

This chapter covers the rough framing and finish carpentry for doors and windows. Before putting the exterior covering on the outside walls of a building, prepare the door and window openings for the frames.

DOORS

Before the exterior covering is put on the outside walls, the door openings are prepared for the frames. Square off uneven pieces of sheathing and wrap heavy building paper around the sides and top of the door opening. Since the sill must be worked into a portion of the rough flooring, no paper is put on the floor. Position the paper at a point even with the inside portion of the stud to a point about 6 inches on the sheathed walls, and tack it down with small nails.

NOTE: Rough openings are usually made 2 1/2 inches larger each way than the size of the door to be hung. (For example, a 2-foot 8-inch by 6-foot 8-inch door would need a rough opening of 2 feet 10 1/2 space allows for the jambs, the wedging, and the clearance space for the door to swing.

TYPES OF DOORS

Doors, both exterior and interior, are classified as job-built or mill-built. This classification is further broken down as batten, panel, and flush doors (Figure 8-1).

NOTE: No hinged interior door should open or swing against a natural entry, swing into hallways, or be obstructed by other swinging doors.

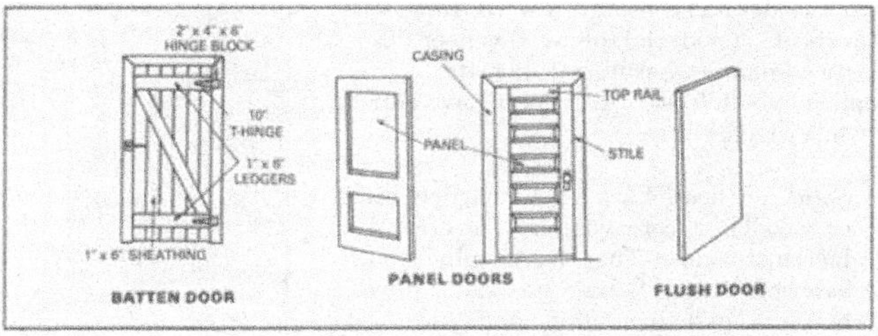

Figure 8-1. Types of doors

Job-Built Doors

The batten door is the most commonly used and most easily constructed type of job-built door. It can be constructed in several ways, such as—

- Using diagonal boards nailed together in two layers, at right angles to each other. This type of door is often used as the core for metal-sheathed fire doors.
- Using vertical boards that are tongue-and-grooved or shiplapped. The door is held rigid by two to four cross pieces, called *ledgers,* which may or may not be diagonally braced. If two additional pieces forming the sides of the door and corresponding to the ledgers are used, these are called *frames*.

In hasty construction *(on-site prefabrication),* the carpenter makes a batten door from several 2 x 6 boards with ledgers and braces, as follows:

- Nail the ledgers with their edges 6 inches from the ends of the door boards.
- Place a diagonal board between the ledgers. It begins at the top-ledger end, opposite the hinge side of the door, and runs to the lower ledger, diagonally across the door. On an outside door, use roofing felt on the weather side to cover the boards.
- Nail wooden laths around the edges and across the middle of the door to hold the roofing felt in place.

NOTE: When these doors are hung, 1/4 inch of clearance should be left around the door to allow for expansion.

- Fasten T-strap hinges to the door ledgers and the hinge blocks on the door casing or post.

Mill-Built Doors

The usual exterior door is the panel type (Figure 8-2). It consists of stiles, rails, and filler panels. Two frequently used interior doors are the flush and the panel types (Figure 8-2).

Panel Doors. Panel doors consist of vertical members called stiles and horizontal members called rails. Stiles and rails form the framework into which panels are inserted. Additional vertical and horizontal members called muntins are used to divide the door into any number of panels. Panels may be solid wood, plywood, particleboard or louvered or have glass inserts.

Flush Doors. Flush doors have flat surfaces on both sides and consist of a wood frame with thin sheets of material (plywood veneer, plastic laminates, hardboard, or metal) applied to both faces. Flush doors have either a solid or hollow core.

- *Solid-core* doors have a solid particle board or woodblock core which is covered with layers of veneer. They are usually used as exterior doors. Solid-core doors provide better sound insulation and have less tendency to warp.
- *Hollow-core* doors have a lightweight core made of various materials that are covered with layers of veneer. They are usually used as interior doors and are less expensive to produce.

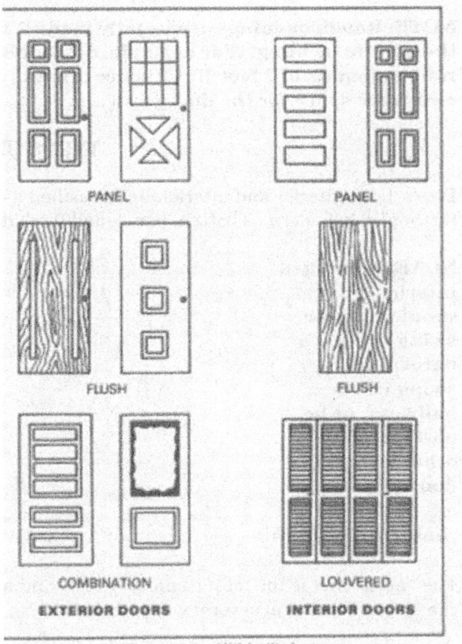

Figure 8-2. Mill-built doors

Specialty Doors

Specialty doors include double doors, sliding doors, and folding doors.

DOOR FRAMES

Door frames are made of the following parts: the head casing, the jambs (head and two sides), and the sill (on exterior doors only). (The principal parts of a door frame are shown in Figure 8-3.) Doors and frames may be fabricated in the shop and installed separately; they may also be Remanufactured (prehung), purchased ready for installation.

Door-frame layout calculations begin with the size of the door (height, width, and thickness), as given on the door schedule. Construction information for door frames is usually given in detail drawings like those shown in Figure 8-4. In the type of frame shown in Figure 8-4, the door jambs (linings of the framing of door opening are rabbeted to depths of 1/2 inch. The rabbet prevents the door from swinging through the frames. A strip of wood may be used instead a rabbet. The door stop also serves to weather proof the door. Most project drawings call for rabbeted exterior door jambs.

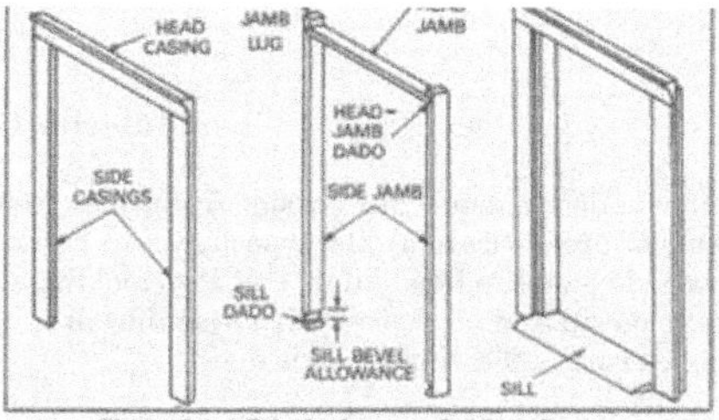

Figure 8-3. Principal parts of a door frame

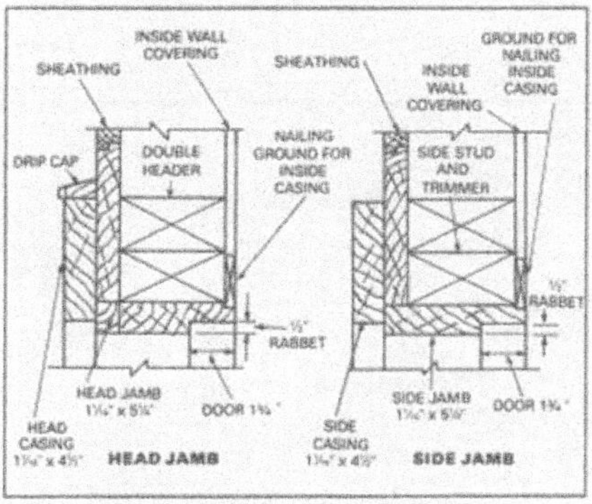

Figure 8-4. Typical door-frame details

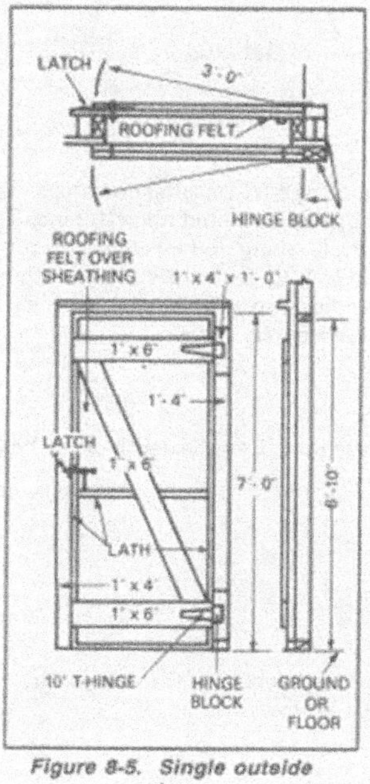

Figure 8-5. Single outside door

Exterior Door Frames

Exterior door frames are made up of two side jambs, a head jamb, a sill, and a stop. They are constructed in several ways. In hasty construction (on-site prefabrication), the frames will be as shown in Figure 8-5. This type requires no frame construction because the studs on each side of the opening act as a frame. Studs are normally placed 16 inches apart on center. Extra studs are added at the sides of door and window openings. Headers are usually used at the top and bottom of such openings.

The siding is applied to the outside wall before exterior doors are hung. The casing is then nailed to the sides of the opening. It is set back the width of the stud. A 3/4- x 3/4-inch piece is nailed over the door but set back the width of the stud; it supports the drip cap. Hinge blocks are nailed to the casing where the hinges are to be placed. The door frame is now ready for the door to be hung.

On an outside door, the outside casings and the sill are considered parts of the door frame. A prefabricated outside door frame—delivered to the site assembled—looks like the righthand view of Figure 8-3, page 8-3. It usually has the door installed, and the entire unit slides between studs.

Interior Door Frames

Interior door frames, like outside frames, are constructed in several ways. In hasty construction (on-site prefabrication), the type shown in Figure 8-6 is used. Interior door frames are made up of two side jambs, a head jamb, and stop moldings which the door closes against. Interior door frames have no sill and no casing, otherwise they are the same as the exterior frames. Figure 8-6 shows the elevation of a single inside door.

NOTE: Both outside and inside door frames may be modified to suit climatic conditions.

Door Jambs

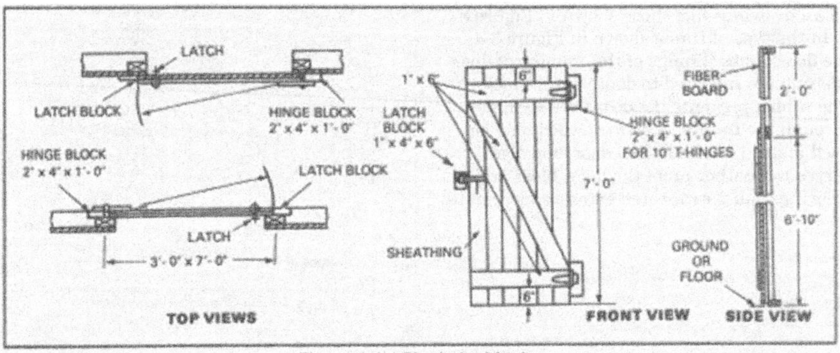

Figure 8-6. Single inside door

Door jambs (Figure 8-7) are the linings of the framing in door openings. The casing and stops are nailed to the door jambs, and the door is hung from them. Door openings should allow 1/2 inch between the frame and the jamb (Figure 8-8, page 8-6) to permit plumbing and leveling of jambs. Inside jambs are made of 3/4-inch stock; outside jambs are made of 1 3/8-inch stock. The width of the stock varies with the thickness of the walls. Inside jambs are built up with 3/8- x 1 3/8-inch stops nailed to the jamb. Outside jambs are usually rabbeted to receive the door.

Jambs are made and set as follows:

Step 1. Cut the side jambs of an entrance door to the height of the door, less the depth of the head jamb rabbet (if any), plus the—

- Diagonal thickness of the sill, plus the sill bevel allowance.
- Thickness of the threshold, if any.
- Thickness of the head jamb.
- Height of the side-jamb lugs.

Step 2. Cut the head jamb to the width of the door, less the combined depths of the side-jamb rabbets (if any), plus the combined depths of the head-jamb dadoes (grooves).

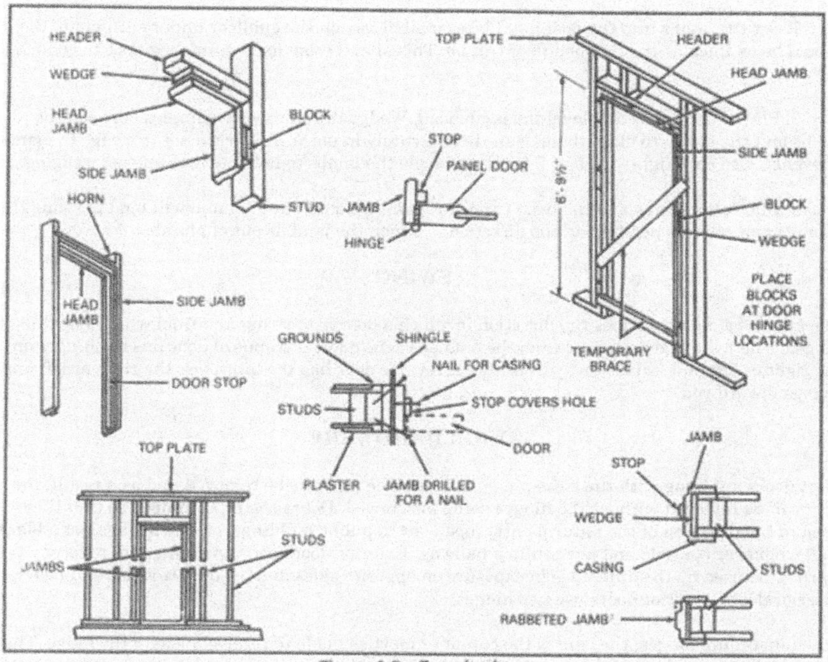

Figure 8-7. Door jambs

NOTE: Regardless of how carefully rough openings are made, be sure to plumb the jambs and level the heads when jambs are set.

Step 3. Level the floor across the opening to determine any variation in floor heights at the point where the jambs rest on the floor.

Step 4. Cut the head jamb with both ends square. Allow the width of the door plus the depth of both dadoes and a 3/16-inch door clearance.

Step 5. From the lower edge of the dado, measure a distance equal to the height of the door plus the clearance required under it. Mark it and cut it square. On the opposite jamb, do the same. Make additions or subtractions on this side for floor variations, if any.

Step 6. Nail the side jambs and jamb heads together with 8d common nails, through the dado into the head jamb.

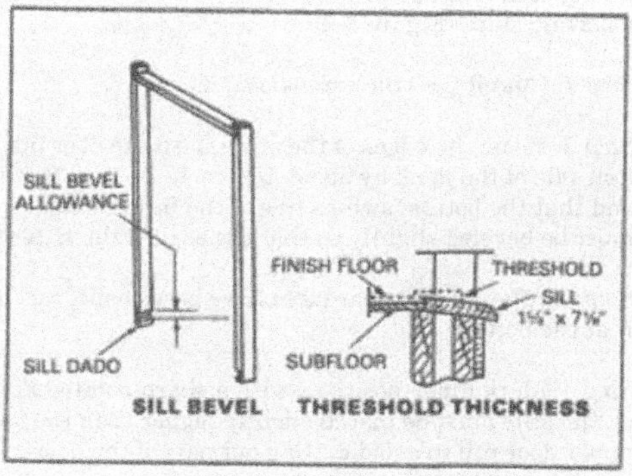

Figure 8-8. Jam allowances

Step 7. Set the jambs into the opening. Place small blocks on the subfloor under each jamb. Blocks should be as thick as the finished floor will be. This allows room for the finished floor to go under the door.

Step 8. Plumb the jambs and level the jamb head. Wedge the sides with shingles between the jambs and the studs, to align them. Nail them securely in place. Take care not to wedge the jamb unevenly. Use a straightedge 5 or 6 feet long inside the jambs to help prevent uneven wedging.

Step 9. Check the jambs and the head carefully. Jambs placed out of plumb will tend to swing the door open or shut, depending on the direction in which the jamb is out of plumb.

SWING

The *hand* of a door describes the direction in which a door is to swing and from which side it is hinged. The hand is determined from the outside of the door. A standard door has the hinges on the right or left and swings away from you. A reverse door has the hinges on the right or left and swings toward you.

DOOR HARDWARE

Most doors are hung with the loose-pin *butt hinge.* The pin may be removed and as a result, the door can be removed without the hinges being unscrewed. Doors should be hinged so that they open in the direction of the natural entry, open out in public buildings, and swing against a blank wall whenever possible and never into a hallway. Exterior doors use three hinges to reduce warpage caused by the difference in exposure on opposite sides and to support wider and heavier exterior doors. Interior doors use two hinges.

When installing hinges, the *gain is* the cutout or mortise made to receive a leaf of the hinge. The depth is determined by the hinge's thickness, and the width is determined by the hinge's size. *Setback is* the distance that the hinge is placed away from the side of the door, usually 3/16 inch.

The *door closer is* a device that closes a door and controls the speed and closing action of the door. Install the door closer according to the manufacturer's instructions.

DOOR INSTALLATION

Doors, both mill-built and job-built, are installed in the finished door frames as described in the following steps (Figure 8-9):

Step 1. Cut off the stile extensions, if any.

Step 2. Plane the edges of the stiles until the door fits tightly against the hinge side and clears the lock side of the jamb by about 1/16 inch. Be sure that the top fits squarely to the rabbeted recess and that the bottom swings free of the finished floor by about 1/2 inch. The lock stile of the door must be beveled slightly so that the edge of the stile will not strike the edge of the door jamb.

Step 3. After proper clearances have been made, tack the door in position in the frame and wedge it at the bottom.

Step 4. Mark hinge positions with a sharp-pointed knife on the stile and the jamb. Hinge positions on the stile must be placed slightly higher than the lower door rail and slightly lower than the upper door rail to avoid cutting out part of the door-rail tenons that are housed in the stile. Three measurements must be marked:

- The location of the butt on the jamb.
- The location of the butt on the door.
- The thickness of the butt on both the iamb and the door.

Step 5. Door *butts* (or *hinges)* (Figure 8-10) are mortised into the door frames as shown in Figure 8-11, page 8-8. Use three butt hinges on all full-length exterior doors to prevent warping and sagging. Place the butts and mortise them with the utmost accuracy so that the door will open and close properly, and so that the door, when open, will not strike the casing. The butt pin must project more than half its thickness from the casing.

Step 6. Using the butt as a pattern, mark the butt dimension on the door edge and face of the jamb.

Step 7. Cut the marked areas, called *gains,* on the door jambs and door to fit the butts. Use a 1-inch chisel and mallet.

Step 8. Test the gains. The butts must fit snugly and exactly flush with the edge of the door and the face of the jamb.

Step 9. Screw half of each of the butt joints on the door and the other three parts on the jamb. Place the butts so that the pins are inserted from the top when the door is hung.

Step 10. Set the door against the frame so that the two halves of the top butt engage. Insert the top pin. Engage and insert pins in the bottom and center butts.

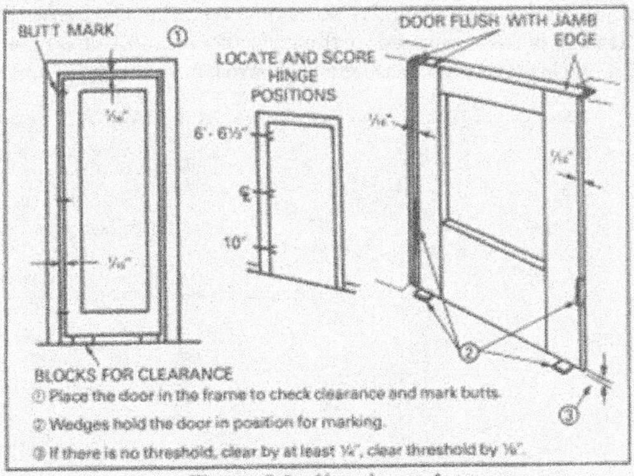

Figure 8-9. Hanging a door

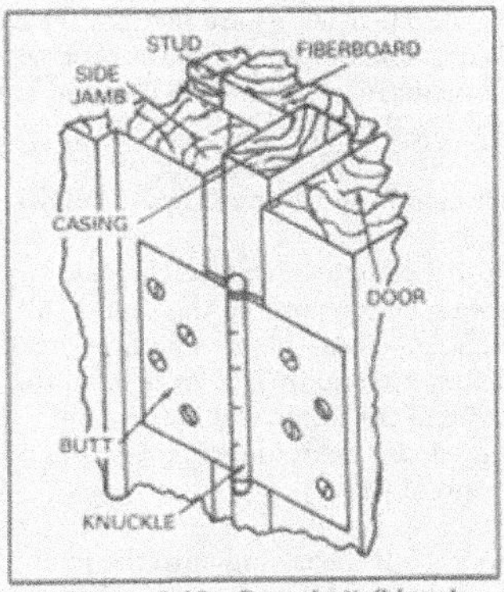

Figure 8-10. Door butt (hinge)

Door Stops

When fitting doors, the stops are usually nailed in place temporarily until the door has been hung. Stops for doors in single-piece jambs are generally 1/2 inch thick and 2 inches wide. They are installed with a butt joint at the junction of the side and head jambs. A 45° bevel cut at the bottom of the stop, about 1 to 1 1/2 inches above the finish floor, will eliminate a dirt pocket and make cleaning or refinishing the floor easier.

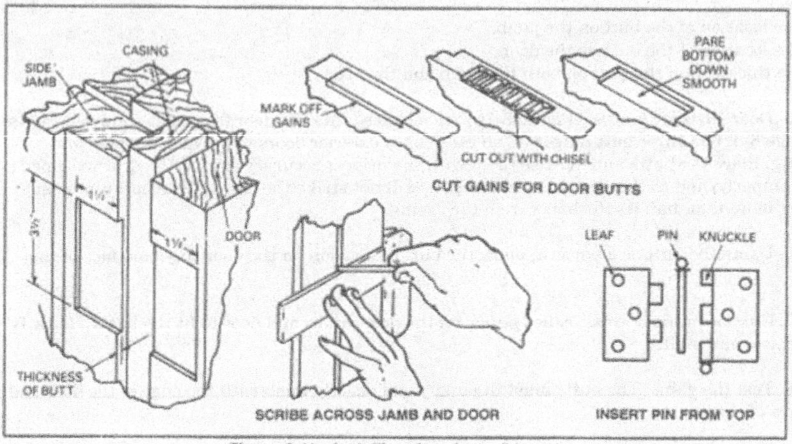

Figure 8-11. Installing door butts (hinges)

Finish Door Trim

Door trim is nailed onto the jambs to provide a finish between the jambs and the wall to cover wedging and spaces between the frame and studs. This trim is called casing. Sizes vary from 1/2 to 3/4 inch thick and from 2 1/2 to 6 inches wide. Most trim has a concave back to fit over uneven plaster. The casing layout depends on the way the side and head casings are to be joined at the corners. The casings are usually set back about 1/4 inch from the faces of the jambs. Care must be taken to make miter joints fit properly. If trim is to be mitered at the top corners, a miter box, a miter square, a hammer, a nail set, and a block plane will be needed. (Door trim and stop are shown in Figure 8-12.)

Door openings are cased up as follows:

Step 1. Leave a margin of 1/4 inch from the edge of the jamb to the casing, all around. Cut one (hinge-side first) of the side casings square and even with the bottom of the jamb. Cut the top or mitered end next, allowing a 1/4-inch margin at the top.

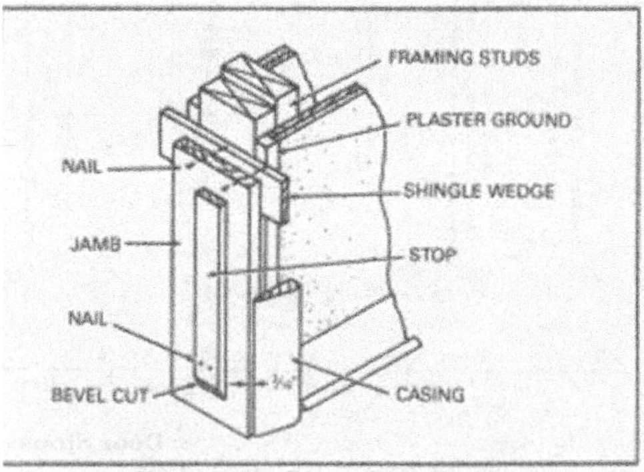

Figure 8-12. Door trim and stop

Step 2. Nail the casing onto the jamb, even with the 1/4-inch margin line. Start at the top and work toward the bottom. Use 4d finishing nails along the jamb side and 6d or 8d case nails along the outer edge of the casings. The nails along the outer edge will need to be long enough to go through the casing and into the studs. Set all nailheads about 1/8 inch below the surface of the wood with a nail set.

Step 3. Apply the casing for the other side and then the head casing.

LOCK INSTALLATION

Two types of locks used in TO construction are the *cylinder* and *tubular* locks. *Cylinder locks* are sturdy, heavy-duty locks designed for installation in exterior doors. They provide high security. *globular locks* are light-duty locks. They are used for interior doors on bathrooms, bedrooms, passages, and closets. Since door locks differ, use lock-set installation instructions, or perform the following steps:

Step 1. After placing the hinges in position, mark off the position of the lock on the lock stile, 36 inches from the floor level.

Step 2. Hold the case of the mortised lock on the face of the lock stile. With a sharp knife, mark off the area to be removed from the edge of the stile that is to house the entire case.

Step 3. Mark the position of the door-knob hub and the position of the key.

Step 4. Mark the position of the strike plate on the jamb.

Step 5. Bore out the wood to house the lock and the strike plate and mortises. (Figure 8-13, page 8-10, shows the installation of the lock and the strike plate.)

Step 6. Clean and install the lock set. The strike plate should be flush or slightly below the face of the door jamb.

Panic hardware is another type of lock. It is also known as a *paretic bar* or *fire-exit bolt.* It is often installed on the exit doors of public buildings. Slight pressure on the touch bar will retract the latch bolts at the top and bottom. Install panic hardware according to the manufacturer's instructions.

WINDOWS

The most common types of windows are *double-hung* and *hinged* (or *casement)* windows (Figure 8-14, page 8-11). All windows consist of two parts, the *frame* and the *sash.*

The double-hung window (Figure 8-14) is made of upper and lower sashes that slide vertically past one another. Screens can be located on the outside of a double-hung window without interfering with its operation. Ventilators and window air conditioners may be placed with the window nearly closed. However, for full ventilation of a room, only one-half of the area of the window can be used. Any current of air passing across its face is lost to the room. Its frame construction and operation are more involved than that of casement windows.

Casement windows (out-swinging or in-swinging) may be hinged at the sides, top, or bottom. Casements have the advantage of catching a parallel breeze and slanting it into a room.

- Out-swinging. The casement window that opens out requires the window screen to be located on the inside with a device cut into its frame to operate the casement.
- In-swinging. In-swinging casements, like double-hung windows, are clear of screens, but they are extremely difficult to make watertight, particularly against a driving rainstorm.

WINDOW FRAMES

Window frames are made of four basic parts: the head, the jambs (two), and the sill. (The sash is the framework that holds the glass in the window.) Where openings are provided, cut away the studs and for equivalent strength, double the studs on each side of the opening to form trimmers. Insert a header at the top. If the opening is wide, the header should also be doubled and trussed. At the bottom of the opening, insert the rough sill.

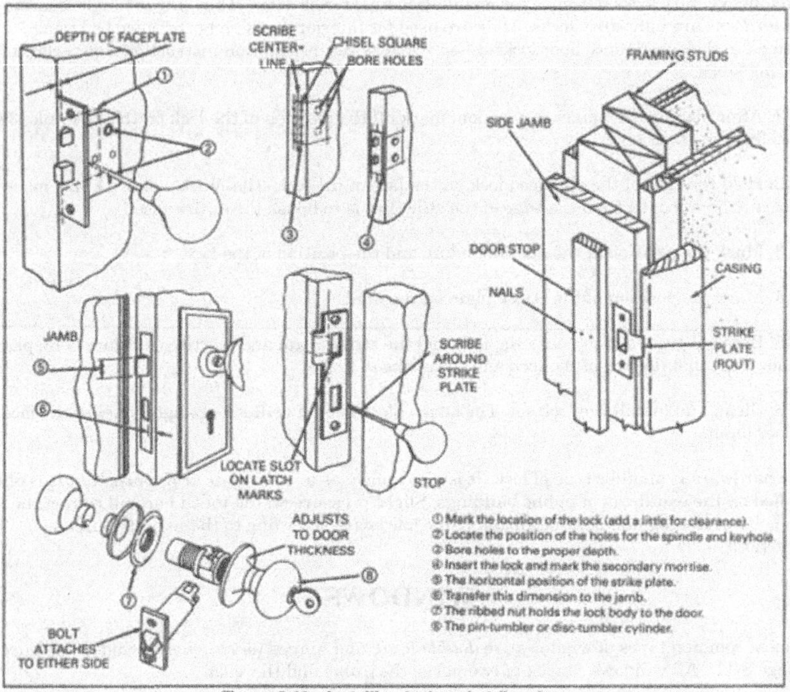

Figure 8-13. Installing lock and strike plate

In hasty construction, millwork window frames are seldom used. Instead, simple openings are left in the walls with the stops all nailed to the stud. The sash may be hinged to the inside or outside of the wall or may be constructed to slide. The sliding sash with overlapping panes is most common in Army construction because it requires little installation time.

Sills have a usual slope of 1 to 5 inches so that they shed water quickly. They are wider than frames, usually extending about 1 1/2 inches beyond the sheathing. They also form a base for the outside finished casing.

WINDOW SASHES

A window is normally composed of an upper and a lower sash. There are two ordinary types of wood sashes: fixed or movable. Fixed sashes are removable only with the aid of a carpenter. Movable sashes may slide up and down in channels in the frame (double-hung), or they may swing in or out and be hinged at the side (casement type).

Sliding sashes are counterbalanced by sash weights that weigh half as much as the sash. Sashes are classified as single or divided, according to the number of pieces of glass (or *lights).*

A sash may be made of 1 x 3 material with reinforced, rolled plastic material, which can be cut to any desired size. For hasty construction of window sashes, perform the following steps:

Step 1. Make two frames with the glass substitute installed on one.

Step 2. Nail the frames together. When the two frames are nailed together, they should be turned so that the joints are not over each other. This staggers the joints and strengthens the sash. Do not make the window sash larger than the available glass substitute. If the sash is too large for the glass substitute to cover, a muntin may be placed in the sash to hold the glass substitute; this should be fastened with corrugated metal fasteners. Where long sashes are made, a muntin should be placed in the center for added strength. Figure 8-15, page 8-12, shows the window frame and sash details.

Step 3. Cut the side pieces to a length equal to the height of the sash, less the width of one piece of material.

Step 4. Cut the top and bottom pieces the same length as the window, less the width of the material.

Step 5. Fasten at the joints with corrugated metal fasteners.

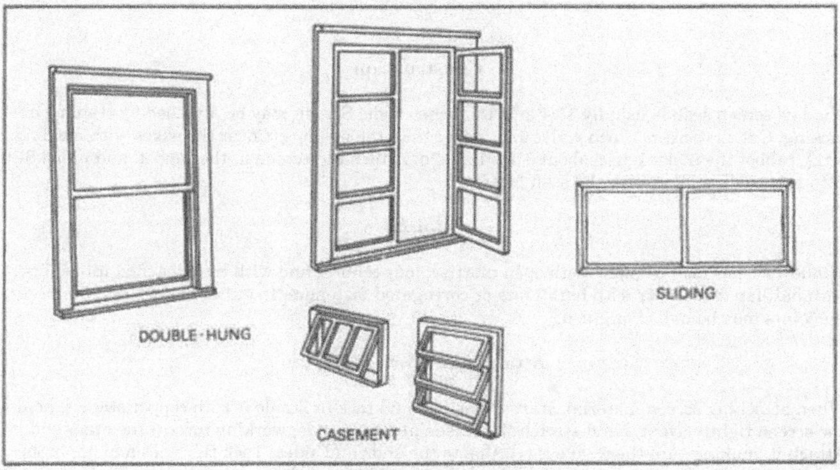

Figure 8-14. Types of windows

ACCESSORIES

The following are a few items that can be added to a structure to enhance efficiency:

WINDOW SCREENS

Screen sash is usually 3/4-inch stock; however, for large windows and doors 1 1/8-inch material is frequently used or 3/4-inch lumber is braced with a horizontal member.

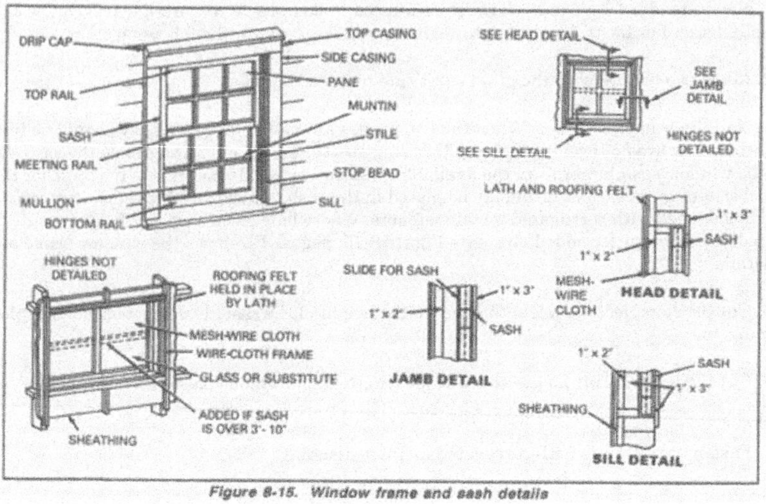

Figure 8-15. Window frame and sash details

Construction

Window screen sash is usually 13/4 or 2 1/4 inches wide. Screen may be attached by stapling or tacking. Cut the screen 1 inch wider and longer than the opening. Cover the edges with molding. Next, rabbet the inside edges about 3/4 x 1/2 inch. Attach the screen in the rabbet, and nail 3/8- x 1/2-inch molding flush with the sash face.

Joints

Window sashes may be made with open mortise, four tenons, and with rails tenoned into stiles; with half-lap corners; or with butt joints or corrugated fasteners. In either of the first two cases, the joints may be nailed or glued.

Attaching Screen Material

When attaching screen material, start at one end and tack or staple it with copper staples, holding the screen tightly. Next, hand-stretch the screen along the side, working toward the other end. Attach it, making sure the weave is parallel to the ends and sides. Tack the sides and apply the molding. Copper staples should be used for bronze or copper screen and cadmium staples for aluminum screens.

DOOR SCREENS

Door screens are made as shown in Figure 8-16. Two separate frames are made of 1 x 4 material for the sides and top; 1 x 6 material is used for the bottom and middle pieces. (Figure 8-17 shows door screen sizes.) The first frame is made of two side pieces as long as the door. The crosspieces are as wide as the door, less the width of the two side pieces. This frame is put together with corrugated metal fasteners or triangular corner splices; then, the screen wire is applied. The second frame is made with the crosspiece as wide as the door. The side pieces are cut to correspond with the distance between the cross-pieces. The second frame is placed over the first frame and nailed securely. For push-and-pull plates, two short 1 x 4 braces are nailed to the side opposite the hinge side.

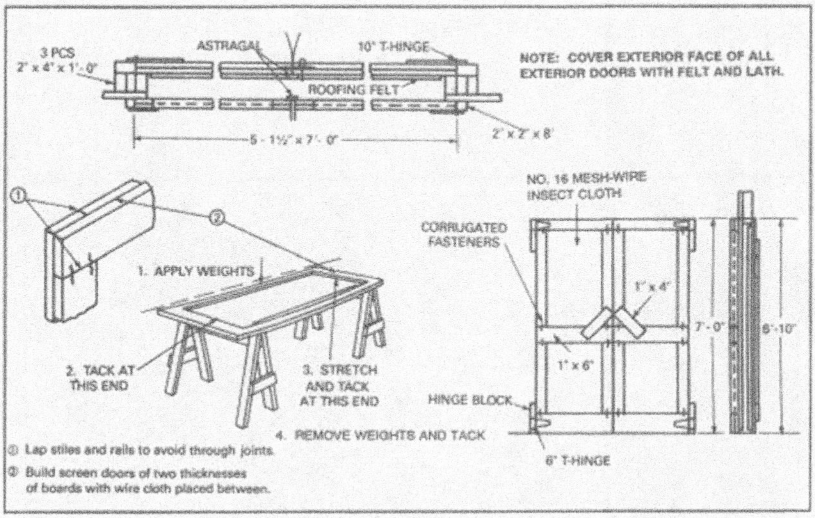

Figure 8-16. Door screen construction

HOODS OR CANOPIES

Hoods or canopies are used in tropical climates to protect the screened opening at the ends of the buildings. They are framed to the end walls with short rafters, which are nailed to the building with knee braces. The rafters are nailed to the wall, their bottom edge flush with the bottom of the end plate. The rafters and braces are made of 2 x 4s nailed with 8d or 10d nails. The sheathing is of the same material as the roof sheathing and is covered with roll roofing. The hood should extend about 2 1/2 or 3 feet from the building. Figure 8-18, page R-14. shows hood or canopy details.

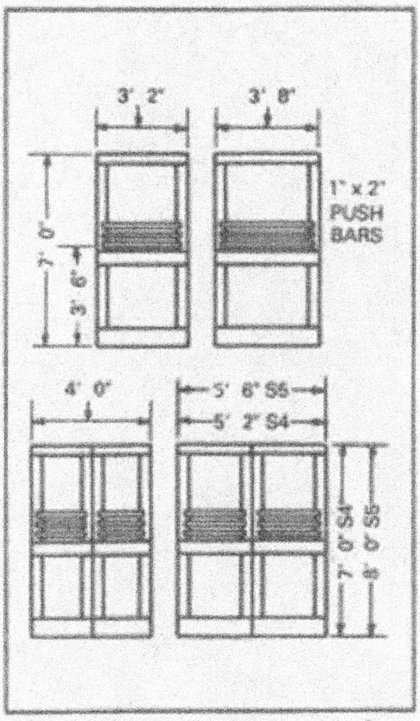

Figure 8-17. Door screen sizes

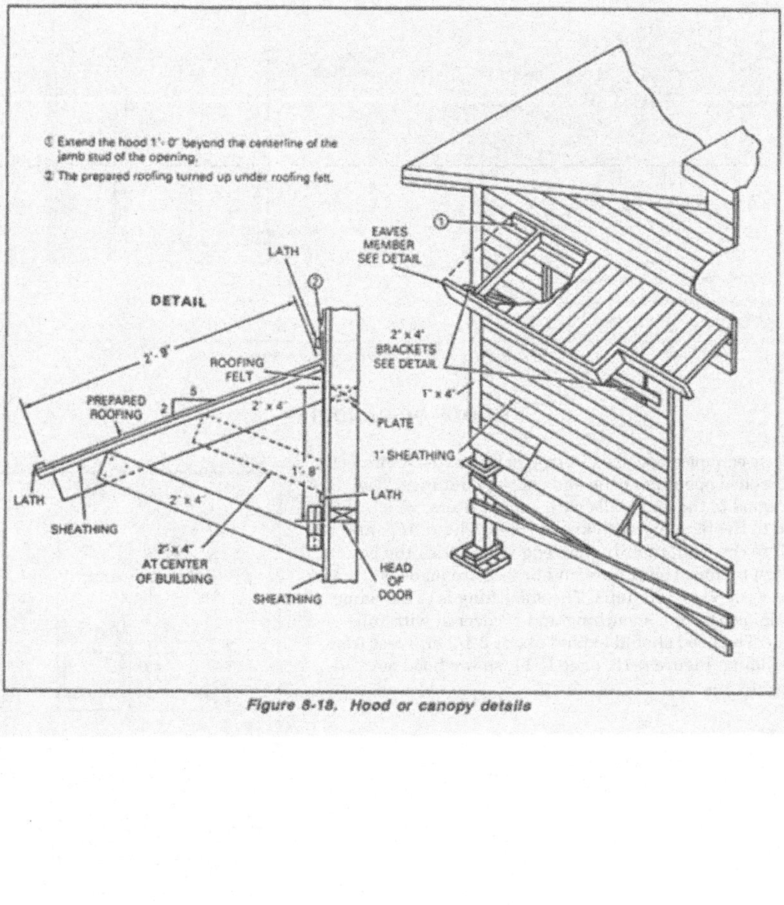

Figure 8-18. Hood or canopy details

FINISH CARPENTRY CHAPTER 9

After the rough framing is complete and a building is weather-tight, carpenters begin the inside finish carpentry. However, finish carpentry may be optional for TO construction. This chapter covers the following interior wall, partition, and ceiling coverings: gypsum board (or sheetrock / wallboard), plywood, and fiberboard (or chipboard). It also covers interior wall and ceiling moldings. (Doors and windows are covered in Chapter 8 and general information, such as floor and wall tile, suspended ceilings, and painting, is covered in Appendix D.)

Over time, "sheetrock" has become the most common term for gypsum board. Also, the term "drywall " is often loosely used to mean gypsum board / sheetrock / wallboard. In this manual we will use ,'sheetrock."

INTERIOR WALL AND PARTITION COVERINGS

In current construction, sheetrock, plywood, and fiberboard are used instead of laths and plaster to cover walls. Sheetrock is normally applied in single (sometimes double) thickness as shown in Figure 9-1, page 9-2.

NOTE: When covering walls and ceilings, always start with the ceiling. After the ceiling is started, begin covering the wall in one corner and work around the room. Make sure that joints break at the center line of a stud or ceiling joist.

PLYWOOD AND FIBERBOARD

Plywood and fiberboard can be used for interior wall coverings; however, plywood is most commonly used. It comes in 4-feet-wide and 5-to 8-feet-long sheets, 1/4 to 3/4 inch thick. It is usually applied vertically from the floor to the ceiling. When plywood is correctly applied (with flush joints), the joints do not need to be concealed. However, to improve wall appearance, joints may be covered with moldings. These may be battens fastened over the joints or applied as *splines* between the panels. Less expensive plywood can be covered with paint or covered in the same way as plastered surfaces. To hang plywood (or fiberboard), see Figure 9-1. Figure 9-2, page 9-3, shows how to fit sheetrock on rough or uneven walls.

SHEETROCK

Sheetrock saves time in construction and has a short drying time as compared to plaster. It is also fire-resistant. It requires moderately low moisture content of framing members. The dry. ing of members will result in *"nail pops,"* which cause the nailhead to form small humps on the surface. Misaligning sheetrock on the studs may cause a wavy, uneven appearance. Wood sheathing will correct misaligned studs on exterior walls.

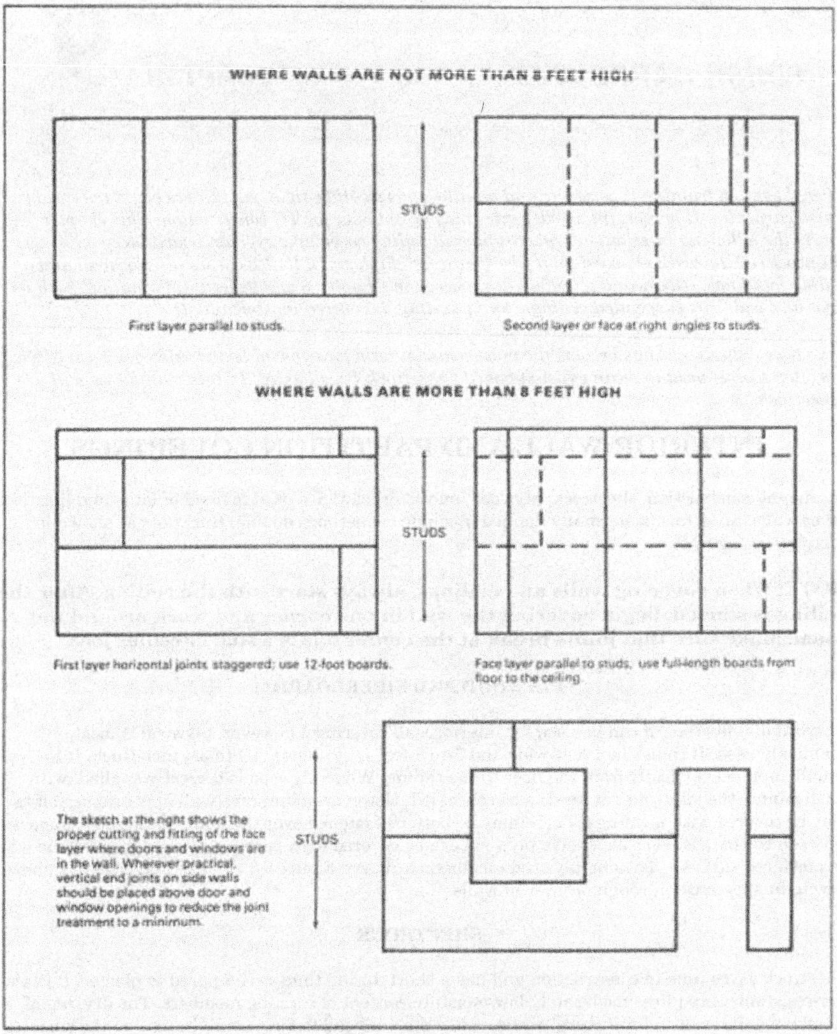

Figure 9-1 Hanging sheetrock and wood panels

Types of Sheetrock

The following are some of the different types of sheetrock used in construction:

- *Gypsum board is* the most commonly used wall and ceiling covering in construction today.
- *Greenboard* or *blueboard is* moisture resistant and is used in bathrooms, laundries, and similar areas.
- *Sound-deadening board is* a sublayer used with other layers of sheetrock (usually type X).
- *Backing board* has gray paper lining both sides. It is used as a base sheet on multilayer applications and is not suitable for finishing and decorating.
- *Foil-backed board* serves as a vapor barrier on exterior walls.
- *Vinyl-surfaced board is* available in a variety of colors. It is attached with special sheetrock finishing nails, screws, or channels and is left exposed with no joint treatment
- *Plasterboard or gypsum lath is* used for a plaster base. It is not compatible with Portland cement plaster.

Sheetrock Dimensions

Sheetrock usually comes in sheets that are 4 x 8, 4 x 9, 4 x l0, 4 x 14, and 4 x 16 feet. Its thickness is 1/4, 3/8, 1/2, or 5/8 inch.

- 1/4 or 3/8 inch is used effectively in renovations to cover existing finish walls with minor irregularities. This thickness is not adequate for single-layer application.
- 1/2 inch is most commonly used. It is adequate for studs or ceiling joists spaced 16 or 24 inches on center.
- 5/8 inch is widely used in multiple, fire-resistant combinations. It is recommended for single-layer walls.

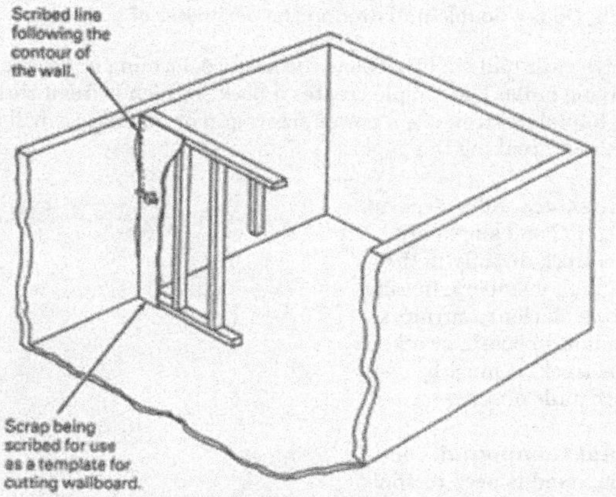

1. Place a piece of scrap material in the angle and scribe (mark) it to indicate the surface peculiarities.
2. Saw the scrap material along the scribed line.
3. Place the scribed strip on the wall panel material to be used. Keep the straight edge of the scrap material parallel with the edge of the panel. Scribe the good sheet of paneling.
4. Saw the sheet along the scribed line.

NOTE: This method can also be used for sheetrock; however, it would be easier to just fill the gaps caused by uneven walls with joint compound.

Figure 9-2. Fitting wall panels to uneven walls

Sheetrock Edges

Sheetrock edges are tapered 1/16 inch thinner than the body of the sheet about 1 1/4 inch on each sheet edge. The shallow channel formed will be brought level with tape and joint compound.

Sheetrock Application

Sheets may be applied either horizontally or vertically; specifications may indicate which method should be used.

Sheetrock Fasteners

Nails used are specially designed with oversized heads for greater holding power and treated to prevent rust and stains. The most common is the annular-ring nail. Other types of nails include the smooth-shank, the diamond-head (used to attach two layers of sheetrock or to attach sheetrock over existing materials), and the predecorated. The predecorated nails have smaller heads, are left exposed, and are colored to match the sheetrock.

Nails. If the sheetrock is single nailed, the nails should be spaced 6 to 7 inches apart on the ceilings and 6 to 8 inches apart on the walls (Figure 9-3). If the sheetrock is double nailed, the centers of the nail pairs are approxi mately 12 inches apart, with each pair 2 to 2 1/2 inches apart and the outer edges 7 inches on center (Figure 9-4). The distance from the edge should be 3/8 to 1/2 inch. Do not double nail around the perimeter of a sheet.

Drive each nail slightly below the surface, forming a "dimple." Be sure not to break the paper when driving nails. The dimple creates a pocket which is filled with joint compound. Screws are made of high-quality steel; use a power screw gun or an electric drill to drive them in just below the surface without breaking the paper.

Adhesives. Adhesives are used to bond single-ply sheetrock directly to the framing members, furring strips, masonry surfaces, insulation board, or other sheetrock. It must be used with nails or screws.

Joint Compound. Joint compound is used to apply tape over joints, to cover nailheads, and to smooth and level the surface. The powdered form is mixed with water to a desired consistency. It is also available ready mixed. This is the most common form and the easiest to work with.

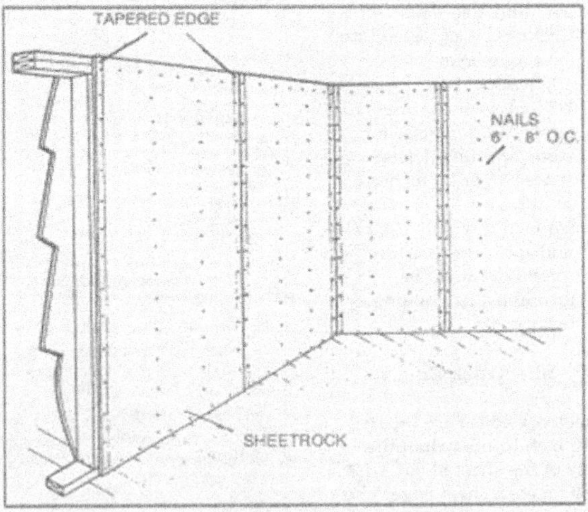

Figure 9-3. Sheetrock applied vertically

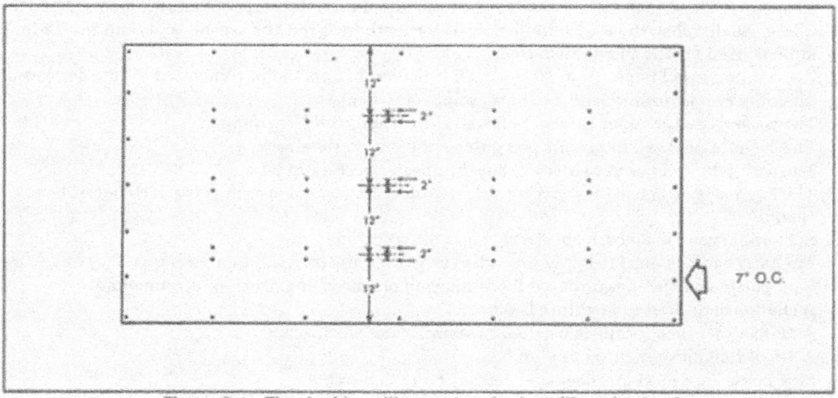

Figure 9-4. The double-nailing system for installing sheetrock

Joint Tape. Joint tape is applied with the first coat of joint compound. It reinforces joints and reduces cracking. The paper type may or may not be perforated. Perforated tape is easier to bed and cover than nonperforated tape. Fiber-mesh tape is self-sticking, which eliminates the need for the first coat of bedding joint compound.

Metal Accessories. Metal accessories include the corner bead and the casing bead. The corner bead is used on all exposed (outside) corners to ensure a clean finish and to protect the outside corners of sheetrock from edge damage It is nailed or screwed 6 inches on center to en sure that it is plumb.

The casing (stop) bead is used where sheetrock sheets butt at wall intersections or wall and exposed ceiling intersections or where otherwise specified. It is matched to the thickness of the sheetrock.

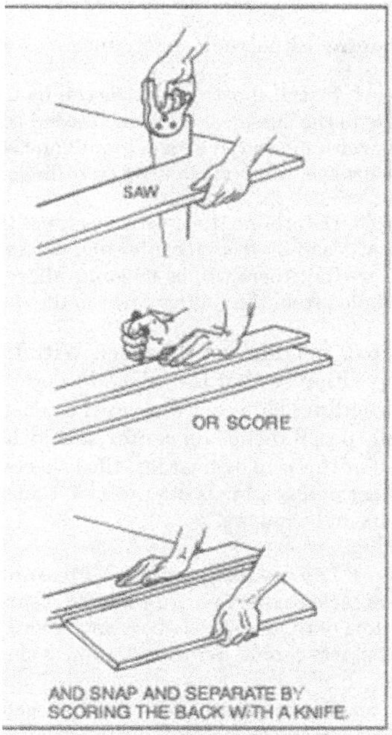

Figure 9-5. Cutting sheetrock

Sheetrock Tools

The following are tools used in the application of sheetrock:

- The sheetrock hammer is used for hammering nails.
- The sheetrock carrier (lifter) is used for carrying or lifting sheetrock.
- Sheetrock knives are used to apply and finish joint

compound. The 4-inch knife is used to bed the tape in the first layer of joint compound and for filling the dimples, the 6-inch knife is used for feathering out the second coat, and the 12-inch knife is used for the third/finish coat.

- The corner trowel flexes from 90° to 103°. It is used to apply joint compound in interior corners.
- The mud pan is used to hold and carry joint compound.
- The corner-bead crimper is used to fasten the corner bead by crimping.
- The T-square is used to lay out and guide a 90° cut on sheetrock.
- The utility knife is used to score or cut the sheetrock (Figure 9-5).
- The keyhole saw is used for cutting out irregular shapes and openings (such as outlet-box openings).
- Surform is used to smooth sheetrock edges after cutting.
- The tape banjo is used to apply tape (dry) or joint compound and tape (wet).
- Sandpaper and sponges are used for feathering or smoothing dried joint compound.
- A chalk line is used to facilitate layout.
- A 16-foot measuring tape is used for measuring the sheetrock.
- A 4-foot hand level is used to plumb.
- Saw horses are used for placing sheetrock on to make cuts.

Sheetrock Installation

There are three steps to installing sheetrock— hanging, finishing, and patching.

Hanging Sheetrock. Apply sheetrock as follows:

Step 1. Install sheetrock on the ceiling first. Measure the distance from the inside edge of the top plate to the outside edge of the second ceiling joist. Measure and cut a piece 48 inches long and to the width measured above. Install and secure the sheet to the ceiling with sheetrock nails. Nail spacing on ceilings is 5 to 7 inches on center.

Step 2. Determine the starting point of the wall. Using a measuring tape, locate a section where the studs are ~ foot on center and where a full sheet could be laid horizontally. Check the layout to ensure that there will be no joints above or below the door or window openings. Sheets will be installed from the ceiling down to the floor, starting at the ceiling.

Step 3. Install the first sheet. With the help of another person, place a sheet of sheetrock in position so that the edges fall on the center of the studs. Place the sheet snug against the ceiling, using a hand level to ensure that it is level. Secure the sheet with sheetrock nails 6 to 8 inches on center, 3/8 inch from the edge. Install succeeding sheets on the top half of the wall against installed sheets, ensuring that joints fall on the center of the studs and proper nail spacing is maintained. Using a utility knife or sheetrock saw, cut out openings for doors and windows.

Step 4. Lay out the receptacles. Measure the distances from an inside corner to both sides of the receptacle box and record them. Measure the distance from the installed sheetrock to the top and bottom of the receptacle box, and record them. Measure and mark these dimensions for the receptacle cutout, allowing 1/16-inch clearance all around.

Step 5. Cut out the opening for the receptacle. With a utility knife, drive a hole within the opening. Using a keyhole saw, cut out the opening. Use a slight undercut bevel so that the back opening is larger than the front opening.

Step 6. Install the prepared sheet. Place the prepared sheet in position, ensuring that the receptacle fits in the opening without breaking the paper. Make adjustments to the opening if necessary. Secure the sheet to the studs with sheetrock nails. Using a Surform, smooth the rough edges of the openings as necessary.

Step 7. Lay out and cut sheets for corner posts. Measure and cut the required number and sizes of sheets to cover corner posts. Scrap pieces of material may be used.

Step 8. Install the corner bead. Using a corner-bead crimper, install the corner bead on the exterior corners of corner posts. Nails may be used if necessary.

Finishing Sheetrock The finishing process consists of covering nailheads and covering seams (covering seams is also referred to as *finishing joints).* Finish sheetrock as follows:

Step 1. Check for improperly recessed nails by running the edge of a sheetrock knife over the nailheads. A clicking sound indicates a nail needing to be recessed.

Step 2. Use a 4-inch knife and mud pan with joint compound to apply a smooth coat of joint compound over the nails. Remove any excess compound.

Step 3. Use the knife and mud pan to apply a heavy coat of joint compound over a sheetrock joint, horizontal or vertical. A heavy coat is enough to ensure a good bond between the tape and sheetrock and to fill in tapered edges. Measure and cut the tape to the length required for a joint. Keeping the tape centered over the joint, start at one end of the joint and work toward the opposite end. Using the knife, press the tape into the compound, removing all excess compound. Work off all excess joint compound, being careful not to wrinkle the tape or leave air bubbles. Continue to tape all the joints in the same manner.

Step 4. Use a 4-inch knife to apply a heavy coat of joint compound over the sheetrock at the inside corner. Measure and cut the tape to the length required for the joint. Fold the tape in half lengthwise, keeping both edges even. Use a corner tape creaser if necessary. Apply the tape at the top and work downward, running the edge of your hand at the center of the tape to ensure that it is in the corner. Using the inside corner tool, press the tape into the compound, working off all excess compound and being careful not to wrinkle the tape or leave air bubbles.

Step 5. Apply the first coat of joint compound over the tape then apply a medium coat of joint compound. Feather the compound with the 6-inch knife to about 2 to 3 inches on each side of the joint. A good job of feathering and smoothing will minimize sanding later.

Step 6. Apply the second coat of joint compound over the tape and nail coverings. The joint compound previously applied must be completely dry. Use the 4-inch knife to apply a thin coat of compound over the nails, removing any excess compound. Using the steps above, apply the second coating to the joints using the 6-inch knife and feathering out 6 to 8 inches on each side of the joint.

Step 7. Apply the third coat of joint compound. The joint compound previously applied must be completely dry. Using the step above, apply the third coat using the 10-inch knife and feathering out 10 to 12 on each side of the joint. Nails should not require a third coat, but it may be applied if necessary.

Step 8. Using a damp sponge or fine sandpaper, sand the surface to a smooth finish, ensuring that there are no voids and that the surface is ready to receive paint.

Patching Sheetrock. There are several different methods of patching sheetrock, depending on the size of the hole.

For small holes, apply fiber-mesh tape directly over the hole. Cut the tape with joint compound and feather the edges. Sand or sponge the area smooth after it has dried.

For fist-size holes, cut out a rectangle around the hole with a keyhole saw. Cut a piece of backing (1 x 2 or 1 x 3) slightly larger than the opening itself. Glue or screw the backing into place. Cut a patch and glue it to the backing using either wallboard adhesive or mastic. Apply tape and coat it with compound. Feather the edges. Sand or sponge the area smooth after it has dried.

For large holes, mark and cut a rectangular section around the damaged area, reaching from the centers of the nearest studs. Cut a patch and screw or nail it to the studs. Apply tape and coat it with compound. Feather the edges. Sand or sponge the area smooth after it has dried.

BASE MOLDING

The interior trim of a building should match or complement the design of the doors, the windows, and the building. Base molding is the trim between the finished wall and the floor. It is available in several widths and forms. Figure 9-6, page 9-8, shows the types of base molding.

Square-edge (or two-piece) baseboard consists of a square-edged baseboard topped with a small base cap. When the wall covering is not straight and true, small base molding will conform more closely to the variations than will a one-piece base alone. This type of baseboard is usually 5/8 x 3 1/4 inches or wider. Installation of square-edged baseboard is shown in Figure 9-7.

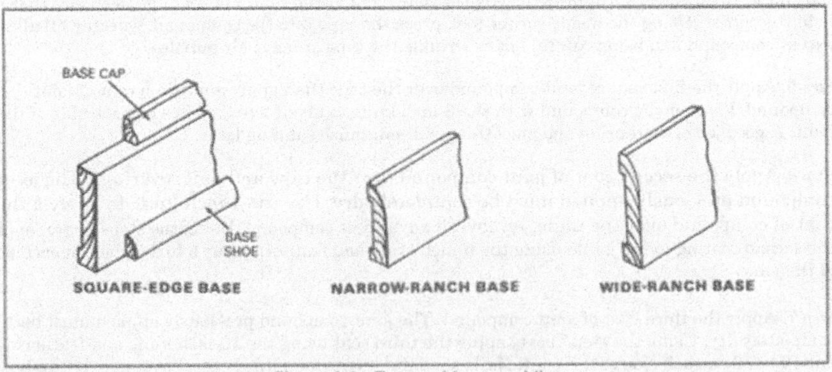

Figure 9-6. Types of base molding

Narrow- and *wide-ranch base* (one-piece baseboard) are 3/4 x 3 1/4 inches or wider and vary from 1/2 x 2 1/4 inches to 1/2 x 3 1/4 inches or wider.

A wood member at the junction of the wall and carpeting serves as a protective bumper; however, wood trim is sometimes eliminated. Most baseboards are finished with a 1/2- x 3/4-inch base shoe. A single-base molding without the shoe is sometimes placed at the wall-floor junction, especially where carpeting might be used.

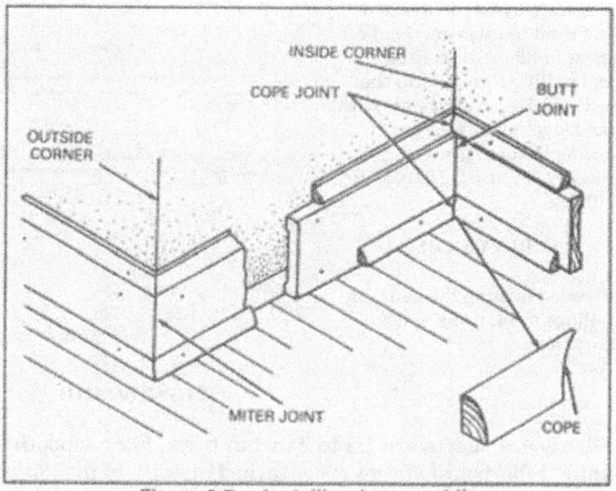

Figure 9-7. Installing base molding

Baseboard should be installed with a butt joint at the inside corners and a mitered joint at the outside corners. (The baseboard installation in Figure 9-7 is done with square-edge baseboard.) It should be nailed to each stud with two 8d finishing nails. Base molding should have a *coped joint* at inside corners and a mitered joint at outside corners. A coped joint is one in which the first piece is square cut against the plaster or base and the second molding is coped. This is done by sawing a 46° miter along the inner line of the miter. The base shoe should be nailed into the subfloor with long, slender nails, but not into the baseboard itself. Then, if there is a small amount of movement in the floor, no opening will occur under the shoe. When several pieces of molding are needed, they should be joined with a lap miter (Figure 9-8). When the face of the base shoe projects beyond the face of the molding, it abuts (Figure 9-9).

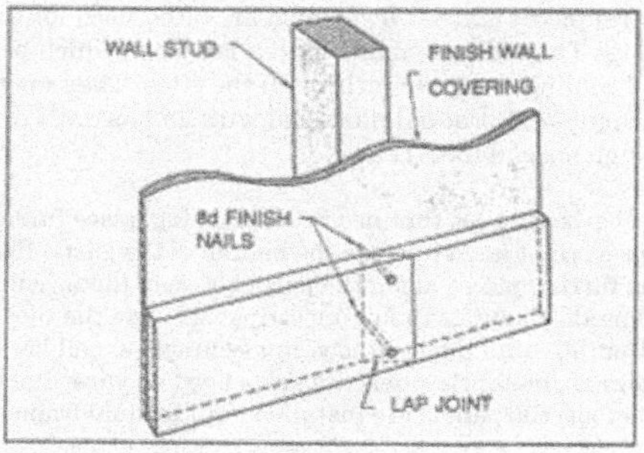

Figure 9-8. Trim lap-miter joint

CEILING COVERINGS

In current construction, sheetrock, plywood, and fiberboard are used instead of laths and plaster to cover ceilings.

SHEETROCK

Cut the panels and treat the joints the same as for walls and partitions, making sure that joints break on the centers of ceiling joists.

A brace may be constructed and used to raise and hold a sheet in place when fitting and nailing the sheet to the ceiling joists. Nail sheets with the lengths going across ceiling joists to prevent sagging (Figure 9-10, page 9-10).

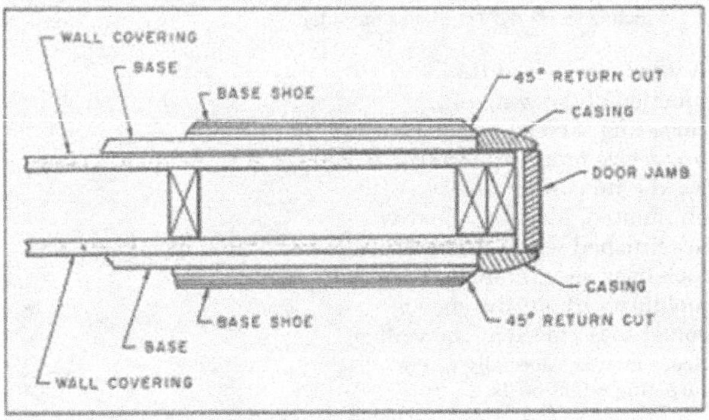

Figure 9-9. Base-shoe trimming

PLYWOOD

Plywood is hung the same on ceilings as on walls and partitions.

FIBERBOARD

Fiberboard sheets are 1/2 to 2 inches thick. For a smooth cut on these sheets, use a utility knife. Fiberboard sheets are attached directly to the joists. To improve ceiling appearance, cover the joints between the sheets with batten strips of wood or fiberboard. Smaller pieces of fiberboard (tiles) require furring strips (wooden strips nailed across joints) (Figure 9-11, page 9-10).

Fiberboard sheets also come in small (rectangular or square) pieces called *tiles,* which are often used for covering ceilings They may be made with a lap joint, which permits blind-nailing or stapling through the edge. They may also be tongue-and-grooved, fastened with 2d box nails driven through special metal clips.

For fiberboard tiles that need solid backing, place furring strips at right angles across the bottom of the joists. Place short furring pieces along the joists between the furring strips. Nail metal channels to furring strips and slide the tiles horizontally into them. In lowering ceilings (usually in older buildings), metal channels are suspended on wire. Some large (2 x 4 foot) tile panels are installed in individual frames.

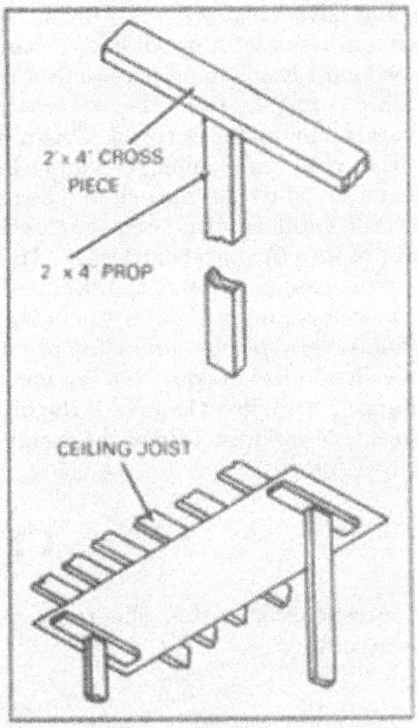

Figure 9-10. Brace for holding ceiling panels

CEILING MOLDINGS

Ceiling moldings are sometimes used at the junction of the wall and the ceiling to finish the sheetrock paneling (sheetrock or wood). Inside corners should be coped joints. This ensures a tight joint even if minor moisture changes occur. Figure 9-12 shows ceiling molding.

For sheetrock walls, a small, simple molding might be best. For large moldings, finish nails should be driven into the upper wallplates and also into the ceiling joist, when possible. (For plastered ceilings, a cutback edge at the outside of the molding will partially conceal any unevenness of the plaster and make painting easier where there are color changes.)

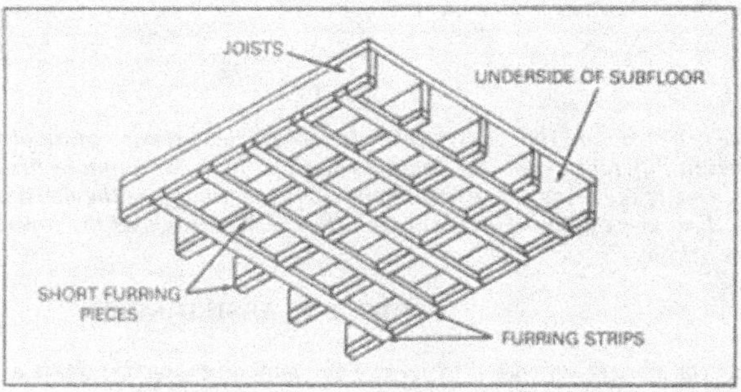

Figure 9-11. Furring strips on ceiling joists

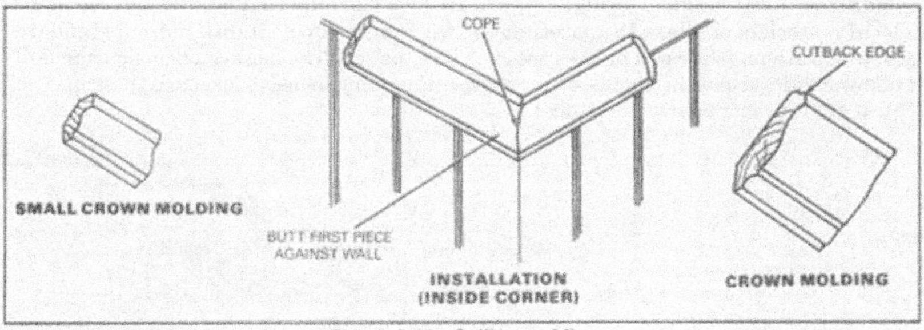

Figure 9-12. Ceiling molding

NONSTANDARD FIXED BRIDGE

CHAPTER 10

A bridge is a structure that carries a roadway over a depression or an obstacle. Bridges may be classified in different ways. Two general classifications, for example, are highway and railroad bridges. One of the bridges most commonly found in the TO is the nonstandard fixed bridge. This chapter discusses the construction of both the substructure and the superstructure of this important military bridge.

BRIDGE CLASSIFICATION

A bridge completely supported by its two end supports (abutments) is called a *single-span* bridge. A bridge having one or more intermediate supports between the abutments is a *multispan* bridge. All supports of a fixed bridge transmit the load directly to the ground.

A *nonstandard fixed highway* bridge (Figure 10-1) is a semipermanent bridge constructed from local materials or Class IV materials drawn from a depot. It differs from standard bridges, which are prefabricated bridges assembled at the site. The most common nonstandard fixed highway bridges are the simple, stringer-type (the stringers being logs) and those made of structural grade timber or structural steel.

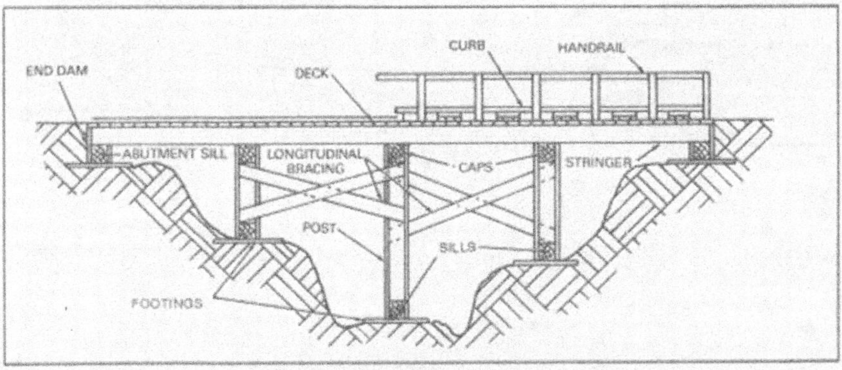

Figure 10-1. Nonstandard fixed highway bridge

A military bridge has two principal parts: the lower part (substructure) and the upper part (superstructure).

SUBSTRUCTURE

The substructure of a bridge supports the superstructure. The substructure consists of

ABUTMENTS

There are two types of end supports or abutments: *footing* and *pile*.

Footing

The footing abutment consists of—

- *Footings.* Footings transmit the load to the ground. They receive the load from the sill and distribute it over a sufficient area to keep the support from sinking into the ground.
- *Sill.* The abutment sill (Figure 10-2) receives the load from the stringers and transmits it to the footings.
- *End dam.* The end dam (or *bulkhead)* is a wall of planks at the end of the bridge to keep the approach road backfill from caving in between the stringers.

Pile

The pile abutment (Figure 10-3) has three main parts:

- Piles driven into the ground, transmitting the load to the soil.
- A cap on top of the piles to receive the load from the stringers.
- Sheeting fastened to the piles to hold the backfill in place.

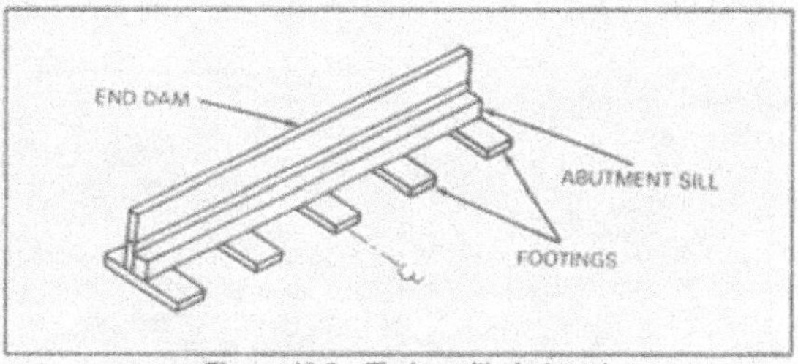

Figure 10-2. Timber sill abutment

INTERMEDIATE SUPPORTS

The following are some of the different types of intermediate supports.

Pile Bent

The pile bent (Figure 10-4) consists of the bent cap, which provides a bearing surface for the stringers and transmits the load to the piles; and the piles, which transmit the load to the soil.

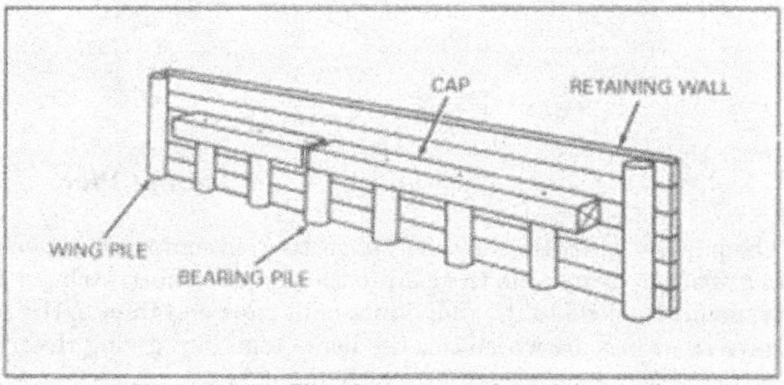

Figure 10-3. Pile abutment and retaining wall

The support for the loads may come either from column action, when the tip of the pile bears on a firm stratum such as rock or hard clay, or from friction between the pile and the soil into which it is driven. In both cases, earth pressure must give some lateral support; transverse bracing is also often used for this purpose.

The pile bent is used for highway bridges only. It is designed to carry both vertical and lateral loads and can be used for spans of up to 50 feet. Its ground-to-ground height is a function of its unbraced length.

Trestle Bent

The trestle bent (Figure 10-4) is like the pile bent except that posts take the place of piles. The posts transmit the load from the cap to the sill, the sill transmits the load to the footings, and the footings transmit the load to the soil. The length of the posts varies according to the height of the bridge above the gap to be spanned. Transverse bracing like that used with the pile bent is provided.

The trestle bent is used for highway bridges only; however, unlike the pile bent, it is designed to carry vertical loads only. It can be used for spans of up to 30 feet and for ground-to-grade heights of up to 12 feet.

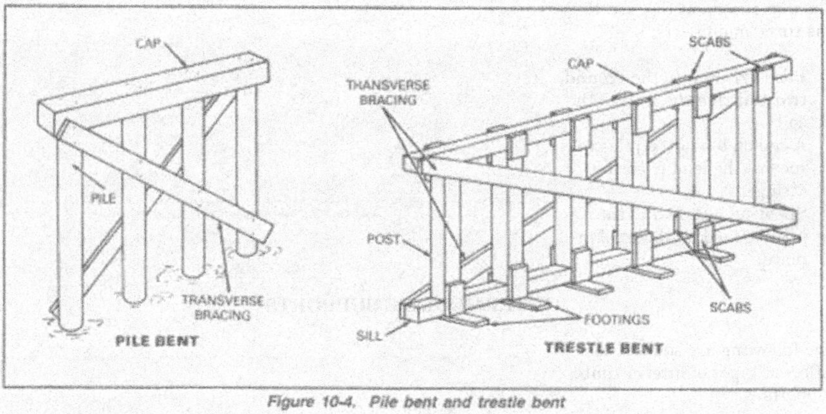

Figure 10-4. Pile bent and trestle bent

Pile-Bent Pier

The pile-bent pier (Figure 10-5, page 10-4) is composed of two or more pile bents with a common cap. The cap transmits the load to the corbels (short, stringer-like members) that, in turn, transmit the load to the individual bent caps and then to the piles and to the soil. Piers usually have cross bracing which ties the bents together, giving them longitudinal rigidity.

The use of multiple bents gives the pile-bent pier great strength. As a result, the pile-bent pier can be used for both highway and railroad bridges. It will carry both vertical and lateral loads, can be used for spans of up to 200 feet, and its ground-to-grade height is governed by its unbraced length.

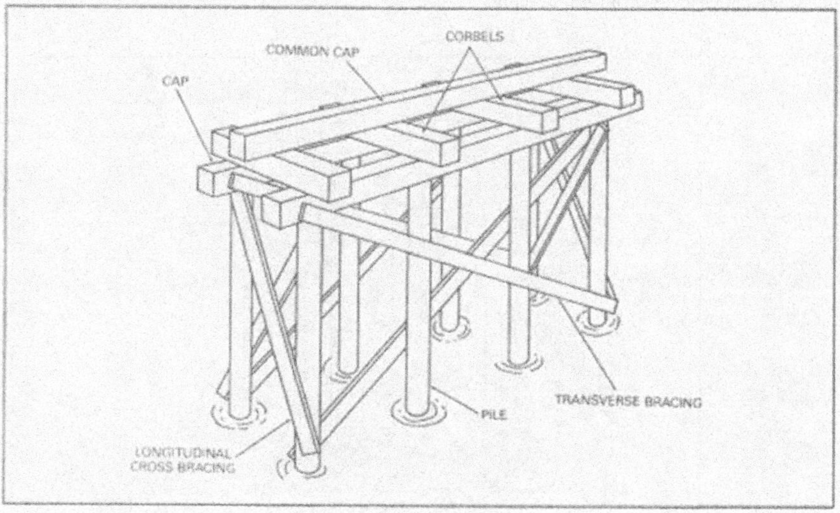

Figure 10-5. Pile-bent pier

Trestle-Bent Pier

The trestle-bent pier (timber-trestle pier) (Figure 10-6, page 10-4) is the same as the pile-bent pier, except that it has sills and footings which transmit the load to the soil.

The trestle-bent pier is used for highway bridges only. It is designed to carry vertical loads only and can be used for spans of up to 60 feet and for ground-to-grade heights of up to 18 feet.

Crib Pier

The crib pier (Figure 10-7, page 10-5) is quite different from pile and trestle piers. It is composed of logs or dimensioned timber fitted together in log-cabin style and is usually filled with rock or other stable fill material. The crib pier should be made so that it needs no exterior bracing. As an expedient, crib piers may be built to the height of the stringers, eliminating the trestle bents.

The crib pier is used for highway bridges only. It is for vertical loads only and can be used for a span of up to 50 feet and a ground-to-grade height of up to 12 feet.

BRACING

Bracing consists of longitudinal bracing, transverse bracing, and diaphragms.

Longitudinal bracing (Figure 10-8) is used to stabilize the bridge centerline.

Transverse bracing (Figure 10-8) provides stability at right angles to the centerline. It is sometimes called sway or lateral bracing.

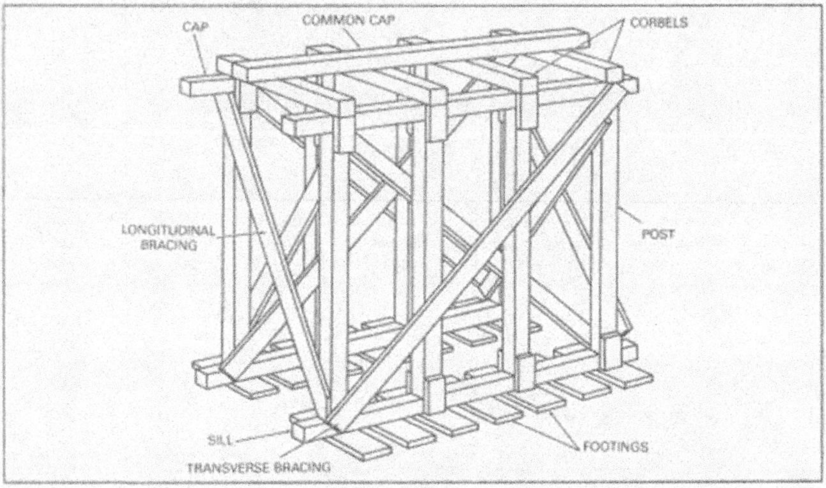

Figure 10-6. Trestle-bent pier

Diaphragms are braces between stringers to prevent them from deflecting laterally (buckling) under load. In spacing these diaphragms, the ratio of distance between diaphragms to the width of the top of stringer (L/b ratio) should not exceed 30 for timber.

L = distance between diaphragms
b = width of top of stringer

Example. If the stringer is 6 inches wide—

Figure 10-7. Crib pier

L = 30 *L* = 180 *inches* (15 *feet)*

In this example, diaphragms should be used every 15 feet between stringers 6 inches wide.

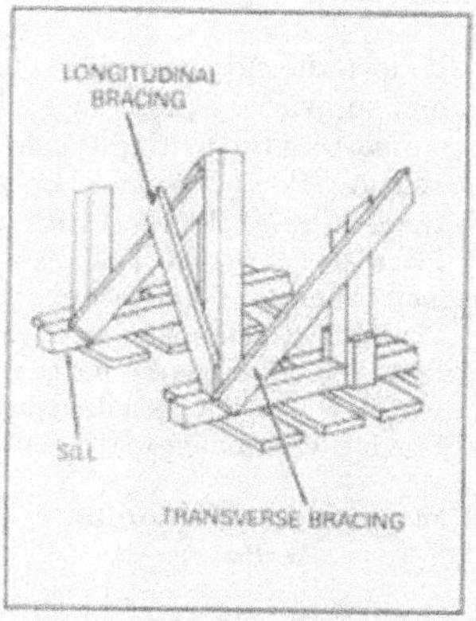

Figure 10-8. Bracing

CONSTRUCTION PROCEDURES

The following paragraphs contain construction procedures for a trestle-bent bridge. This includes laying out the centerline and constructing abutments, retaining walls, and trestle bents.

Layout of Centerline

The first task in constructing a trestle-bent bridge is laying out the centerline (Figure 10-9, page 10-6):

Step 1. Stretch a line or tape representing the centerline across the stream or ravine.

Step 2. Attach the line to stakes driven into the ground at least 15 feet behind the proposed location of the abutment sills. For defiles wider than 100 feet, use intermediate stakes as needed to prevent sag.

Step 3. Place the line at the level of the intended top of the flooring or at some known distance above or below it.

Construction of Abutments

Saving time in abutment construction is especially important on short bridges. Abutment and approach preparation often requires as much time as the rest of the bridge. Use the simplest abutment possible; often a timber sill with timber footings is adequate (page 10-2).

The end dam is installed after the stringers and planks.

After the centerline is fixed—

Step 1. Place the abutment sill at approximately its correct location under the tape. See that it is at right angles to the centerline by using a line from the centerline stake 15 feet behind the sill to each end of the sill. Both distances must be the same.

Step 2. Once the sill is properly

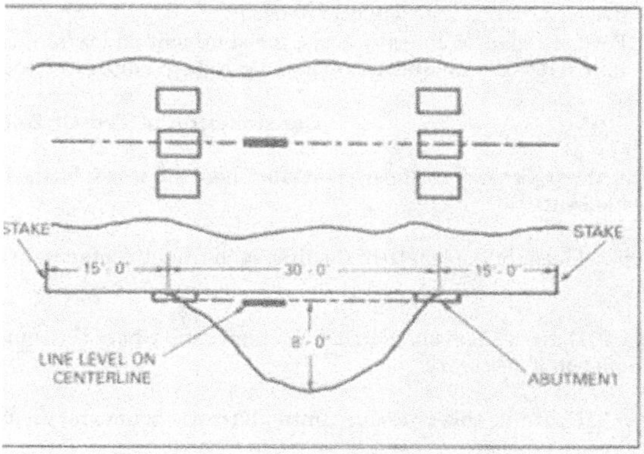

Figure 10-9. Laying out a centerline

located, mark its position and remove it to construct the foundation.

- Remove the earth as needed to provide a level surface for footings. The sill must be level and supported equally by each footing when installed. Make sure that the surface supporting the footings is about 2 inches higher than its final position to allow for settling. Do not dig too deeply. If this is done by mistake, do not backfill with earth. Instead, raise the level with planking.
- Place the two outside footings so that their outer edges are under the ends of the sill. Place the long dimension of the footings parallel to the bridge centerline.
- Place the remaining footings, equally spaced, between and in line with the outside footings.
- Place the sill on the footing centerline so that the load is in the middle of each footing. Place the sill with the largest dimension vertical.
- Provide for drainage of the abutment area.

Construction of Retaining Walls

Retaining walls and revetments, when needed, are part of the abutment construction. The simplest type of retaining wall is built of planks or logs supported by piles or posts. (Figure 10-10 shows an abutment and retaining wall; Figure 10-11 shows retaining-wall details.)

Figure 10-10. Abutment and retaining wall

- Use wing walls to prevent the earth from washing out behind the retaining wall.
- Drive piles or posts 4 feet into the ground.
- Fasten anchor cables from the top of the piles to a deadman behind the retaining wall or to the wing-wall end. These deadmen and anchors can be eliminated if two or three rows of piles, driven as far as they will go, are used.
- For long spans and heavy loads, the abutment and retaining wall are often constructed as a unit. This may also be necessary where steep banks and poor soil conditions exist.

Construction of Trestle Bents

After the position of the near-shore abutment sill is established, locate the position of the first trestle bent:

Step 1. Measure the length of the first span from the abutment sill along the centerline (Figure 10-12).

Step 2. Drive a small stake under the centerline where the center of the trestle bent is to be. Use a plumb bob if necessary.

Step 3. Continue this procedure until all trestle bents and the far-shore abutment sill are located.

Step 4. Excavate and place footings under the trestle bent the same as for the abutment (page 10-

5). Outside footings under the trestle sill are centered under the outside posts of the bent.

- Measure the vertical distance from the centerline down to the top of the footings.
- If the centerline was placed at the intended top of the flooring, this distance minus the thickness of the tread, deck, and stringers gives the height of the trestle bent.
- If steel stringers are to be used, allow also for the thickness of the nailing strips.

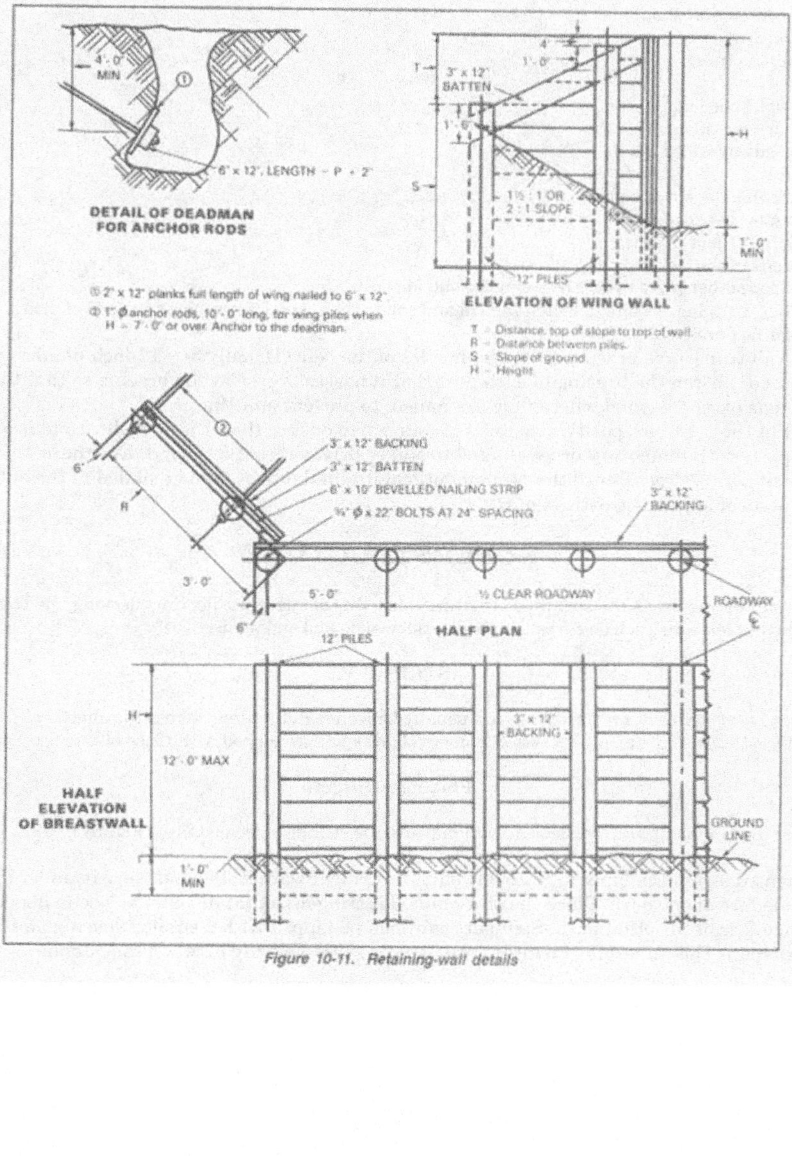

Figure 10-11. Retaining-wall details

Step 5. To obtain the correct height of the trestle-bent posts, subtract the thickness of the cap and sill from the height of the trestle bent.

Additional Construction Procedures

There are some additional procedures that should be followed when constructing a substructure.

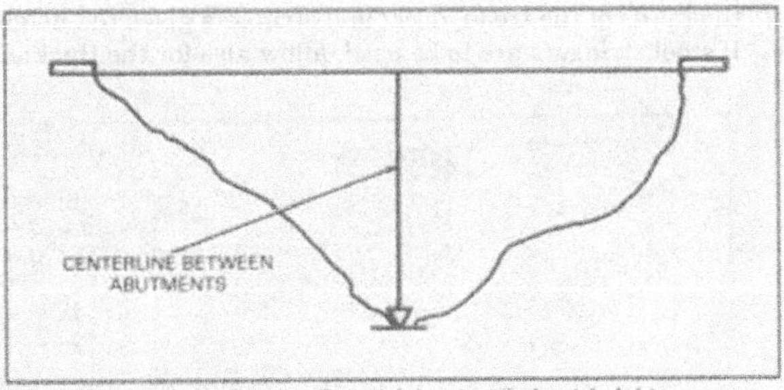

Figure 10-12. Determining trestle-bent height

- Make the length of the cap and sill equal to the roadway width plus 2 feet.
- Center the outside posts under the roadway edges 1 foot from the ends of the cap and sill. Space other posts evenly between the outside posts.
- Use driftpins or bolts to fasten the sill and cap to the posts. Use scabbing instead of driftpins for fast erection.
- Nail transverse bracing across both sides of the bent. Usually 3- x 12-inch planks are used. Fasten the bracing to each post that it passes over. Cut the bracing so that the ends extend beyond where they are nailed, to prevent splitting.
- Put the bent into position, using a plumb bob to ensure that it is straight. Hold it in place with temporary braces nailed to stakes driven into the ground. Use these temporary braces until the permanent longitudinal bracing can be nailed to the outside posts of adjacent trestle bents.

SUPERSTRUCTURE

The superstructure is the spanning structure consisting of stringers, flooring (decking and tread), and other features such as curbs, handrails, sidewalks, and end dams.

STRINGERS

When wood stringers are used, they are usually long enough to extend across the abutment sills and trestle caps on which they rest. Stringers of one span are lapped with those of the next span.

Placing Stringers

After the abutment and trestle bents are in place, the stringers are installed (Figure 10-13).

When stringers are lapped, place one outside stringer so that its inside face is under the inside face of one curb. Place the other outside stringer so that its outside face is under the inside face of the other curb. Stringers can then be lapped with a similar spacing on the next span. The remaining stringers are usually spaced evenly between the outside stringers. On some narrow one-lane bridges, stringers may be grouped closer together under the vehicle tracks.

When stringers are butted, or continuous across the span, place the outside faces of both outside stringers under the inside faces of the curbs.

Fastening Stringers

Fasten stringers (Figure 10-14) as follows:

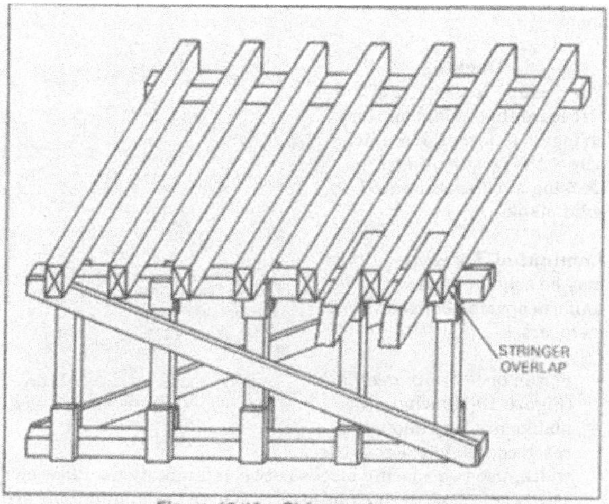

Figure 10-13. Stringer placement

Wood Stringers. Fasten wood stringers by driving nails diagonally through the side of the stringer into the cap or by driftbolts. When using driftbolts, bore a hole, smaller in diameter and 3 inches shorter than the driftbolt, through the stringer and into the cap.

Steel Stringers. Fasten steel stringers by—

- Driving railroad spikes into the cap beside the flange.
- Driving 60d nails partially into the cap and bending them over the bottom flange.
- Driving nails or driftbolts through prebored holes in the bottom flange.

NOTE: When steel stringers are not fastened through their flanges, frequent inspection is necessary to be sure the stringers have not shifted. Fasten wood nail strips to the top flange of steel stringers to provide a means of fastening the flooring.

When a laminated deck (planks placed on edge) is to be installed, the planks may be fastened to steel stringers either by using metal clips provided for the purpose or by driving nails partially into the deck and bending them around the stringer flange (Figure 10-15).

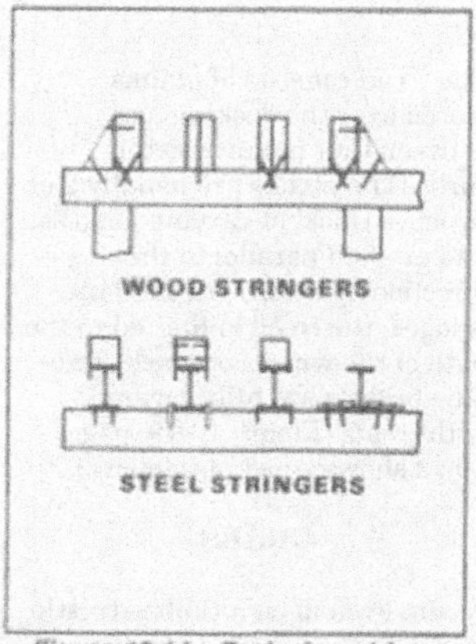

Figure 10-14. Fastening stringers

FLOORING

The flooring system of a typical timber-stringer trestle bridge consists of two main parts: the decking and the tread.

Decking

The decking is the part of the structure that is laid on the stringers to form the roadway across the trestle bridge. Decking may be laminated or solid plank.

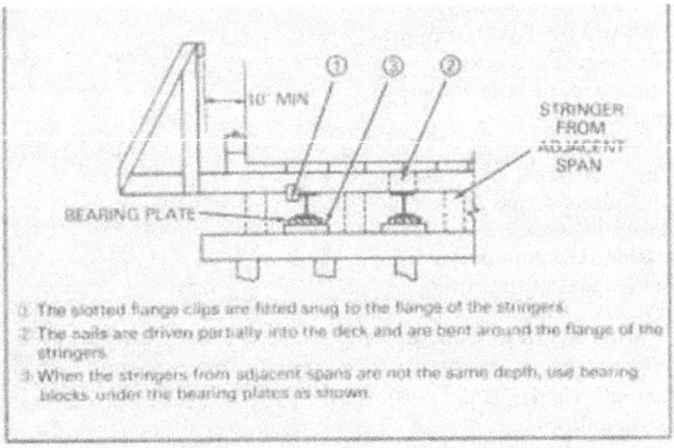

Figure 10-15. Open laminated deck, sectional view

Laminated. Laminated decks may be solid or open with uniform spacing between members.

- For an *open laminated* deck (Figure 10-16) where the planks are long enough to reach completely across the width, use two spacing blocks between laminations. Place spacers on the stringer nearest one-third the length of the lamination. Where the laminations are not long enough (usually true for two-lane bridges), lap the laminations on a **central** stringer. Put a spacer block at each outside stringer.
- For a *solid laminated* deck, place laminations solidly against one another.

Solid Plank. For a solid-plank deck (Figure 10-17), lay planks horizontally, at right angles to the stringers. Leave a 1/4-inch space between planks to allow for swelling when wet.

Extend the decking about 2 feet at approximately 5-foot intervals to support the handrail posts (Figure 10-18, page 10-12).

Tread

The tread consists of planks placed over the decking and between (but not under) the curbs. The planks are usually 2 or 3 inches thick, of varying lengths, and are laid parallel to the direction of traffic. On one-lane bridges, the tread is limited to the path of the wheels or track. Two-lane bridges are fully covered with tread. (Figure 10-19, page 10-12 shows tread placement.)

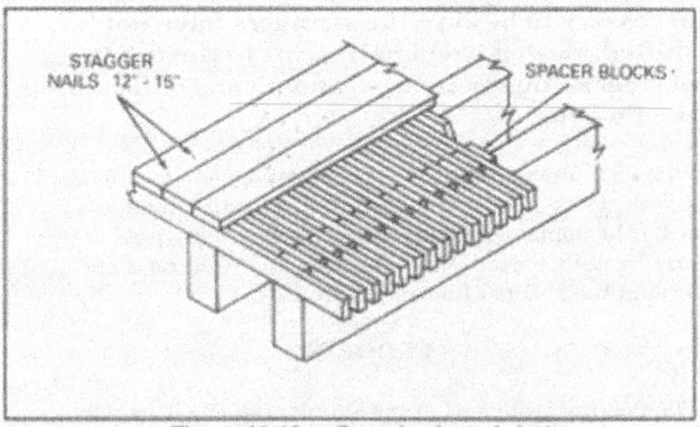

Figure 10-16. Open laminated deck

CURBS

A curb system on a timber-trestle bridge is used to guide traffic on the bridge. When assorted sizes of lumber are available, make curbs of 6- x 6-inch timber supported on 6- x 12- x 30-inch curb risers, spaced on about 5-foot centers. The curb is usually bolted to the decking with 1/2-inch bolts, two per curb riser.

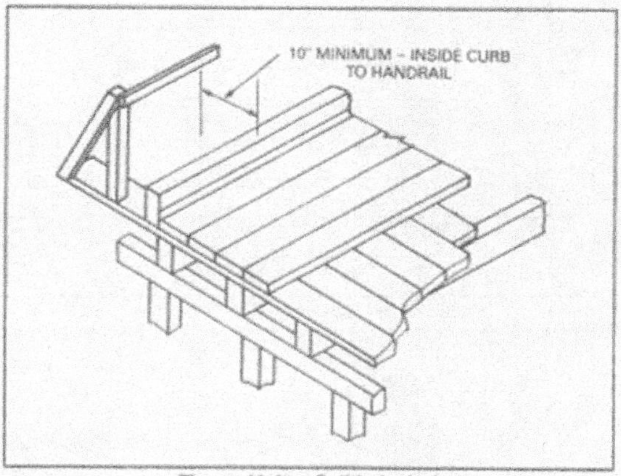

Figure 10-17. Solid-plank deck

HANDRAILS

Handrails (Figure 10-20, page 10-12) mark the bridge route and provide safety for pedestrians crossing the bridge. Make handrails of 2- x 4-inch or larger material, if available. Over a laminated deck, make handrail posts and knee braces of the same material as the deck so that they can be fastened snugly between the laminations, which are extended to receive them.

For solid-plank decks, toenail 4 x 4 posts, or two 2 x 4s nailed together, to the extended planks. Make the posts 42 inches high and space them on 5-foot centers. Place the posts so that the distance from the inside face of the curb to the inside face of the handrail is at least 10 inches.

SIDEWALKS

If sidewalks are necessary, form them by extending the decking an additional 36 inches. Place stiffening members underneath the outside edge. Support them with braces attached to the stringers, where necessary.

END DAM

The end dam is the wall that withstands the earth pressure of the abutment of a bridge (see Figure 10-1, page 10-1). After the stringers and flooring are in place, construct an end dam of flooring planks across the end of the stringers. The end dam should extend across the roadway from the top of the footing to the top of the tread. After placement of the end dam, complete the approach up to the top of the bridge deck. Post the traffic-control and classification signs. The bridge is now ready for traffic.

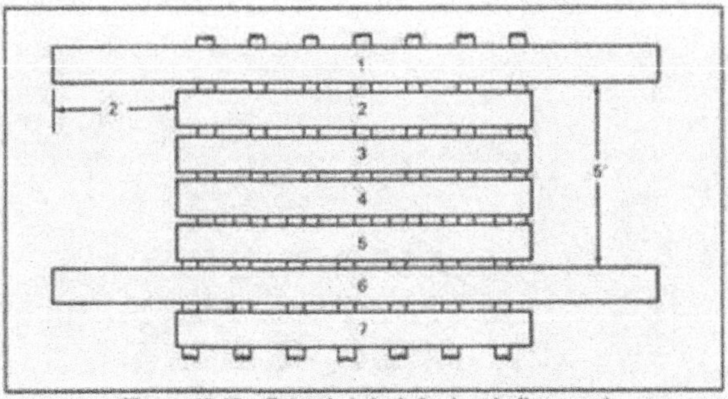

Figure 10-18. Extended deck for handrail support

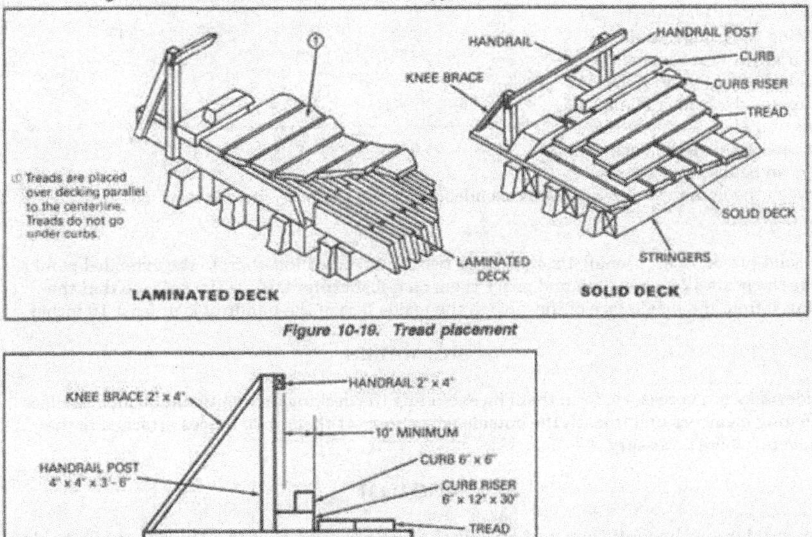

Figure 10-19. Tread placement

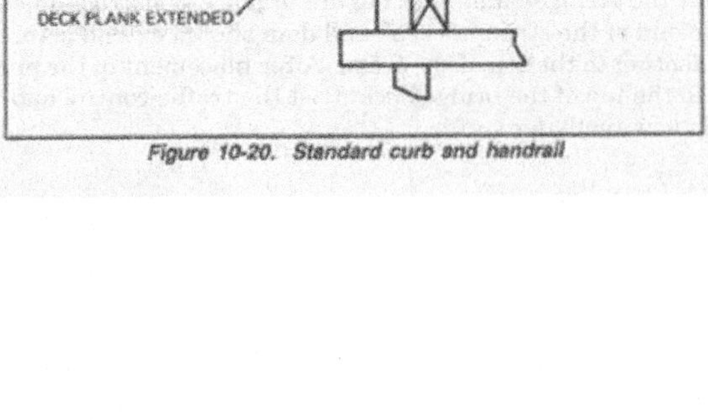

Figure 10-20. Standard curb and handrail

TIMBER-PILE WHARVES CHAPTER 11

Wharves are used for loading and unloading ships. This chapter describes how a carpenter constructs a timber-pile wharf. The topics covered include—

- *Layout and installation of piles for pile-wharf construction.*
- *Construction of a wharf superstructure.*
- *Installation of docking hardware.*

(For more detailed information on timber-pile wharves, see FM 5-480.)

TYPES OF WHARVES

Wharf is an overall term that applies to any waterfront structure designed to make it possible for vessels to lie alongside the shore for loading and unloading. The term *wharf is* confined in practice to the T- and U-type *marginal* wharves (Figure 11-1, page 11-2). A marginal wharf usually consists of a timber or steel superstructure supported by a series of timber, steel, or concrete pile bents.

The other structures shown in Figure 11-1 are called *piers,* except the *quay. A* quay is a reinforced landing place made toward the sea or at the side of a harbor. All structures shown in Figure 11-1 may consist of fill supported by bulkheads.

TYPES OF PILES

To protect a wharf against normal wear and tear, three types of piles are used: bearing, fender, and mooring piles. The types of piles are discussed in the following paragraphs:

BEARING PILES

Bearing piles support the wharf or pier framework and decking. The piles should be straight and measure at least 6 inches across the top, 18 inches across the butt (bottom), and from 60 to 80 feet in length. Pile length varies according to the depth of the water and condition of the bottom. Bearing piles should be spaced from center to center 6 to 10 feet apart in one direction and 5 feet apart in the other direction.

FENDER PILES

The force of a moving ship coming in direct contact with bearing piles is enough to collapse an unprotected wharf. To protect and absorb the initial shock, fender piles are placed about 2 1/4 feet out from the centerline of the outside row of bearing piles. These piles are placed about 18 feet apart and along the sides where ships dock.

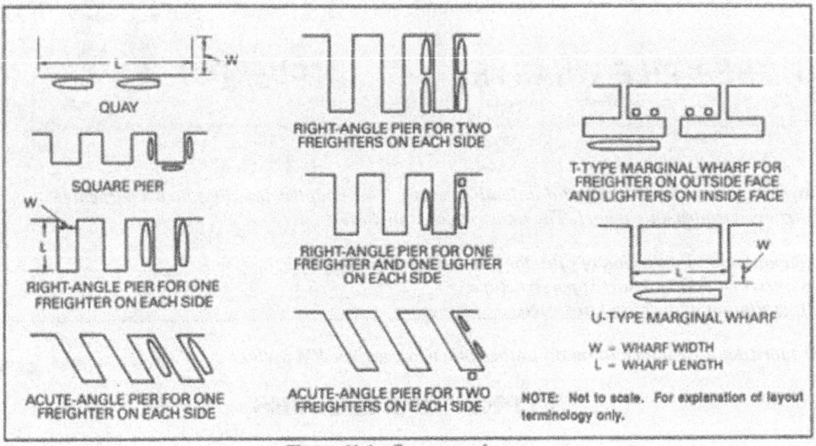

Figure 11-1. Common wharves

MOORING PILES

Mooring piles are aligned with the outside row of bearing piles and are spaced about 30 feet apart. This type of pile is braced along the outside row of bearing piles and usually extends to about 4 feet above the floor (or deck) of the platform. The 4-foot extension provides ample space to secure mooring lines.

NOTE: Timber piling must be treated with creosote or some other preservative Compound to protect it from fungi and marine borer attacks.

INSTALLATION OF PILES

Pile-driving equipment and the methods of driving and pulling piles are covered in FM 5-480. The equipment is operated by a special crew, but the carpenter is present during the pile driving to direct the alignment of the piles.

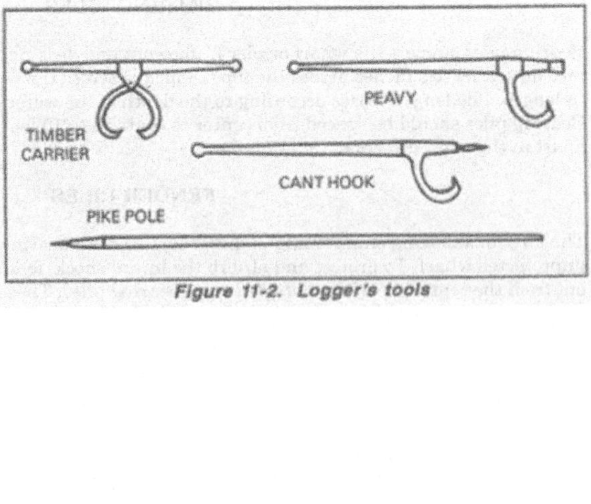

Figure 11-2. Logger's tools

USING SPECIAL TOOLS

Since most of the heavy

timbers used to build waterfront structures cannot be manhandled, special tools, known as logger's tools (Figure 11-2), are used to move and place these timbers. They consist of—

- *Peavy and cant hooks.* Lever-type tools, used mainly to roll timbers.
- *Timber carriers.* Two-man tools, used mainly to pick up and carry timbers.
- *Pike poles.* Used to hold or steady timbers while they are being placed.
- *Cranes.* Normally, two men are assigned to a crane: the operator and the helper. The helper drives the crane carrier (truck) and hooks and unhooks loads. Using standard signals, the helper tells the operator when to lift and lower the load and where to position it. After the heavy timbers have been moved and placed, the carpenter's level is used to level them properly.

NOTE: Although the crane cannot be considered a special tool, it is mentioned here because of its use to raise and lower heavy timber.

STRAIGHTENING PILES

Piles should be straightened as soon as any misalignment is noticed. The desired accuracy of alignment varies with each job; however, if a pile is more than a few inches out of plumb, it should be set true. The greater the penetration along the wrong line, the more difficult to get the pile back into plumb. To realign piles, use one of the following:

- A block and tackle (Figure 11-3), with the impact of the hammer jarring the pile back into line.
- A jet (Figure 11-4, page 11-4), either alone or in conjunction with a block and tackle.
- A block and tackle and an alignment frame (Figure 11-5, page 11-4) to pull the piles In a bent into proper spacing and to align them after they have been driven.

When a floating pile driver is used, a frame (template) for positioning piles may be fastened to the hull. A floating

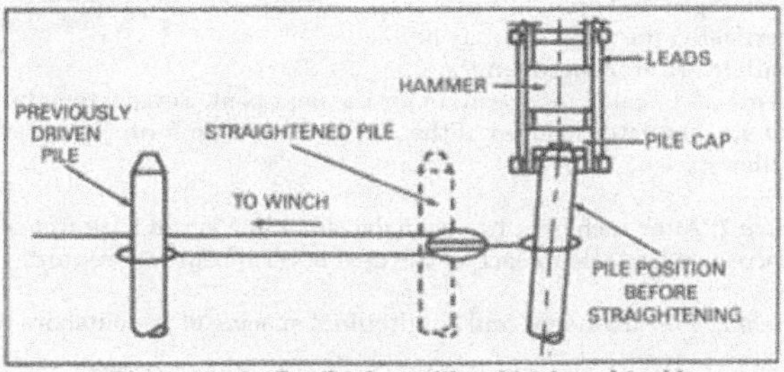

Figure 11-3. Realigning with a block and tackle

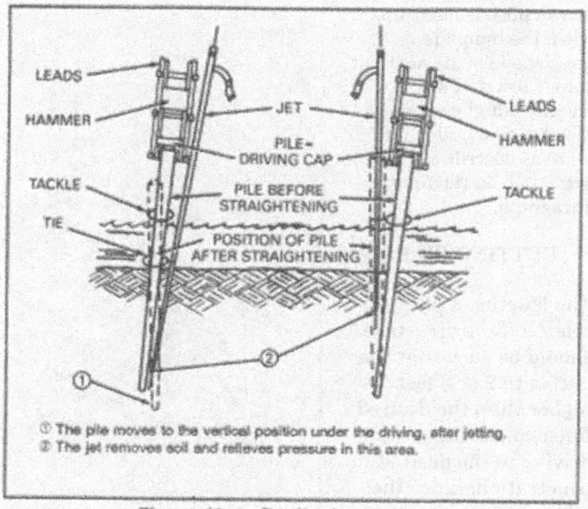

Figure 11-4. Realigning with a jet

template (Figure 11-6, page 11-6) is sometimes used to position the piles in each bent. The spacing of battens is such that the centerline between them is on the pile line desired. Battens are placed far enough apart so that, as the pile is driven, the larger diameter butt end will not bind on the template and carry it underwater.

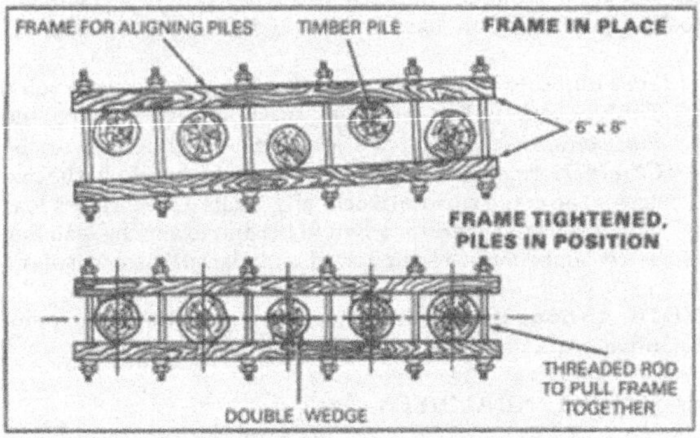

Figure 11-5. Realigning with a frame

A chain or collar allows the template to rise and fall with the tide. If the ends of the battens are hinged and brought up vertically, the template may be withdrawn from between the bents and floated into position for the next bent. Several templates may be used for a bent; or a single template is moved, if the pile spacing is uniform. The position of the piles is controlled as follows:

Step 1. After each bent has been driven, a line is run back from each pile in the outer bent to the corresponding pile in each of the next several bents shoreward.

Step 2. The alignment and longitudinal spacing of the outshore bent are verified.

Step 3. Any deviation in position of previously driven piles is made up when the template is positioned for the next bent. Piles that are slightly out of position may later be pulled into place as described previously in the first paragraph.

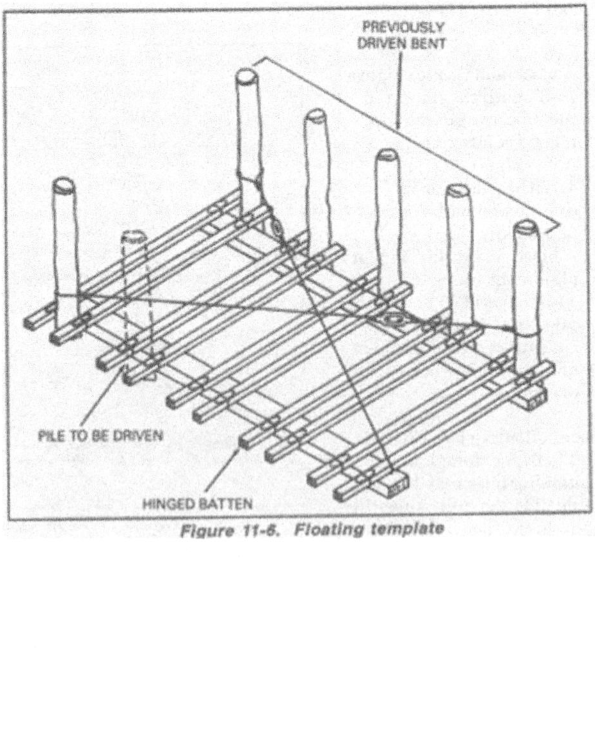

Figure 11-6. Floating template

CUTTING PILES

The lengths of pile selected for a structure should be such that the butts are 2 or 3 feet higher than the desired finished elevation after driving to the desired penetration. Since the pile capping should bear evenly on every pile in the bent, trimming should be carried out accurately by nailing saw guides across all piles in the bent. Figure 11-7 shows cutting piles.

CAPPING TIMBER PILES

Caps are large timbers placed on top of the timber-bearing piles to support the superstructure. The pile capping is fastened as shown in Figure 11-8, page 11-6, and as follows:

Step 1. After the piles have been cut, the cap is put in place; a hole for a driftpin is bored through the cap into the top of each pile; and the driftpins are driven into the holes.

Step 2. At a joint between pile cap timbers, a splice scab is bolted across the joint to each side of the pile cap.

Step 3. The working platform, aligning cables, or spacing frame may then be removed, since the driftpins will hold the piles in the proper position.

BRACING PILES

Bents are braced as shown in Figure 11-8 and as follows:

Step 1. Bolt diagonal timbers to each pile with the bracing running in one diagonal direction on one side of the bent and in the opposite diagonal direction on the other side.

Step 2. If the piles in a bent differ a lot in diameter at the point of bracing, make one of the following corrections:

- Large piles may be flattened down with an ax (hewed or dapped).
- Small piles may be blocked out with filler pieces.
- The flexibility of the braces may be used to pull them tight against the piles.

(Figure 11-9, page 11-6, shows transverse bracing.)

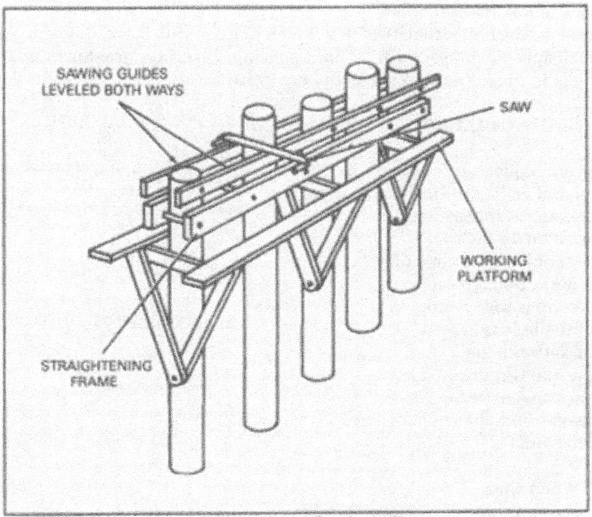

Figure 11-7. Cutting piles

WHARF SUPERSTRUCTURE

After the timber pile bents have been aligned, braced, and capped, the construction of the wharf superstructure is begun. Building the superstructure consists of installation of the stringers, the decking, and the curbs or stringpieces; and erection of the fender systems. Figure 11-10 shows stringers and decking in place.

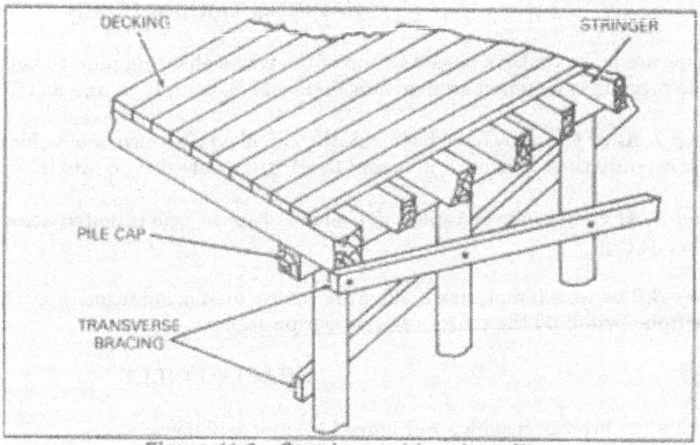

Figure 11-8. Capping and bracing piles

STRINGERS

Stringer positions are measured from the centerline of the wharf. The stringers are toenailed to the pile caps with two 3/8- x 10-inch spikes at each bearing point. The ends of the stringers overlap to provide complete bearing on the pile caps. Spacer blocks are toenailed between stringers with two 60d nails.

DECKING

Standard decking consists of 4- x 8-inch planks, which are spiked to each stringer with two 5/16- x 7-inch spikes, and set with 1/4-inch spacing. Openings greater than 1/4 inch may be used between planks in areas that are subject to heavy rains.

STRINGPIECES

The stringpiece (or curb) is placed on 2- x 10-inch blocking, 24 inches long, spaced on 48-inch centers along the edge of the deck. Stringpiece bolts are countersunk and the hole is seated with bituminous material. (Figure 11-11 shows a wharf-edge cross-section of a timber-pile wharf.)

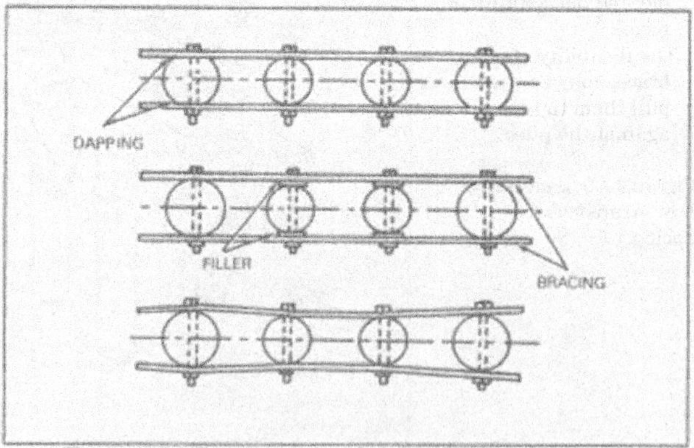

Figure 11-9. Transverse bracing for different size piles

- When the stringpieces are

parallel to the direction of the wharf stringers, the stringpieces are bolted through the blocking, the decking, and the stringer end pieces.
- When the stringpieces are *perpendicular* to the direction of the stringer, they are bolted through the blocking, the decking, alternate stringers, and the pile cap.

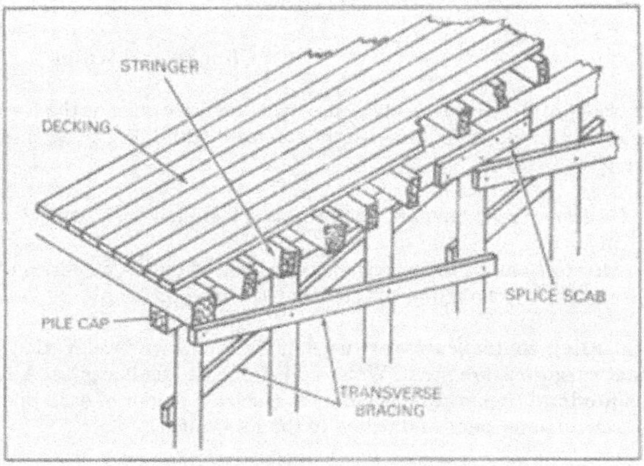

Figure 11-10. Stringers and decking in place

FENDER PILES AND CHOCKS

Timber is the most suitable material for wharf fenders in TO construction. Fender piles serve the following purposes:

- They cushion a wharf from the impact of ships and protect the outer row of bearing piles from damage.
- They protect the hulls of craft from undue abrasion.

The 3- or 4-foot extension of a fender pile above the deck level of a wharf supplements wharf-mooring hardware, but is not used for warping a ship into or out of the berth.

Since fender piles are not part of the structural support of the wharf, they are easier to replace than bearing piles.

Protection of Fender Piles

Protective devices that lengthen the life of fender piles are—

- A heavy timber wearing ribbon, which may easily be replaced. It is sometimes installed along a line of fender piles at the elevation receiving the heaviest abrasion.
- Floating logs or camels (floating fenders).
- Rope wrappings, particularly on corner fenders.

Fender Piles for Quays

Structures that are almost completely rigid, such as solid-fill quays, sometimes have fender piles backed up with heavy springs to give a combination of yield and resistance.

Installation of Fender Piles

Fender piles are driven at a slight batter (angle). Usually 1 to 12 fender piles are used along the outside edge of all rows of bearing piles, except on the extreme inshore wharf sections. Every third fender pile may extend 3 to 4 feet above the curb. The others are cut off flush with the top of the curb.

Chocks and Wales

Chocks are timber braces placed between fender piles at the level of the stringpiece or pile cap to hold them in position and give them lateral stability. Chock ends should be firmly seated against the piles.

- On *timber-pile* wharves, each chock is fastened with two bolts through the stringer endpiece or pile cap.
- On *steel-pile* wharves, each chock is bolted by 12- x 12-inch blocks driftpinned to the ends of the stringers or bolted to the ends of the wharf pile cap.

Wales (horizontal beams) are used at mean low water elevation when tidal currents are swift or tidal variations are great. Wales add rigidity to fender piles. A 12- x 12-inch continuous longitudinal timber wale is bolted to the back fender of each pile. Timber chocks are placed between fender piles and bolted to the line wales.

PILE CLUSTERS AND CORNER FENDERS

Pile clusters, whether at the faces or corners of wharves or acting as dolphins (isolated pile clusters), must combine beam strength, rigidity, and stability against horizontal stresses. Therefore, the individual piles that make up the cluster must be joined so that the cluster will act as a unit.

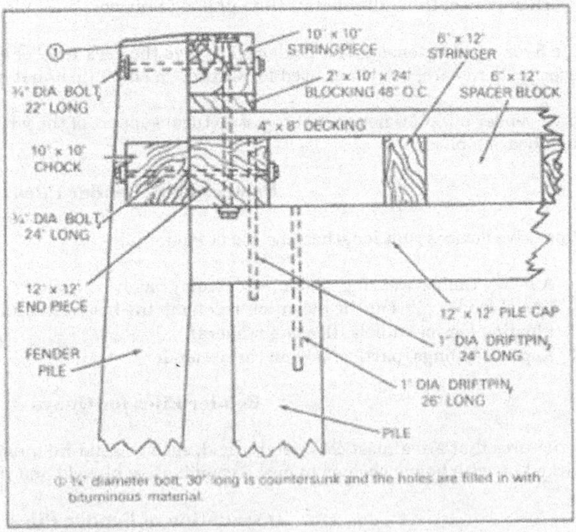

Figure 11-11. Wharf-edge cross-section of a timber-pile wharf

Mooring Piles

Clusters of three or more piles are used to supplement or replace wharf-mooring hardware. The top of the cluster is lashed together.

Mooring piles are placed at intervals along the wharf face when bollards and other mooring hardware are not avail able. A maximum of three piles of each cluster extends 3 feet or more above the wharf deck.

Corner Fenders

Piles clustered at exposed corners of the wharf, bolted and lashed together, are provided so that a ship may use the corner to pivot when warping in and out of the berth. The wharf is strongly reinforced at the corners with layers of diagonal planking laid one across the other. This reinforcing is backed up with diagonal batter piles.

The standard corner-fender cluster is made up of 10 piles battered for adequate spacing at the points. Timber connectors may be used in conjunction with the bolts to tie the piles more firmly into a single rigid member. To avoid undue abrasion to ship hulls and to outside pile surfaces, heavy rope mats may be lashed to the clusters at the level of contact. To supplement mooring hardware, the corner piles extend 3 to 4 feet above deck level.

Deck Reinforcing on Wood-Pile Wharves. Before stringers are set, wooden piles battered inward are driven to support a cap, set diagonally across each corner, and bolted to the bottom face of the other caps. Another piece of cap timber is set to act as a strut between the fender cluster and the diagonal cap.

The space between the cluster and the diagonal cap is then floored over with two layers of plank each 6 inches thick, laid diagonally (and transversely to each other) to fill the thickness between the cap timbers. To complete the reinforcement, stringers are set close and spiked together over the outer half of each corner panel.

Deck Reinforcing on Steel-Pile Wharves. In steel-pile marginal wharves and piers with corner fenders, the deck in each corner panel is similarly reinforced with timber. Wood piles battered inward carry a diagonal cap timber bolted to the bottom flanges of the steel-pile caps. The diagonal cap is strutted against the fender cluster, the diagonal layers of plank are applied, and the stringers are set close and spiked together, as described above for wood-pile wharves.

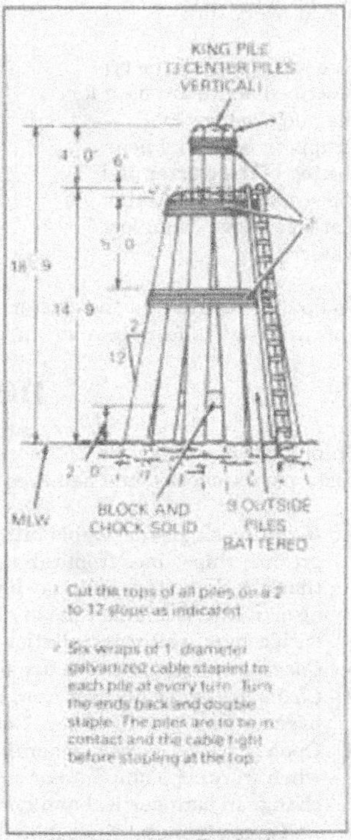

Figure 11-12. Typical pile dolphin

FLOATING LOG FENDERS (CAMELS)

Floating logs are used to absorb part of the impact shock when a ship is berthed. They protect the surface of fender piles while a ship is tied up. The simplest type of fender is a single line of floating logs, each secured by two or more lengths of 1/2-inch galvanized chain fastened to 3/4-inch eyebolts in the fender log and the wharf pile. Some arrangement, such as loose steel collars around the wharf piles, is provided to allow the logs to rise and fall with the tide.

Floating clusters of logs or strongly constructed rafts are called *camels*. In addition to absorbing impact shock and protecting fender piles from the sliding friction of a ship moving in the berth, camels may be required to breast a ship off the face of the wharf into deeper water.

PILE-MOORING DOLPHINS

Dolphins (Figure 11-12 and Figure 11-13, page 11-10) are isolated clusters of piles to which a ship may be moored. The center of the cluster, called a *king pile,* may be a single pile or a cluster driven vertically and wrapped to act as a unit. The other piles are driven in one or more concentric rings around the king pile, each battered towards the center. The king pile is normally left somewhat longer than the others for use as a mooring post.

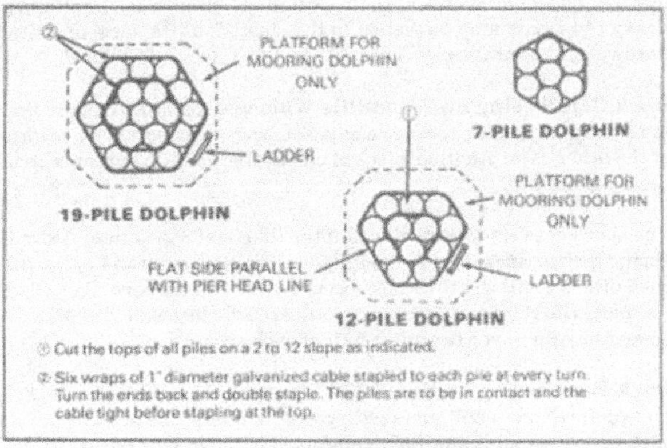

Figure 11-13. Typical dolphin plan views

When composed of a cluster, the king pile is wrapped with at least six turns of 1-inch diameter galvanized wire rope, stapled to each pile at every turn.

Two wrappings of the type described above are used for the pile cluster. One wrapping is located near the top of the cluster and the second about 2/3 the distance above mean low water.

To further ensure that the cluster will act as a unit, the piles are chocked and bolted together approximately 2 feet above mean low water.

DOCKING HARDWARE

Ships tie up to wharves with lines fastened to mooring fittings such as bollards, corner mooring posts, cleats, chocks, and pad eyes (Figure 11-14).

- *Bollards*, single- or double-bitt, are steel or cast-iron posts to which large ships are tied. They prevent ships' lines from riding up off the post. Bollards may have waist diameters smaller than ton diameters and may have caps, or projecting, rounded horns. Double-bitt bollards are also known as double bitts or double steamship bitts. Bollard bodies may be hollow for filling with concrete after installation They are usually designed to take line pulls of about 35 tons.
- *Corner mooring posts* are usually designed to take pulls of up to 60 tons.
- *Cleats* are generally cast iron, with arms extending horizontally from a relatively low body. The base may be open or closed. Cleats are used for securing smaller ships, tugs, and workboats.
- Open or closed *chocks,* generally made of cast iron, are used for directing and snubbing lines when working a ship into or out of its berth. A closed chock must be used when there is a change in both vertical and horizontal directions of a line.
- *Pad eyes* are metal rings mounted vertically on a plate and intended to receive a ship's line spliced with thimble and shackle. They are used only for securing small craft.

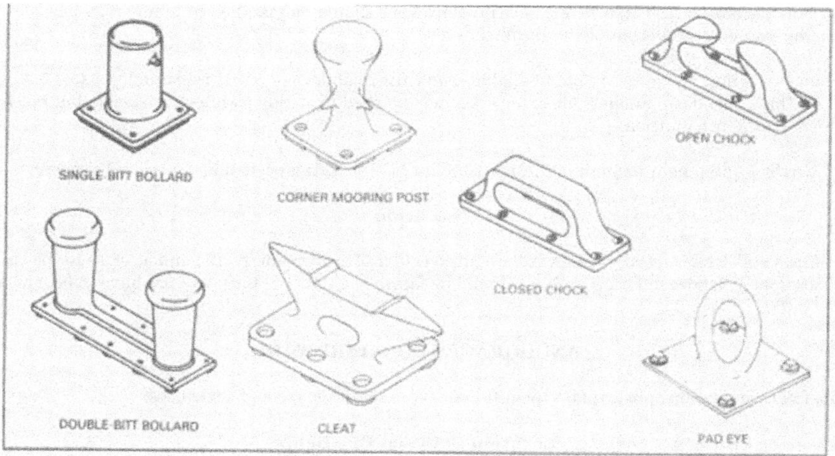

Figure 11-14. Mooring hardware

HARDWARE INSTALLATION

Proper installation requires that the vertical and horizontal stress on any structural unit on which mooring hardware is attached be partially transferred to the wharf structure. This is done by increasing the number and size of stringers under the hardware installation, and by providing anchorage for mooring hardware bolts that will transfer the stress through the pile cap of one or more bents to several piles.

Stringer Reinforcement

The number and size of stringers are increased at the location of major hardware items. When base widths of hardware are greater than 12 inches, but less than 24 inches, at least two 12- x 12-inch stringers are needed. For base widths greater than 24 inches, but less than 36 inches, three 12 x 12-inch stringers are needed; and so forth. Stringers are laid close together and are spiked to each other and at each bearing point. Mooring hardware bolts pass through stringers, filler blocks, and anchorage timbers.

Standard Installation

Standard wharf structures use the following mooring hardware:

- Pier, 90 x 500 feet—six large double-bitt bollards on each side on 100-foot centers and five 42-inch cleats on each side centered between bollards.
- Offshore marginal wharf, 60 x 500 feet— six large double-bitt bollards and five 42-inch cleats spaced as above on the outshore side only.
- Lighterage quay 35 x 500 feet—eleven 42-inch cleats on 50-foot centers.

Nonstandard Installation

For nonstandard wharf structures, mooring hardware should be installed in numbers, types, and spacing approximately that of standard wharves.

When cleats and pad eyes are not available, every third fender pile must be extended 3 to 4 feet above the wharf deck. Fender-pile extensions may be used to steady a ship in the berth, but not to winch a ship into position.

On berths located near enough to the shore, bollards or mooring posts may be located onshore.

Location

Bollards and other mooring hardware are placed clear of cranes and traffic and as close to the curb as possible. Onshore mooring anchors should be located so that the lines will not have to be moved for traffic.

ANCHORAGES FOR HARDWARE

The following paragraphs explain the different types of hardware and their uses:

Location Between Pile Bents

To provide anchorage for heavy items of mooring hardware located between pile bents, a grillwork of 12 x 12 timbers is bolted underneath the pile cap (Figure 11-15). Each of the four piles directly affected by the upward pull on the grillwork is strapped to the pile cap with 3- x 3/8-inch steel strapping. The straps are spiked to piles and pile caps. Filler blocks of 12 x 12 timbers are centered to receive the mooring hardware bolts.

Location on Pile Bents

Mooring hardware is also located directly over the outside bearing pile of a bent as shown in Figure 11-16. Mooring hardware with 22- to 26-inch bolt centers is anchored as follows:

- Two 12 x 12 by approximately 20-foot-long timbers are bolted under the pile cap over which the hardware is located and to both sides of three piles of the bent.
- 12 x 12-inch filler timbers approximately 4 feet long are bolted to the wharf pile cap under the hardware bolt location.
- Each of the three piles directly affected by the upward pull on the

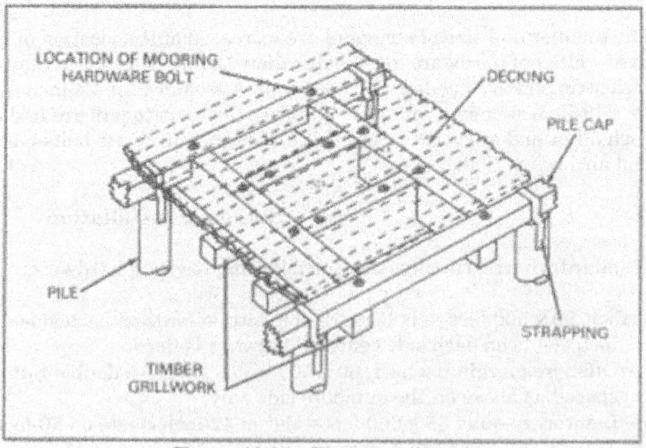

Figure 11-15. Timber grillwork

grillwork is strapped to the pile cap with steel strappings as described above (see Figure 11-15).

Items of mooring hardware with bolt centers greater than 26 inches require using timber wider than 12 inches or doubling the number of timbers, or locating the hardware between bents using the timber grillwork anchorage described previously.

Bracing

The wharf structure is longitudinally braced at the location of bollard installations. Diagonal bracing is done from just below the pile caps to about low-water level at the location of each bollard. The cross bracing is bolted to each pile.

Installation of Light Items

Light items of mooring hardware with bolt centers less than 8 inches, such as cleats, chocks, and pad eyes, are bolted through the stringpiece, blocking, decking, and stringer end piece.

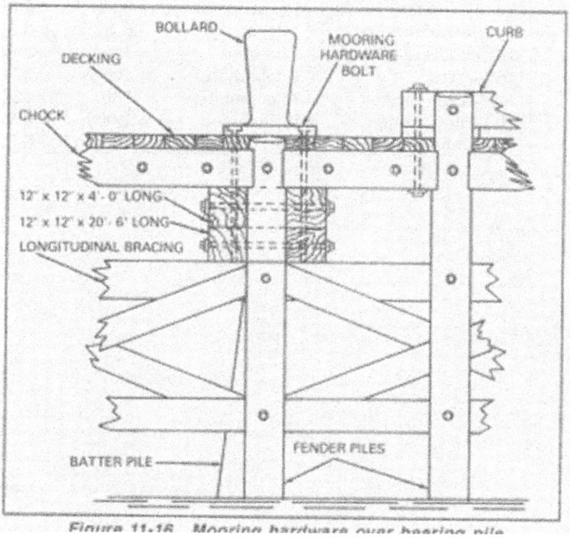

Figure 11-16 Mooring hardware over bearing pile

CONVERSION TABLES

APPENDIX A

Metric to US	US to Metric
LENGTH	
1 mm = 0.04 in (0.03937 in) 1 cm = 0.3937 in 1 m = 3.281 ft or 1.094 yd = 39.3707 in 1 km = 0.621 statute-mile 1 km = 0.5399 nautical mile (nmi)	1 yd = 91.44 cm 1 ft = 30.48 cm 1 in = 2.54 cm 7/8 in = 2.22 cm (22.22 mm) 3/4 in = 1.90 cm (19.05 mm) 5/8 in = 1.59 cm (15.88 mm) 1/2 in = 1.27 cm (12.70 mm) 3/8 in = 0.98 cm (9.52 mm) 1/4 in = 0.64 cm (6.35 mm) 1/8 in = 0.32 cm (3.18)
AREA	
1 sq cm = 0.155 sq in 1 sq m = 10.76 sq ft or 1.196 sq yd 1 sq km = 0.386 sq miles 1 hectare (ha) = 2.47 acres	1 sq in = 6.45 sq cm 1 sq ft = 0.0929 sq m 1 sq yd = 0.836 sq m 1 sq mi = 2.59 sq km 1 acre = 0.405 ha 1 acre = 43560 sq ft

LINEAR MEASURE

Metric Denomination		Meter	US Equivalent
	1 millimeter	= 0.001	= 0.0394 in
10 millimeters	= 1 centimeter	= 0.01	= 0.3937 in
10 centimeters	= 1 decimeter	= 0.1	= 3.937 in
10 decimeters	= 1 meter	= 1.	= 39.3707 in = 3.28 ft
10 meters	= 1 dekameter	= 10.	= 32.809 ft
10 dekameters	= 1 hectometer	= 100.	= 328.09 ft
10 hectometers	= 1 kilometer	= 1,000.	= 0.62138 mi
10 kilometers	= 1 myriameter	= 10,000.	= 6.2138 mi

CARPENTRY ABBREVIATIONS AND SYMBOLS

APPENDIX B

ABBREVIATIONS

Carpenters use the following abbreviations in connection with lumber:

ad	air-dried
al	all length
av	average
avw	average width
avl	average length
bd	board
bf	board foot
bdl	bundle
bev	beveled
bm	board (foot) measure
btr	better
clg	ceiling
CL	centerline
clr	clear
CM	center-matched; that is, tongue-and-groove joints are made along the center of the edge of the piece.
com	common
csg	casing
ctg	crating
cu ft	cubic foot
D&CM	dressed (one or two sides) and center-matched
D&M	dressed and matched; that is, dressed one or two sides and tongue and grooved on the edges. The match may be center or standard.
ds	drop siding
D&SM	dressed (one or two sides) and standard-matched
D2S&CM	dressed two sides and center-matched
D2S&SM	dressed two sides and standard-matched
0	diameter
dim	dimension
e	edge
FAS	firsts and seconds, a combined grade of the two upper grades of hardwoods
fbk	flat back
fbm	feet board measure (board feet)
fcty	factory (lumber)
fg	flat grain

flg	flooring
fok	free of knots
frm	framing
ft	foot or feet
hdl	handle (stock)
hdwd	hardwood
hrt	heart
hrtwd	heartwood
in	inch or inches
kd	kiln-dried
kd	knocked down
lbr	lumber
lgr	longer
lgth	length
lin ft	linear foot; that is, 12 inches
lr	log run
lr mco	log run, mill culls out
m	thousand
mfbm	thousand (feet) board measure
mco	mill culls out
merch	merchantable
mr	mill run
mfsm	thousand (feet) surface measure
mw	mixed width
no	number
1s & 2s	ones and twos, a combined grade of the hardwood grades of firsts and seconds
ord	order
p	planed
pat	pattern
pky	pecky
pln	plain, as in plain sawed
pn	partition
qtd	quartered (with reference to hardwoods)
rd	round
rdm	random
res	resawed
rfg	roofing
rfrs	roofers
rip	ripped
rl	random length
rw	random width
S&E	surfaced one side and one edge
S2S&M	surfaced two sides and standard- or center-matched
S2S&SM	surfaced two sides and standard-matched
sap	sapwood
S1E	surfaced one edge
S1S1E	surfaced one side and one edge
S1S2E	surfaced one side and two edges
S2E	surfaced two edges
S4S	surfaced four sides
S&CM	surfaced one or two sides and center-matched

S&M	surfaced and matched; that is, surfaced one or two sides and tongue and grooved on the edges. The match may be center or standard.
S&SM	surfaced one or two sides and standard matched
S2S&CM	surfaced two sides and center matched
sb	standard bead
sd	seasoned
sdg	siding
Sel	select
sesd	square-edge siding
sf	surface foot; that is, an area of 1 square foot
sftwd	softwood
shd	shipping dry
ship	shiplap
sm	standard matched
snd	sap no defect
snd	sound
sq	square
sq e	square edge
sq e&s	square edge and sound
sgrs	squares
std	standard
stk	stock
sw	sound wormy
t&g	tongue and groove
tb&s	top, bottom, and sides
TO	theater of operations
tbrs	timbers
vg	vertical grain
wal	wider, all length
wdr	wider
wt	weight
wth	width

SYMBOLS

Symbols commonly used in carpentry are given below. For additional information on the various symbols used in construction plans and blueprints, refer to TM 5-704.

Architectural

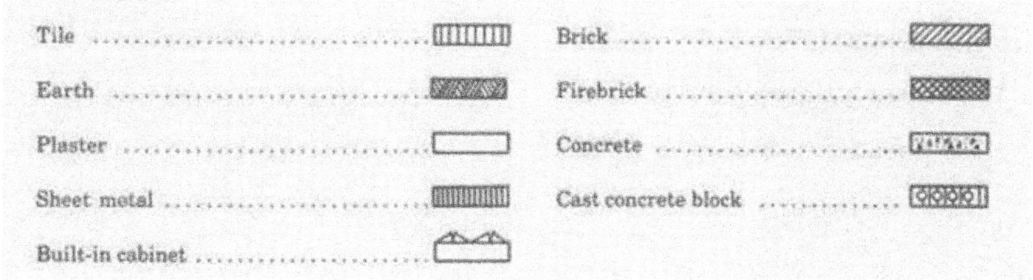

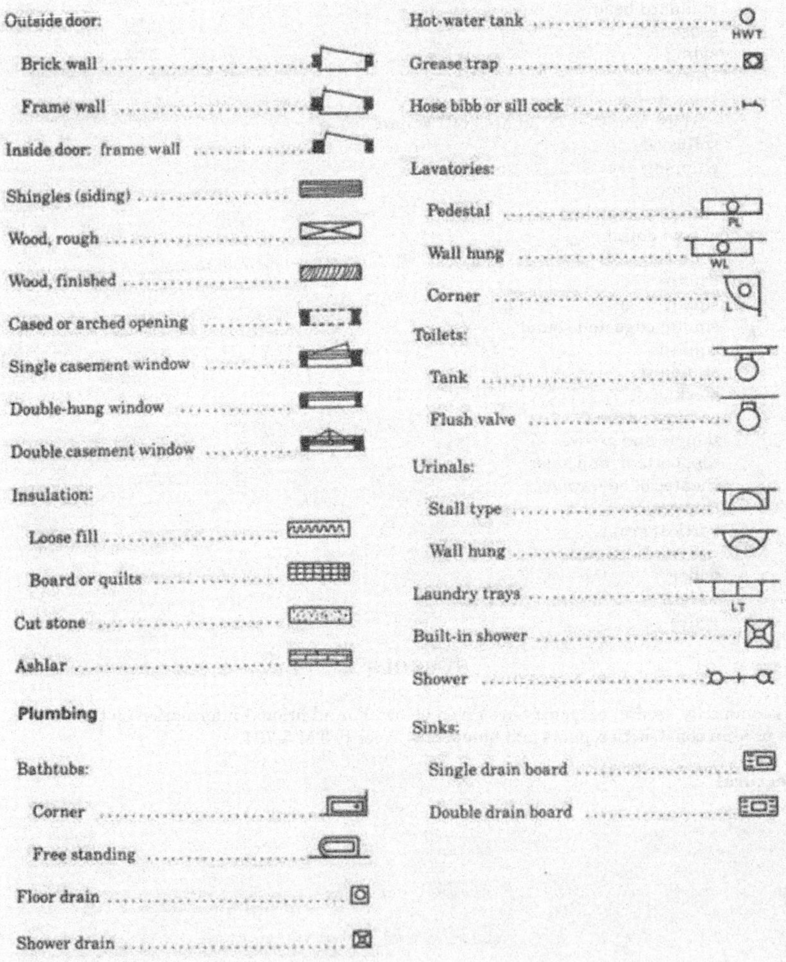

Outside door:
Brick wall
Frame wall
Inside door: frame wall
Shingles (siding)
Wood, rough
Wood, finished
Cased or arched opening
Single casement window
Double-hung window
Double casement window
Insulation:
Loose fill
Board or quilts
Cut stone
Ashlar
Plumbing
Bathtubs:
Corner
Free standing
Floor drain
Shower drain
Hot-water tank
HWT
Grease trap
Hose bibb or sill cock
Lavatories:
Pedestal
PL
Wall hung
WL
Corner
L
Toilets:
Tank
Flush valve
Urinals:
Stall type
Wall hung
Laundry trays
LT
Built-in shower
Shower
Sinks:
Single drain board
Double drain board

Electrical

Pull switch	
Single-pole switch	S_1
Double-pole switch	S_2
Triple-pole switch	S_3
Buzzer	
Floor outlet	
Bell	
Drop cord	
Ceiling outlet	
Wall bracket	
Single convenience outlet	
Double convenience outlet	
Motor	

MANPOWER ESTIMATES — APPENDIX C

This chapter contains tables (C-1 through C-14, pages C-1 through C-8) which may be used in preparing manpower estimates for carpentry work. The tables do not include provisions for loading and hauling materials to the job site.

All tables presume average working conditions in terms of weather, skill, crew size, accessibility, and the availability of equipment.

Table C-1. Rough framing

Work element description	Unit	Man-hours/unit
Beams (3 - 2" x 8')	1,000-bd-ft measure	30
Floor joists, sills	1,000-bd-ft measure	25
Bridging	100 pr	5
Wall frames, plates	1,000-bd-ft measure	45
Furring, including plugging	1,000 lin ft	32
Blocking	1,000-bd-ft measure	20
Grounds for plaster	1,000 lin ft	48
Door bucks	ea	3
Ceiling joists	1,000-bd-ft measure	25
Rafters	1,000-bd-ft measure	45

Trusses (span feet)	Each	Man-hours assembly	Man-hours placement
20		1.5	1
30		2.0	8
40		12.0	8
50		20.0	6*
60		24.0	6*
80		32.0	6*

Typical crew: 1 leader and 8 workers. Minimum crew: 1 leader and 2 workers.

*Assumes the use of an organizational crane, 1 operator, 1 oiler, and 2 or 3 workers on guylines.

Table C-2. Sheathing and siding

Work element description	Unit	Man-hours/unit
Wall sheathing	1,000 sq ft	
Building paper		8
Tongue and groove		24
Plywood		16
Fiberboard		16
Roof decking	1,000 sq ft	
Tongue and groove		32
Plywood		20
Siding	1,000 sq ft	
Plywood		16
Corrugated asbestos		32
Drop siding		32
Narrow bevel		48
Shingles		40

Typical crew: 1 leader and 8 workers.

Table C-3. Insulation

Work element description	Unit	Man-hours/unit
Thermal	1,000 sq ft	
Board		
Floor*		32
Wall		8
Ceiling		24
Roof		24
Rock wool		
Loose		16
Batts		12
Foil alone		8
Rigid foam		24
Acoustic	1,000 sq ft	
Strip		24
Quilt		8

Typical crew: 1 leader and 4 workers.

*Install vermin shield.

Table C-4. Finish carpentry

Work element description	Unit	Man-hours/unit
Walls	1,000 sq ft	
Plywood		32.0
Plasterboard (includes tape)		48.0
Ceilings	1,000 sq ft	
Wood		48.0
Plasterboard (includes tape*)		64.0
Cemented tile		32.0
Panel with suspension		72.0
Baseboard (2 member)	1,000 lin ft	72.0
Molding (chair)	1,000 lin ft	48.0
Door frame, trim	ea	2.5
Sliding door with pocket	ea	8.0
Window frame, trim	ea	3.0
Installing prefab closets	ea	16.0
Setting kitchen cabinets	ea	1.5
Shelving	1,000 sq ft	64.0
Chalkboard (complete)	1,000 sq ft	110.0
Stairs		
Closed stringer, built on job	story	16.0
Closed stringer, prefab	story	8.0
Open stringer	story	24.0

Typical crew: 1 leader and 3 to 8 workers.
For small rooms, increase time required for wall- and ceiling-board installation by 30 to 50 percent.

*Includes furring strips when necessary.

Table C-5. Door installation

Work element description	Unit	Man-hours/unit
WOOD DOORS AND FRAMES:		
Door frames and trim		
Single exterior	ea	3
Double exterior	ea	3
Single interior	ea	3
Double interior	ea	4
Sliding door frame	ea	4
Door: fit, hang, and lock		
Single exterior	ea	5
Double exterior	ea	8
Single interior	ea	5
Double interior	ea	7
Screen doors	ea	2
METAL DOORS:		
Single	ea	6
Double	ea	9
MISCELLANEOUS DOORS COMPLETE WITH TRIM AND HARDWARE:		
Rolling, manual-operated	ea	29
Rolling, motor-operated	ea	36
Sliding, manual-operated	ea	20
Sliding, motor-operated	ea	25
Sliding, fire	ea	18
Garage doors		
Wood 16' x 7'	ea	8
Aluminum 16' x 7'	ea	10
Scuttles	ea	10
CAULKING	1,000 lin ft	5

Includes jambs, stops, casings, and weather stripping.
Does not include sills or thresholds.
On wood doors, if power planes, hinge butt routers, and lock mortises are used, deduct 25 percent from installation time.

Table C-6. Flooring

Work element description	Unit	Man-hours/unit
Wood floors	1,000 sq ft	
Subfloor		
Tongue and groove		24
Plywood		16
Finish floor		
Softwood		24
Hardwood		32
Soft tile	1,000 sq ft	
Cemented		24
Nailed		32
Linoleum	1,000 sq ft	32

Typical flooring crew: 1 leader and 4 workers.

Table C-7. Window installation

Work element description	Unit	Man-hours/unit
Wood windows		
Double hung	ea	4
Casement, single	ea	4
Fixed wood sash	ea	3
Jalousie	ea	2
Skylights	ea	8
Louvers	ea	5
Screens	ea	2
Venetian blinds	ea	2
Metal windows		
Double hung	ea	2
Casement	ea	2
Commercial projected	ea	2
Skylights	ea	9
Weather stripping	ea	3
Caulking	1,000 lin ft	16

Suggested crew size: 2 to 6 workers.

Installation includes drilling fasteners, expansion sills, installing plugs, toggle blocking, hinges, locks, and other hardware.

For special panic-device doors, add three hours for single doors and four hours for double doors.

Table C-8. Built-up roofing, insulation, and flashing

Work element description	Unit	Man-hours/unit
Roofing	1,000 sq ft	
2 ply		12
3 ply		20
4 ply		25
5 ply		30
Insulation	1,000 sq ft	25
Flashing	1,000 sq ft	60
Typical crew: 1 leader and 6 workers.		

Table C-9. Roll roofing

Work element description	Unit	Man-hours/unit
Paper (plain) and felt	1,000 sq ft	7
Asphaltic aluminum (including primer)	1,000 sq ft	16
Canvas (including 2 coats of paint)	1,000 sq ft	25
Typical crew: 1 leader and 6 workers.		

Table C-10. Shingle roofing

Work element description	Unit	Man-hours/unit
Wood	1,000 sq ft	35
Slate	1,000 sq ft	55
Metal	1,000 sq ft	50
Asphalt	1,000 sq ft	30
Typical crew: 1 leader and 4 workers.		

Table C-11. Metal, asbestos-cement, and tile roofing (pitch at least 3 inches per foot)

Work element description	Unit	Man-hours/unit
Metal (corrugated and V-crimp)	1,000 sq ft	
Wood purlins		18
Metal purlins		36
Sheet (seamed)		60
Asbestos-cement	1,000 sq ft	
Wood purlins		35
Metal purlins		45
Tile	1,000 sq ft	
Clay		55
Metal		60

Typical crew: 1 leader and 5 workers.

Table C-12. Pile bracing and capping

Work element description	Unit	Man-hours/unit
Bracing*	ea	
Horizontal		1.0
Diagonal		0.8
Capping	1,000 lin ft	
Wood		100.0
Steel		150.0
Concrete		200.0

Typical crew: 1 leader and 6 workers.

*Table based on 4-inch x 10-inch x 4-foot bracing members.

Table C-13. Pier framing

Work element description	Unit	Man-hours/unit
Stringers	1,000-bd-ft measure	40
Bridging	1,000 lin ft	40
4-inch deck	1,000 sq ft	20
2-inch wearing surface	1,000 sq ft	16
Bull rail	1,000 lin ft	60
Bumper	1,000 lin ft	36

Typical crew: 1 leader and 10 workers.

Table C-14. Deck hardware

Work element description	Unit	Man-hours/unit
Bits	ea	3
Bollards	ea	4
Chocks	ea	3
Cleats	ea	2
Pad eyes	ea	1

Typical crew: 1 leader and 4 workers.

GENERAL INFORMATION — APPENDIX D

FLOOR AND WALL TILE

The following paragraphs include information pertaining to the various types of tile and their installation procedures.

The number of tiles needed is calculated by performing the following procedures:

First, calculate the square feet of the area to be tiled. If you are using 12-inch-square tiles, the total floor area (in square feet) equals the total number of tiles needed, plus an additional 10 percent waste factor. If another size of tiles is being used, multiply the area by 144 to convert to square inches. Then divide that number by the area (square inches) of the tiles to find the required amount (include a 10 percent waste factor).

Example: You are using tiles 9 x 9 inches. To tile a floor 12 feet long and 9 feet wide—

Multiply the room dimensions to find the area: *12 feet x 9 feet = 108 square feet*

Multiply the area by 144: *108 x 144 = 15,552 square inches*

Calculate the area of the tile: 9 *inches x 9 inches = 81 square inches*

Divide the room area (square inches) by the tile area (square inches): *15,552 divided by 81 = 192 tiles*

Add 10 percent waste factor: *192 + 19 = 211 tiles required*

RESILIENT FLOOR TILE

Resilient floor tile is durable, easily maintained, comfortable and attractive, and low cost. It is made of rubber, vinyl, linoleum, and asphalt. Common sizes of this tile are either 9 x 9 inches or 12 x 12 inches.

A notched trowel (used for spreading adhesive) and a tile cutter are required for installation. To lay out and install resilient floor tile, perform the following procedures:

Locate the center of the end walls of the room. Establish a main centerline by snapping a chalk line between these two points. Lay out another centerline at right angles to the main centerline. This line may be established using a framing square or the triangulation method. With the centerline established, make a trial layout of the tiles along the centerlines. Measure the distance between the wall and the last tile. If this measurement is less than 1/2 tile, move the centerline half the width of the tile closer to the wall. This adjustment will eliminate the need to install border tiles that are too narrow. Since the original centerline is moved exactly half the tile size, the border tile will remain uniform on opposite sides of the room. Check the layout along the other centerline in the same way.

Spread adhesive over one quarter of the total area, starting with the quarter farthest from the door and working toward the door. Ensure that the floor surface is clean before you spread the adhesive. Spread up to the chalk lines but do not cover them. Be sure to use a notched trowel with the notch depth recommended by the manufacturer of the adhesive. Allow the adhesive to take an initial set before setting the first tile. The time required will vary, depending on the type of adhesive used.

Start laying the tiles at the center of the room. Make sure the edges of the tiles are aligned with the chalk line. Lay rows by width, stair-stepping additional rows and ensuring that the tiles are tight against one another in a cross-grained pattern unless otherwise specified. After all of the full tiles have been laid, install the border or edge tiles around the room. To lay out a border tile, place a loose tile over the last tile in the outside row with the grains running in opposite directions (if using a cross-grained pattern). Then, take another tile and place it in position against the wall and mark a pencil line on the first tile. Cut the tile along the marked line.

After all the tiles have been installed, remove any excess adhesive using a cleaner or solvent and procedures approved by the manufacturer.

CERAMIC AND OTHER SPECIALTY TILES

This tile is used extensively where sanitation, stain resistance, easy cleaning, and low maintenance are desired. Types of tile include ceramic, mosaic, paver, quarry, brick-veneer, cement-bodied, marble, and other stone tiles. These can be used for both interior and exterior flooring. Tile is used on both walls and floors. Field tile is regular tile placed on all courses in the main field of an installation. Trim tile is a specially shaped tile used to border and complete the main field of tile; it is available in a wide variety of shapes, sizes, and colors to match field tile.

Tiles come with two types of finishes—glazed and unglazed. Glazed tiles are coated with a glaze before firing to give the tile color and to preserve its surface. They may be fired to a smooth or textured finish. Glazed tiles are most commonly used for walls but may also be applied to floors and countertops. They are used mainly for interiors. Unglazed tiles are fired without a glaze coating. They derive their color from the clay from which they are made. Adhesives used are Thinset or Organic Mastic. Thinset is a powdered cement-based product that is mixed with either water, a latex or acrylic additive, or epoxy. It is very versatile. Organic Mastic is premixed in a solvent or latex base. It may deteriorate if exposed to heat or water.

Grout is a powder made from sand and cement and is used to seal the cracks between the tiles. It is mixed with either water or, to increase durability, an additive. It is available in a variety of colors.

The following tools and equipment are required for installing ceramic and specialty tiles:

A striking tool is used to compact the grout into the joints.

- *A beating block* is a board used to even the tile surface after it has been set.
- A square-notched trowel is used to spread adhesive.
- *A pointing trowel is* used to spread adhesive in tight spots.
- *A tile cutter is* used to score the tile surface so that it can be snapped by applying pressure to the score.
- *A fine file or tile stone is* used to smooth rough edges after cutting tile.
- *A time nipper is* used to clip tile and cut irregular openings.
- *A squeegee or sponge is* used to remove excess grout from the tile surface.
- A sponge float or rubber-faced trowel is used to spread grout over the surface.
- *An electric tile saw is* similar to a mason saw. It is used to make clean, accurate cuts.

To lay out and install ceramic and specialty tiles, perform the following procedures:

Check the area to be tiled to determine if it is square. If the area is slightly out of square, minor changes in the layout can accommodate these conditions. If the area is seriously out of square, the process stops for any required structural repairs or surface preparation. If the framing problems are serious, it may not be possible to tile the area.

Draw the layout on paper. Layout depends greatly on the pattern desired and the type, size, and shape of the tile being used. Use as many full tiles and as few cut tiles as possible. Place cut tile away from visual focal points (doorways, thresholds, and so forth); tiles should be set symmetrically for a more attractive finish and appearance.

Place reference lines on the floor or wall. Once the layout has been established on paper, transfer it to the floor or wall. A reference line should be snapped to mark the rows of cut tiles around the perimeter. A grid of reference lines should be snapped to enclose all full tiles in sections no larger than 3 square feet.

To install tiles, first spread adhesive over a small area or section (3 x 3 feet), making sure to spread it just up to the lines so that the lines will still be visible. Align the first tile against a 90° intersection in the grid and press it gently into the adhesive. After each course of tile is applied, use a beating block to level the surface. After all the tiles are set, allow the adhesive to set the required time, according to manufacturer's instructions. Prepare the grout and spread it over the tile surface, ensuring that the joints are filled. When the grout begins to dry, clean the tile with a damp sponge. After the grout has dried, wipe off the haze with a clean rag or towel. After the grout has completely dried and hardened (approximately 72 hours), a grout sealer may be applied.

SUSPENDED CEILINGS

Suspended ceilings are primarily designed for acoustical control; however, ceilings are also lowered to save on heating and air conditioning expenses; finish off exposed joints; and cover damaged plaster.

ACOUSTICAL TILE

Acoustical tile absorbs sound, reduces noise, reflects light, and resists flame. Its thickness ranges from 3/16 to 3/4 inch; its width from 12 to 30 inches; and its length from 12 to 60 inches. The most common size panels used are 2 x 2 feet and 2 x 4 feet.

GRID-SYSTEM COMPONENTS

The grid-system components used in suspended ceilings include the following: the *main tee* (12-foot lengths), the *cross tee* (2- and 4-foot lengths), the *wall angle* (10-foot lengths), the *splice plate* (available in aluminum only), *suspending devices,* and *suspending wire.*

Suspending devices include screw eyelets; suspending hooks and nails; 8d common nails or larger, driven into wood joists and bent into a U-shape; and an approved Hilti fastener for concrete or steel.

Suspending wire includes 16-gauge anneal wire placed at 4-foot intervals and attached to suspending devices at the ceiling and to the main tees in the grid system.

INSTALLATION

First, lay out the grid pattern. This is based on the ceiling's length and width at the new ceiling height. If the ceiling's length or width is not divisible by 2 feet, increase to the next higher dimension divisible by 2 feet. For example, if a ceiling measures 12 feet 7 inches x 10 feet 4 inches, the dimensions should be increased to 14 feet x 12 feet for layout purposes. Draw the layout on paper. Make sure that the main tees run perpendicular to the joists. Position the main tees so that the border panels at the room's edges are equal and as large as possible. Draw in cross tees so that the border panels at the room's ends are equal and as large as possible.- Determine the number of pieces of wall angle by dividing the perimeter by 10 and adding 1 additional piece for any fraction. Determine the number of main tees and cross tees by counting them on the grid pattern layout.

Next, establish the ceiling height. Mark a line around the entire room at the desired height to serve as a reference line. There must be a minimum of 2 inches between the new ceiling and the existing ceiling. Ensure that this line is level and marked continuously so that it meets at intersecting corners. Next, install the wall angle. Secure the wall angle along the reference line, ensuring that it is level.

Install the suspension wire. Suspension wires are required every 4 feet along the main tees and on each side of all splices. Attach the wires to the suspending devices. The wires should be cut at least 2 feet longer than the distance between the old and new ceiling. Now, install the main tees. Main tees need to be laid out from the center to ensure that the slots line up with the cross-tee locations. Cut them where appropriate. Tees 12 feet long or less are installed by resting the ends on opposite wall angles and inserting the suspension wire. Tees over 12 feet long must be cut to ensure that the cross tees will not intersect the main tee at a splice joint. Rest the cut end on the wall angle and attach suspension wires along the tee. Make necessary splices and continue attaching suspension wires along the tee until the tee rests on the opposite wall angle. Ensure that the main tees are level and secured before continuing.

Install the cross tees. Cut and install border tees on one side of the room. Install the remaining cross tees according to the grid-pattern layout. At opposite wall angles, install the remaining border tiles. Finally, install the acoustical panels. Install the full-size panels first. Handle panels with care and ensure that the surfaces are kept clean from hand prints and smudges. If you are working on a large project, work from several cartons to avoid a noticeable change of uniformity. Cut and install the border panels.

PAINTING CEILINGS AND WALLS

The following tools and equipment are required for painting:

- Paint brushes, wall, 2 to 4 inches wide.
- Paint roller with cover.
- Paint pan.
- Stepladder.
- Paddle (stir stick).
- Rags.
- Paint, latex, flat.
- Bucket of water.

Prepare the paint for application. Remove the cover from the paint container. Remove any film layer from the top of the paint. Using the paddle (stir stick), mix the paint thoroughly, in a figure-8 motion. Scrape off and break up any unsettled matter on the bottom or lower sides of the container. Pour the paint into the paint pan until it is 2/3 full.

CEILING

Brush a narrow strip of paint around the perimeter of the ceiling along the inside edges where the wall and the ceiling meet. Using a roller, paint the remaining portion of the ceiling. Cross roll to ensure complete paint coverage without voids.

WALLS

Brush a narrow strip of paint along the inside corners of the wall and corner post. Cut in around all trim and baseboards with a trim brush. Using a roller, paint the remaining portion of the wall and corner post. The corner post may be painted with a brush. When the first coat has completely dried, apply a second coat in the same way. Ensure that the entire surface is covered and without voids.

CLEANUP

Clean paint spots from painted surfaces. Use the appropriate solvent and a clean rag. Repaint spots if necessary. Pour excess paint back into the container. Thoroughly clean brushes and rollers.

GLOSSARY

AFM Air Force Manual
anchor irons of special form used to connect timbers or masonry
anchor bolts bolts which fasten columns, girders, or other members to concrete or masonry
apron a plain or molded finish piece below the stool of a window; covers the rough edges of plaster
attn attention

backing the bevel on the top edge of a hip rafter that allows the roofing board to fit the top of the rafter without leaving a triangular space between it and the lower side of the roof covering
balloon frame the lightest and most economical form of construction. The studding and corner posts are set up in continuous lengths from the first-floor line or sill to the roof plate.
baluster a small pillar or column used to support a rail
balustrade a series of balusters connected by a rail; generally used for porches, balconies, and such
band a low, flat molding
base the bottom of a column; the finish of a room at the junction of the walls and floor
base molding the molding on the top of a baseboard
batten (cleat) a narrow strip of board fastening several pieces together
batten door a door made of sheathing and reinforced with strips of board nailed crosswise
batter board a temporary framework for locating the corners when laying a foundation
batter pile a pile driven at an angle to brace a structure against lateral thrust
bd ft board feet
beam an inclusive term for joists, girders, rafters, and purlins
bedding a filling of mortar, putty, or other substance in order to secure a firm bearing
bed molding a molding used to cover the joint between the plancier and frieze (horizontal decorative band around the wall of a room); also used as a base molding on heavy work and sometimes as a member of a cornice
belt course a horizontal board across or around a building; usually made of a flat member and a molding
bent a single, vertical framework consisting of horizontal and vertical members supporting the deck of a bridge or pier
bevel board (pitch board) a board used to lay out bevels in framing a roof or stairway
bird's mouth a cutout near the bottom of a rafter; fits over the rafter plate
blind-nailing driving nails so that the holes are concealed
board lumber less than 2 inches thick
board foot the equivalent of a board 1 foot square and 1 inch thick
boarding in the process of nailing boards on the outside studding of a house
bollard a steel or cast-iron post to which large ships are tied

BOM bill of materials

braces pieces fitted and firmly fastened to two others at any angle; used to strengthen the angle thus treated

bracket a projecting support for shelves or other structures

break joints to arrange joints so that they do not come directly under or over the joints of adjoining pieces, such as in shingling and siding

bridging pieces fitted in pairs from the bottom of one floor joist to the top of adjacent joints and crossed to distribute the floor load; it can also be pieces of width equal to the joists and fitted neatly between them

building paper cheap, thick paper used to insulate a building before the siding or roofing is put on; sometimes placed between double floors

built-up member a single structural component made from several pieces fastened together

built-up timber a timber made of several pieces fastened together, forming one piece of larger dimension

butt joint a joint made by fastening two end pieces together without overlapping

camber the angle cut in a piece of wood (such as a joist) to match an adjacent member (such as a rafter)

carriages the supports, or the steps and risers, of a flight of stairs

casement a window in which the sash opens on hinges

casing the trimming around a door or window opening, outside or inside; the finished lumber around a post or beam

CB circuit breaker

ceiling narrow, matched boards; sheathing of the surfaces that enclose the upper side of a room

center-hung sash a sash hung on its centers so that it swings on a horizontal axis

chamfer a beveled surface cut on the corner of a piece of wood

checks splits or cracks in a board, ordinarily caused by seasoning

chock heavy timber fitted between fender piles along the wheel guard of a pier or wharf

chord the principal member of a truss on either the top or bottom

clamp a mechanical device used to hold two or more pieces together

clapboards a special form of outside covering of a house; a type of siding

cleats metal arms extending horizontally from a relatively low base, used for securing small ships, tugs, and work boats

cm centimeter(s)

collar tie usually a 1 x 4 or a 1 x 6 fastened in a horizontal position to secure the upper ends of rafters together; placed on every other rafter system

column a square, rectangular, or cylindrical support for roofs, ceilings, and so forth

combination frame a combination of the principal features of the full and balloon frames

common rafter a rafter which runs square with the plate and extends to the ridge

cone concrete

concrete an artificial building material made by mixing cement and sand with gravel, broken stone, or other aggregate, plus sufficient water to cause the cement to set and bind the entire mass

conductors pipes for conducting water from a roof to the ground or to a receptacle or drain; downspout

const construction

cornice the molded project that finishes the top of the wall of a building

counterflashings strips of metal used to prevent water from entering the top edge of the vertical side of a roof flashing; allows expansion and contraction without danger of breaking the flashing

cripple rafter rafters that cut between valley and hip rafters

cross brace bracing with two intersecting diagonals; also refers to the top surface of a piece of wood that has been cut at an angle to match an adjacent member

d penny
DA Department of the Army
dado a rectangular, square-based groove cut across the full width of a board and commonly used to form joints; a rectangular groove cut. Dados are used in sliding doors and window frames.
dado joint a joint formed by two intersecting boards, meeting at right angles, and with the end of one notched into the other for half the second board's thickness
deadening construction intended to prevent the passage of sound
dead load a permanent load on a building or other structure, which includes the weight of its structural members and the fixed loads they carry
decking the heavy plank floor of a pier or bridge
df double flow
dn down
diam diameter
diagonal inclined member of a truss or bracing system used for stiffening and for wind-bracing
double-hung window a window with an upper and a lower sash, each with sash cords and weights
dovetail joint a joint made by cutting pins the shape of dovetails, which fit between dovetails onto another piece
drawboard joint a mortise-and-tenon joint with holes bored so that when a pin is driven through, the joint becomes tighter
drip the projection of a window sill or water table to allow the water to drain clear of the side of the house below it

ea each
elev elevation
end-lap joint a joint formed at the corner where two boards lap. Each board is cut away to half its thickness and they are halved to a distance equal to their width so that they are flush when fitted together

F Fahrenheit
fascia a flat member of a cornice or other finish; generally, the board of the cornice to which the gutter is fastened
fender pile the outside row of piles that protects a pier or wharf from damage by ships
fishplates metal plates, fastened to each side of a butt splice for support
fished joint an end butt splice strengthened by pieces nailed on the sides
filler a piece used to fill the space between two surfaces
fin finished
flashing the material used and the process of making watertight the roof intersections and other exposed places on the outside of the house
flr floor
flue the opening in a chimney through which smoke passes
flush even, adjacent structural surfaces, or those in the same plane
flush joint a mortar joint formed by cutting extra mortar away from the wall
FM field manual
footing an enlargement at the lower end of a wall, pier, or column that distributes the load
footing form a wooden or steel structure placed around the footing to hold the concrete to the desired shape and size

foundation that part of a building or wall that supports the superstructure

frame the timber skeleton of a building and the surrounding or enclosing woodwork of windows, doors, and so forth
frame box a window frame with boxes to contain the sash weights
framing the rough timber structure of a building, including interior and exterior walls, floor, ceiling, and roof
ft foot, feet
furring narrow strips of board nailed upon the walls and ceiling to form a straight surface upon which to lay the laths or other finish

gable the vertical, triangular end of a building from the eaves to the apex of the roof
gal gallon
gauge a tool used by carpenters to strike a line parallel to the edge of a board
gambrel a symmetrical roof with two different pitches or slopes on each side
girder a timber used to support wall beams or joists
girt (ribboned) the horizontal member of the walls of a full or combination frame house that supports the floor joists or is flush with the top of the joists
GL ground level
glued joint a joint held together with glue
grade the horizontal ground level of a building or structure
groove a long, hollow channel cut by a tool into which a piece fits or in which it works. Two special types of grooves are the dado and the housing.
ground a strip of wood assisting the plasterer in making a straight wall; also, a place to which the finish of the room may be nailed

ha hectare
half-lap joint two pieces joined by cutting away half the thickness of each so that they fit flush into each other
halved joint a joint made by cutting half the wood away from each piece to bring the side flush
hanger a vertical-tension member supporting a load
header a short joist into which the common joists are framed around or over an opening
head jamb (yoke) the top horizontal member of a door or window frame
headroom the clear space between floor and ceiling, as in a stairway
heel of a rafter the end or foot that rests on the wall plate
hip rafter rafters extending from the outside angle of the plates toward the apex of the roof
hip roof a roof that slopes up toward the center from all sides, necessitating a hip rafter at each corner
housed joint a joint in which one piece is grooved to receive the other
housing a groove cut at any angle with the grain and part way across the piece; housings are used for framing stair risers and treads in stringers
hypotenuse the length of a rafter

in inch(es)

jack rafter a short rafter framing between the wall plates; a hip rafter
jamb the side piece or post of an opening; sometimes applied to the door frame
joists timbers supporting the floorboards

kerf the cut made by a saw
km kilometer(s)

knee brace a corner brace, fastened at an angle from the wall stud to the rafter, stiffening a wood or steel frame to prevent angular movement

kraftpaper strong, brown paper used for a variety of building needs

lap joint a joint of two pieces lapping over each other
laths narrow strips which support plastering
lattice crossed wood, iron plate, or bars
lb pound(s)
ledgers the support for the second-floor joists of a balloon-frame house or for similar uses; ribband
level a term describing the position of a line or plane when parallel to the surface of still water; a tool used for testing horizontal and vertical surfaces and for determining differences of elevation
lgth length
lin ft linear feet
lintel (cap) a horizontal structural member spanning an opening and supporting a wall load
lip molding a molding with a lip which overlaps the piece against which the backing of the molding rests
live load the weight a building must bear due to the combination of furniture, other movable objects, and the people who occupy the building
lookout the end of a rafter, or the construction that extends beyond the sides of a house to support the eaves; also the projecting timbers at the gables, supporting the verge boards
louver a type of window (usually in peaks of gables and the tops of towers) provided with horizontal slats, which exclude rain and snow and allow ventilation
lumber sawed parts of a log such as boards, planks, scantling, and timber

m meter(s)
matching (tongue-and-groove) the method used in cutting the edges of a board to make a tongue on one edge and a groove on the other
max maximum
meeting rail or check rail the bottom rail of the upper sash of a double-hung window
member a single piece complete in itself, within a structure
mi mile(s)
min minimum
miter the joint formed by two abutting pieces meeting at an angle
mm millimeter(s)
mortise the hold, generally rectangular, for a tenon or any such hole cut into or through a piece by a chisel
mortised joint a joint made by cutting a hole or mortise in one piece—and a tenon, or piece to fit the hole, upon the other
mullion the construction between the openings of a window frame to accommodate two or more windows
muntin the vertical member between two pieces of the same panel work; the vertical sash bars separating the different panels of glass

N North
NCOIC noncommissioned officer in charge
newel the principal post at the foot of a staircase; also the central support of a winding flight of stairs
nmi nautical miles
no. number

nosing the part of a stair tread that projects over the riser or any similar projection; a term applied to the rounded edge of a board

O.C. on center

p pile
pad eyes metal rings vertically mounted on a plate; used for tying small vessels
parting bead the strip or bead that separates the upper and lower sashes of a window (also called a parting strip)
partition a permanent interior wall that divides a building into rooms
partition, nonbearing a dividing wall that separates areas of a structure but does not provide support for the room, overhead partitions, or floor joists
pc pieces
picture molding a molding shaped to form a support for picture hooks, often placed on the wall at some distance from the ceiling to form the lower edge of the frieze
pier timber, concrete, or masonry supports for girders, posts, or arches; intermediate supports for adjacent ends of two bridge spans; or a structure extending outward from the shore into the water, used as a dock for ships
piers masonry supports set independent of the main foundation
pilaster a portion of a square column, usually set within or against a wall
piles long posts driven into swampy soil or wherever it is difficult to secure a firm foundation upon which to lay the footing course of masonry or other timbers
piling large timbers or poles driven into the ground or the bed of a stream to make a firm foundation
pitch inclination or slope, as for roofs or stairs; the rise divided by the span
pitch board a board sawed to the exact shape formed by the stair tread, riser, and slope of the stairs and used to lay out the carriage and stringers
plan a horizontal, geometrical drawing of a building showing the walls, doors, windows, stairs, chimneys, columns, and other structural components
plank a wide piece of sawed timber, usually 1 1/2 to 4 1/2 inches thick and 6 inches or more wide
plaster a mixture of lime, hair, and sand (or of lime, cement, and sand) used to cover outside and inside wall surfaces
plate the top, horizontal piece of the walls of a frame building upon which the roof rests
plate (or seat) cut the cut at the bottom end of a rafter that rests upon the top of the plate
plot plan (site plan) a drawing showing all necessary property lines, contours, building lines, building locations, existing or new buildings, and utility easements
plow to cut a groove running in the same direction as the grain of the wood
plumb cut any cut made in a vertical plane; the vertical cut at the top end of a rafter
ply a term used to denote a layer or thickness of building or roofing paper (such as two-ply or three-ply)
porch an exterior, ornamental entranceway
post a timber set on end to support a wall, girder, or other members of the structure pr pair
prefab prefabricated
pulley stile the member of a window frame that contains the pulleys and between which the edges of the sash slide
purlin a timber supporting several rafters at one or more points, or the roof sheeting directly
PX post exchange

qty quantity

R riser
rabbet a corner cut out of an edge of a piece of wood
rabbet joint two pieces of timber, rabbeted and fit together

rafters beams that slope from the ridge of a roof to the eaves making the main body of the roof's framework
rail the horizontal members of a balustrade or panel work
rake the trim of a building extending in an oblique line, as rake, dado, or molding
rake molding the cornice on the gable edge of a pitched roof, the members of which are made to fit those of the molding of the horizontal eaves
return the continuation of a molding or finish of any kind in a different direction
ridge the top edge or corner formed by two intersecting roof surfaces
rise the vertical distance through which anything rises, as the rise of a roof or stair
riser the vertical board between two treads of a flight of stairs
RL random length
rm room
roof decking the layer of wood or plywood applied directly to the rafters, under the shingles
roofing the material put on a roof to make it weatherproof
rub joint a glued joint made by carefully fitting the edges together, spreading glue between them, and rubbing the pieces back and forth until they adhere
rubble roughly broken quarry stone
rubble masonry uncut stone, used for such things as rough work, foundations, and backing run the horizontal length of a piece, such as a rafter, when it is in position

S switch
saddle board the finish of the ridge of a pitched roof house; sometimes called *comb board*
sash the framework that holds the glass in a window
sash lift a metal hook, bar, or plate attached to a sash to enable lifting and lowering of the window
sawing, plain lumber sawed regardless of the grain, the log simply squared and sawed to the desired thickness
scab a short piece of lumber used to splice or prevent movement of two other pieces
scaffolding or staging a temporary structure or platform enabling workmen to reach high places
scale a short measurement used as a proportionate part of a larger dimension; the scale of a drawing is frequently expressed as 1/4 inch = 1 foot
scantling lumber with a cross section ranging from 2 by 4 inches to 4 by 4 inches.
scarfed joint a timber spliced by cutting various shapes of shoulders or jogs which fit each other
scarfing a joint between two pieces of wood allowing them to be spliced lengthwise
scotia a hollow molding used as part of a cornice and often under the nosing of a stair tread
scribing marking a piece of wood to provide for fitting one of its surfaces to the irregular surface of another
section a drawing showing the type, arrangement, and proportions of the various parts of a structure. It is assumed that the structure is cut by a plane, and the section is the view gained by looking in one direction.
shakes imperfections in timber caused by high winds or imperfect growth conditions
sheathing wallboards or roofing boards; generally applied to narrow boards laid with a space between them, according to the length of a shingle exposed to the weather
sheathing paper (building paper) the paper used under siding or shingles to insulate the house
shingles thin, oblong pieces of wood or other material, thinner at one end, used for covering roofs or walls
siding the outside finish between the casings

sills the horizontal timbers of a house which either rest upon the masonry foundation or, in the absence of such, form the foundation
sizing working material to the desired size; a coating of glue, shellac, or other substance applied to a surface to prepare it for painting or other finishing
sleeper a timber laid on the ground to support a floor joist
sole the horizontal member on which the studs bear
spac spacing
span the distance between the bearings of a timber or arch
specifications written or printed directions regarding the details of a building or other construction
spirit level an instrument for measuring the exactness of the horizontal or vertical position of a building or structure member
splice joining of two similar members in a straight line
sq cm square centimeter(s)
sq ft square foot, feet
sq in square inch(es)
sq km square kilometer(s)
sq m square meter(s)
sq mi square mile(s)
sq yd square yard(s)
square a tool used by carpenters to obtain accuracy; a term applied to a surface area of 100 square feet
stairs, box those built between walls, usually having no support except the walls
standing finish term applied to the finish of the openings and the base and all other interior finish work
stringer a long, horizontal timber in a structure supporting a floor
stucco a fine plaster used for interior decoration and fine work; also used for rough outside wall coverings
stud an upright beam in a building framework
studding the framework of a partition or the wall of a house; usually referred to as 2 x 4s
subfloor a wood floor laid over the floor joists on which the finished floor is laid

threshold the beveled piece over which the door swings; sometimes called a carpet strip
tie beam (collar beam) a beam so situated that it ties the principal rafters of a roof together and prevents them from thrusting the plate out of line
timber lumber with a cross section more than 4 by 6 inches, such as posts, sills, and girders
tin shingle a small piece of tin used for flashing and repairing a shingle roof
TM technical manual
TO theater of operations
toenailing driving a nail, spike, or brad at an angle into the end of one piece of wood to fasten it to a second piece; avoids having the nails show above the surface
top plate a piece of lumber supporting the ends of rafters
to the weather a term applied to the projection of shingles or siding beyond the course above
TRADOC United States Army Training and Doctrine Command
tread the horizontal part of a step
trim a term sometimes applied to exterior or interior finished woodwork and the finish around openings
trimmer the beam or floor joist to which a header is framed
trimming putting the inside and outside finish hardware on a building

truss structural framework of triangular units, used in place of rafters, for supporting loads over long spans

US United States
USAES United States Army Engineer School

valley rafter rafters extending from an inside angle of the plates toward the ridge or centerline of the house
valleys the internal angle formed by the two slopes of a roof
verge boards the boards that serve as the eaves' finish on the gable end of a building
vestibule an entrance to a house; usually enclosed

W watts
WH wall hung
wainscoting matched boarding or panel work covering the lower portion of a wall
wale a horizontal beam
wash the slant on a sill, capping, and so forth, to allow water to run off easily
water table the finish at the bottom of a house that carries water away from the foundation
wharf a structure that provides berthing space for vessels, to facilitate loading and discharge of cargo
wind a term used to describe the surface of a board when twisted (winding) or when resting upon two diagonally opposite corners, if laid upon a perfectly flat surface
wooden brick a piece of seasoned wood, made the size of a brick, and laid where necessary to provide a nailing hold in masonry walls

REFERENCES

SOURCES USED

These are the sources quoted or paraphrased in this publication.

FM 5-480. *Port Construction and Repair.* 12 December 1990.
FM 5-742. *Concrete and Masonry.* 14 March 1985.
TM 6-302-1. *Army Facilities Components System: Design.* 28 September 1973.
TM 5-302-2. *Army Facilities Components System: Design.* 28 September 1973.
TM 5-704. *Construction Print Reading in the Field* (AFM 85-27). 2 January 1969.

DOCUMENTS NEEDED

These documents must be available to the intended users of this publication.

DA Form 2028. *Recommended Changes to Publications and Blank Forms.* 1 February 1974.
DA Form 2702. *Bill of Materials.* 1 July 1963.

READINGS RECOMMENDED

These readings contain relevant supplemental information.

FM 5-412. *Project Management.* 13 June 1994.
TM 5-303. *Army Facilities Components System: Logistic Data and Bills of Materiels.* 1 June 1986.

INDEX

Field Manual
No. 5-10

Headquarters
Department of the Army
Washington, DC, 3 October 1995

Combat Engineer Platoon

TABLE OF CONTENTS

Page

PREFACE

Field Manual (FM) 5-10 is designed as a companion manual for *FM 5-34* and is structured using the following Battlefield Operating Systems (BOS) as a framework:

- Intelligence.
- Maneuver.
- Mobility and survivability.
- Fire support (FS).
- Air defense (AD).
- Combat service support (CSS).
- Command and control (C^2).

Each BOS is not discussed in detail. Instead, they are used to outline considerations for how the engineer platoon supports the total combined-arms (CA) effort. The bulk of the platoon's missions concentrates on mobility, countermobility, and survivability operations in support of the mobility and survivability BOS. However, the platoon leader and sergeant must have an understanding of the role their platoon plays in supporting all the BOS. More importantly, the platoon leader and sergeant must understand the impact of all the BOS on platoon planning, preparation, and execution.

This FM also has three appendices: *Appendix A* covers basic formations, movement techniques, and hand-and-arm signals; *Appendix B* contains checklists for security, precombat, and packing; and *Appendix C* covers case studies for mounted and dismounted breaching.

The proponent of this publication is Headquarters (HQ), United States (US) Army Engineer School. Send comments and recommendations on *Department of the Army (DA) Form 2028* directly to Commander, US Army Engineer School, ATTN: ATSE-T-PD-PM, Fort Leonard Wood, Missouri 65473-6650.

Unless this publication states otherwise, masculine nouns and pronouns do not refer exclusively to men.

CHAPTER 1
Intelligence

Intelligence is the key for preparing for the next battle. The analysis of intelligence, in detail, determines the impact of the enemy, weather, and terrain on operations. It is a continuous process that supports the planning for and execution of all operations. It is imperative that the platoon leader understand the intelligence process and his part in that process.

THREAT

The threat facing the US Army has gone through a quantum change with the breakup of the Warsaw Pact and the former Soviet Union. However, the majority of the forces we will face in mid- to high-intensity conflicts will use the "Soviet" model for their operations. *Figures 1-1 and 1-2, pages 1-2 and 1-3,* provide templates for the Soviet-style platoon and company defense as well as movement formations. A detailed analysis of threat mine-warfare capabilities and tactics can be found in *FM 90-13-1, Appendix A, and FM 20-32, Appendix C.*

RECONNAISSANCE

Reconnaissance is critical to the engineer. During offensive operations, the engineer must see the battlefield to plan breaching operations. History has proven that when a unit conducts CA rehearsals and obstacle reconnaissance it has greater success. The platoon leader should organize his platoon based on the reconnaissance mission the commander gives him. This frequently means providing expert engineers to assist the task force (TF) scouts. The platoon leader may also organize all or part of his platoon for a reconnaissance mission if there is a critical requirement for engineer-specific intelligence. If at all possible, the engineers who conduct the reconnaissance should be trained in this task as well as cross trained with the maneuver unit's scout platoon. The platoon leader must ensure that his designated reconnaissance element links up with the scout platoon.

TYPES

The general types of engineer reconnaissance missions are—

- Technical, which—
  - Is conducted to collect engineer-specific technical data.

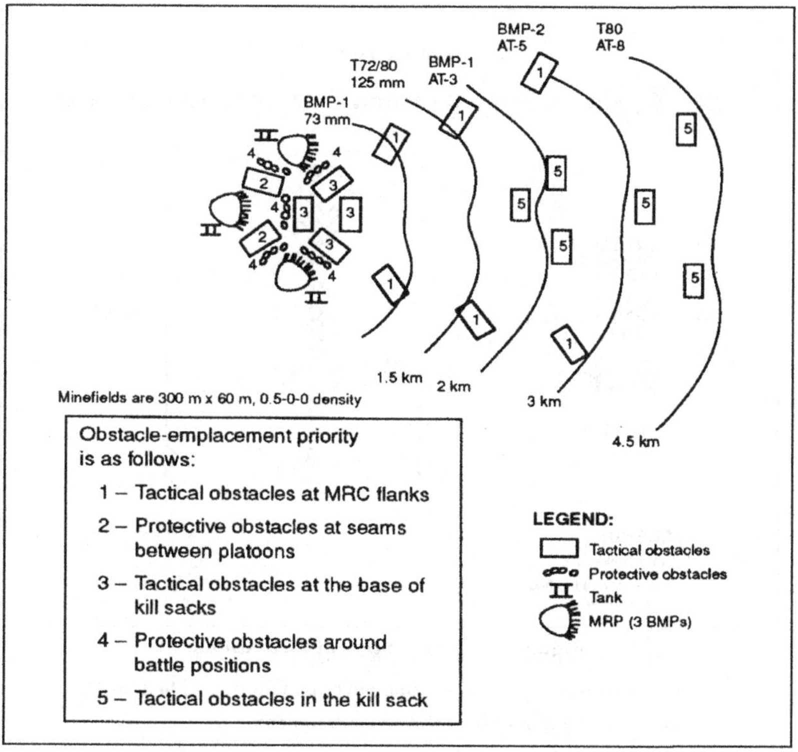

Figure 1-1. Motorized rifle company (MRC) defense template

- Is conducted to perform a specified task from higher HQ in a Level 1 threat environment.
- Involves (almost always) engineers only and is usually conducted behind the forward line of own troops/forward edge of the battle area (FLOT/FEBA).

- Tactical, which—
  - Is conducted to collect engineer-specific information requirements on an enemy point, area target, or attack axis or a route.

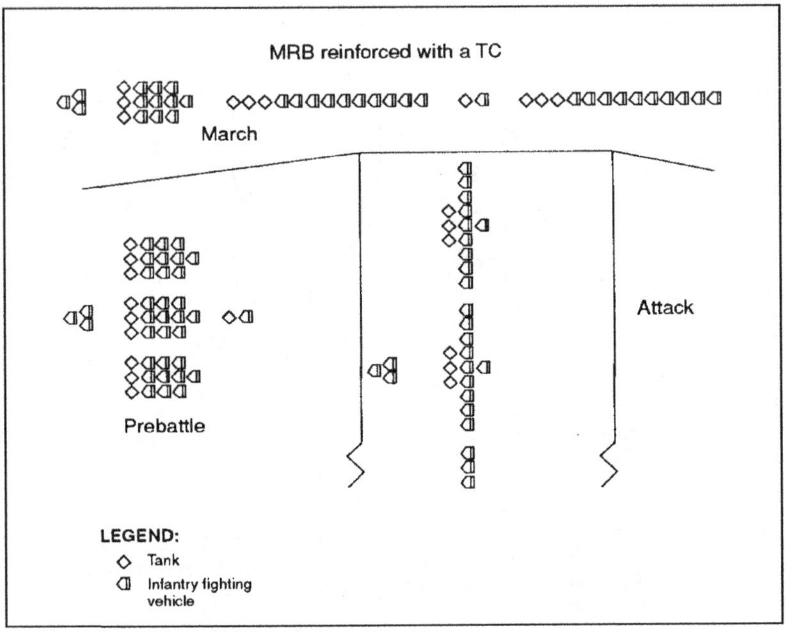

Figure 1-2. Typical threat formations

- Is conducted to support the maneuver plan as an implied task.
- Is conducted in a Level 3 threat environment.
- Requires engineer participation as part of a CA effort.
- Is usually conducted forward of the FLOT/FEBA.

PLANNING

Reconnaissance is a CA effort. The engineers either augment the existing reconnaissance element or conduct their own reconnaissance as part of the consolidated reconnaissance and surveillance (R&S) plan. Engineer reconnaissance must be focused on a specific target, and the reconnaissance element should be given specific intelligence requirements.

In heavy forces, the engineer company commander provides the reconnaissance target and does the majority of the engineer reconnaissance planning. However, in light forces, the engineer platoon leader does the engineer reconnaissance planning.

Platoon reconnaissance planning should include—

- A warning order (WO) based on brigade and TF planning, which includes—
  - Tentative target locations.
  - Type of reconnaissance or target (see *Figure 1-3).*
  - Specific information, equipment, and coordination requirements.
- The movement of reconnaissance assets to effect the linkup.
- The maximum time for leaders to prepare for the reconnaissance mission.
- The assignment of detailed reconnaissance routes.
- The actions to be taken on contact.

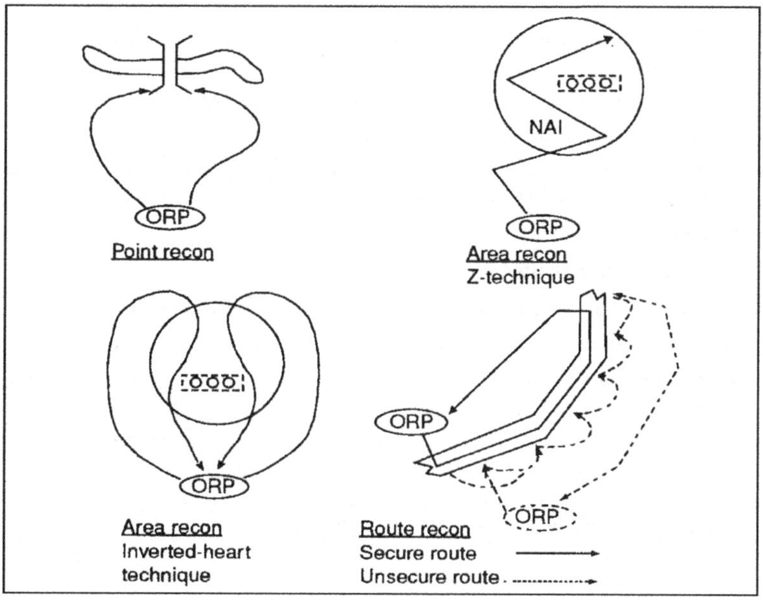

Figure 1-3. Recon techniques

- Consolidation instructions.
- Specific reporting requirements.

MOUNTED AND DISMOUNTED PATROLLING

Patrolling is a CA operation. Engineers are normally part of a larger unit's patrol. When patrolling mounted—

- Position all reconnaissance assets forward.
- Use aggressive and continuous patrols.
- Use area and zone reconnaissance.
- Allow enough time for a thorough reconnaissance.
- Use stealth, not combat power, for reconnaissance.
- Modify formations to overwatch the point element.
- Use terrain driving techniques; use draws and folds in the terrain to mask movement.
- Use night, fog, dust, and storms to conceal movement.
- Reduce speeds to avoid dust signatures.
- Travel at the highest tactical speeds, and place artillery fires on expected enemy positions to avoid long-range fires.
- Keep vehicles dispersed.

When patrolling dismounted—

- Reconnoiter all danger areas, choke points, and critical terrain features.
- Expect contact with enemy dismounted scouts.
- Use darkness, storms, and periods of limited visibility to infiltrate enemy positions.
- Use binoculars or night-vision devices at night with the illumination from ambient light (ensure that laser protection is maintained).
- Use artillery illumination and spotting rounds on reconnaissance targets to assist in land navigation and orientation.

CHAPTER 2
Maneuver

Engineer support to maneuver units requires detailed planning. Engineers may be part of the maneuver unit's formations and, if so, will provide the combat multiplier of engineer effort. *FM 90-13-1* is the primary doctrinal manual for breach planning. *FMs 20-32 and 90-7* are the primary doctrinal manuals for obstacle planning and construction. These references provide the technical information required to plan engineer support for offensive and defensive maneuvers.

The platoon leader should have an understanding of the tactical decision-making process, troop-leading procedures (TLP), and orders preparation. *Figure 2-1, page 2-2,* shows the relationship between TLP, the estimate process, and the intelligence preparation of the battlefield (IPB) process.

In addition, the platoon leader must advise the maneuver commander on the military aspects of the terrain since he is the terrain expert. Normally, the platoon leader only has a map to prepare his terrain analysis. However, he should request terrain-analysis products from his parent engineer company. In both light and armored units, the engineer company has access to terrain-analysis and terrain-visualization products or the ability to develop them. These products greatly aid platoon-level planning.

OFFENSE

The platoon leader is the key engineer expert to his supported unit. He must make several key decisions during the offense. He recommends where critical breach assets are located and which ones are best based on the situation. For example, the supporting engineer platoon leader, in an armored unit, would recommend where and how the M1 tank's plow would be used; or a light platoon leader could recommend to the TF commander a covert breach based on his terrain analysis. In both cases, the engineer platoon leader assists the maneuver commander with his planning. For more information on breaching, see *FM 90-13-1*. The following considerations are specific to the offense:

- Using a highly mobile engineer force, well forward and integrated into maneuver formations, to maintain the momentum of the attack.

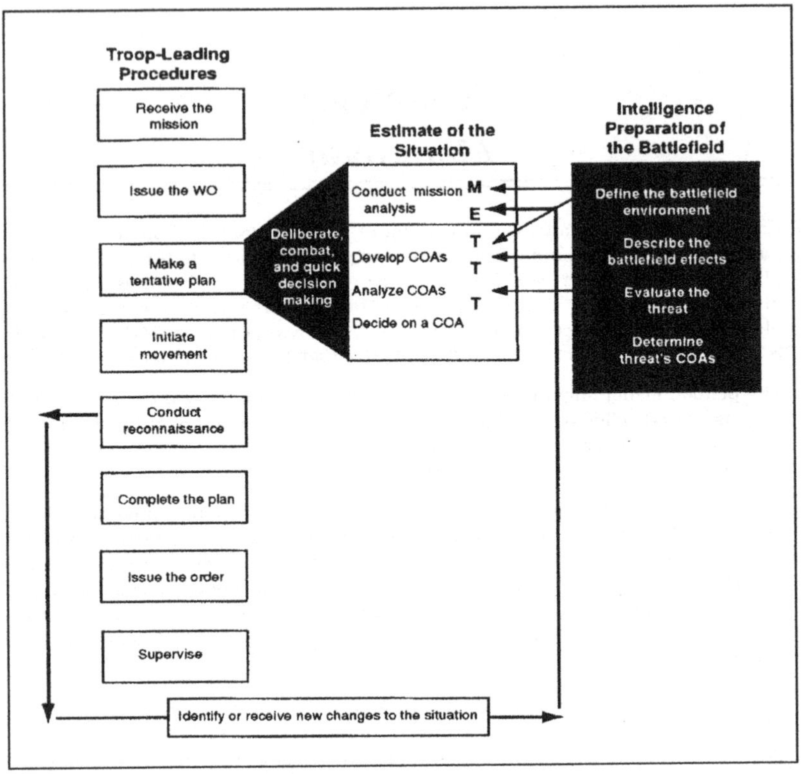

Figure 2-1. Relationship between TLP, the estimate process, and the IPB process

- Arranging for lane turnover between the forward breaching force and follow-on engineers for lane improvement and obstacle clearance.
- Ensuring that marking material and the standing operating procedure (SOP) for lane marking are coordinated between passing units.
- Ensuring that critical engineer equipment, such as armored vehicle-launched bridges (AVLBs), follow-on tactical bridging, lift capability for

line-charge reloading, and lane-marking materials to replenish marking systems, is available.

- Planning situational-obstacle capability for flank security. Scatterable mine (SCATMINE) systems work well in this application.
- Planning tactical-obstacle capability for protection against counterattacks (CATKs). Artillery- and aircraft-delivered family of scatterable mines (FASCAM) are planned for quick emplacement. Ground FASCAM systems are planned for prolonged defenses.
- Ensuring that Class IV/Class V barrier materials are available for a transition to the defense.
- Removing curbs from the AVLB if M1 tanks with rollers are to cross. The roller's dog bone should be raised.

DEFENSE

As with the offense, the platoon leader recommends to the maneuver commander where engineer assets are best used. It is critical that the platoon leader be an integral part of obstacle planning and integration. For more information on obstacle integration, see *FM 90-7*. The platoon leader should also coordinate and develop the obstacle-protection plan. It is critical that the obstacles are protected against enemy breaching attempts, especially covert attempts. The platoon leader should coordinate the inspection of the obstacles that his platoon built before the expected enemy attack. This will ensure that the enemy has not covertly breached and, if so, allows for the obstacles to be repaired before the enemy's main attack. The following considerations are specific to the defense:

- Understanding the commander's intent for obstacles.
- Ensuring that adequate transportation assets are available to haul Class IV/Class V supplies forward (see *Table 2-1, page 2-4*).
- Identifying critical engineer tasks early. Terrain preparation requires time for completion. Engineers must not remain idle while planning is in progress. If possible, mortars, air-defense artillery (ADA), fire-support-team (FIST) vehicles, CSS assets, and the tactical operations center (TOC) should be dug in while direct-fire systems are being positioned. Typically, these systems can be fortified between issuing the WO and completing the direct-fire systems' positioning.
- Practicing operations security (OPSEC) measures to prevent premature disclosure of the defense.

Table 2-1. Class IV/Class V haul capacities

Vehicle	Concertina Wire	M15 AT Mine	M21 AT Mine	M16 AP Mine	M14 AP Mine	MOPMS Mine	Flipper Mine	Volcano Mine	MICLIC Reload
HMMWV, M998 2,500 lb 215 cu ft	2	51	27	55	56	15	11	1	*1
2 1/2 -ton truck 5,000 lb 443 cu ft	4	102	55	111	113	30	23	2	1
5-ton truck 10,000 lb 488 cu ft	7	204	109	222	227	61	46	5	3
5-ton dump truck 10,000 lb ***135/291 cu ft	2/4	112/204	32/70	168/222	71/153	23/51	39/46	3/5	2/3
20-ton dump truck 40,000 lb 754 cu ft	11	628	179	888	443	132	184	20	11
HEMTT truck 20,000 lb 540 cu ft	8	408	128	444	317	94	92	10	7
12-ton S&P 24,000 lb 875 cu ft	13	489	208	533	514	148	110	12	9
40-ton lowboy 80,000 lb 1,760 cu ft	27	1,466	419	1,777	1,035	308	368	43	27
M54812,000 lb 529 cu ft	8	244	125	266	272	74	55	6	4
#Mines / wt/lb Cube cu ft	40/ 1,180 64	1/ 49 1.2	4/ 91 4.1	4/ 45 .8	90/ 44 1.9	21/ 162 5.7	40/ 217 3.4	240/ 1,850 37.6	** 2,656 64.8

for concertina = bundles; 1 bundle = 40 rolls * Overloads vehicle **Line charge + rocket ***Without/with sideboards

- Ensuring that the engineer organization for combat allows for rapid transition to the offense. The reserve must always have a designated force of engineers. Obstacle placement must not interfere with spoiling and CATKs.
- Ensuring that engineer units are not held in reserve but remain committed and work on the commander's priority tasks.

PLATOON LEADER'S PLANNING

The platoon leader can use *Tables 2-2 and 2-3, pages 2-6 and 2-7,* as a guideline when planning and conducting offensive and defensive operations.

REHEARSAL PRINCIPLES

Regardless of the event or task to be rehearsed or the type of rehearsal used, certain principles are universal for conducting effective rehearsals. They are—

- Supporting the scheme of maneuver and the commander's intent.
- Providing clear tasks/conditions/standards (T/C/S).
- Conducting multiechelon CA rehearsals.
- Determining key participants.
- Enforcing standards/training to standard.
- Using the after-action-review (AAR) process to provide feedback to all participants.
- Complementing the preparation phase.
- Instilling confidence in the plan and in the leaders.

During the command estimate process, use the rehearsal principles. These principles produce the—

- Participant levels.
- Rehearsal technique.
- Initiation of precombat checks (PCCs), precombat inspections (PCIs), and mission-specific drills.

Proper application of the rehearsal principles enables the unit to progress through a "crawl/walk/run" process that ends with mission success (see *Figure 2-2, page 2-8*).

Table 2-2. TF-engineer offense briefing

Briefing Topic	Main Points to Brief
Engineer task organization	• How the platoon will be task-organized • Special equipment that will be task-organized to subordinate units • Times and locations for linkup
Enemy situation	• Obstacle template • Enemy's obstacle intent • Enemy's expected level of defensive preparation • Confirmed intelligence on enemy obstacles and positions (include an analysis of this intelligence) • Enemy employment of FASCAM • Impact of enemy obstacles on friendly maneuver
OBSTINTEL collection effort	• Engineer R&S plan • NAIs critical to OBSTINTEL • How the OBSTINTEL effort supports the scheme of maneuver • Scout, engineer, and maneuver OBSTINTEL responsibilities and relationships
Scheme of engineer operations	• How the platoon will support the scheme of maneuver from the objective to the LD, to include consolidation • Engineer contingency plans and pertinent decision points
TF and company/team actions at obstacles	• In-stride breaching, to include— — Areas where the unit can expect to in-stride breach — Special equipment required — Special instructions • Deliberate breaching, to include— — Transitioning from attack formations to breaching formations — How the fundamentals of breaching will be synchronized — Key events and control measures synchronizing the breach — Composition of and instructions for the support, breach, and assault forces — Scheme of fires supporting the breach — Obscuration plan — Coordinating instructions, command and signals for the breach • Assault breaching, to include— — Allocation of engineers — Special coordinating instructions
Countermobility support to offensive maneuver	• Planned locations for situational obstacles • Triggers for situational-obstacle employment
Coordinating instructions	• Lane marking, breach traffic control, order of march through the lane, specific breaching signals, and breach C^2
Service support	• What support the platoon is receiving from the supported company/team and the parent company, if different from the SOP • Medical evacuation, supplies, and vehicle recovery • Current equipment status if critical to the mission
Unresolved issues	• Issues that the plan does not adequately address

Table 2-3. TF-engineer defense briefing

Briefing Topic	Main Points to Brief
Engineer task organization	• How the platoon will be task-organized • Special equipment that will be task-organized to subordinate units • Times and locations for linkup
Enemy	• Enemy maneuver formations through the sector, to include— — Most probable enemy COA — Place and time the enemy will change formation — Enemy breaching assets and capabilities (include high-priority targets)
Countermobility support	• Scheme of obstacles, to include— — Obstacle zones and restrictions — Obstacle belts and groups and their effects — Desired effect of obstacle groups on enemy maneuver — How the scheme of obstacles support the direct- and indirect-fire plans — Triggers and decision points for situational and reserve obstacles — Location of Class IV/ClassV supply points • Obstacle status reporting, to include— — Completion of obstacles — Barrier-material status • Obstacle-protection and turnover plans
Mobility support	• Where the lanes need to be to support MSRs, company/team supply routes, and reconnaissance passages • Responsibility for lane siting and closure • Lane marking and contact points • Obstacle restrictions that support the TF's mobility • Engineer-support plan for mobility
Survivability support	• Protective-obstacle plan, to include— — Material that the unit allocates — Any engineer assistance required — Protective-obstacle reporting responsibilities • Construction plan for fighting positions, to include— — Any engineer assistance required to construct individual positions — Allocation of engineer equipment and time and the priority for constructing vehicle positions — Linkup points and times with the company/team • Equipment control plan to ensure that maneuver companies know— — How many positions will be constructed by type — When, where, and how much equipment they will receive — How long they will retain the equipment
Engineer time line	• Engineer time line, to include— — Estimated completion time — Enemy trigger to stop work — Withdrawal route or passage — Engineer consolidation plan — Class VI/V supply-point recover or destruction
Coordinating instructions	• Obstacle-siting responsibilities • Reporting times • Class IV/Class V supply-point location, security plan, and responsibility

Table 2-3. TF-engineer defense briefing (continued)

Briefing Topic	Main Points to Brief
Service support	• Critical equipment and material shortages affecting defensive preparations • Medical evacuation plan for engineers • Barrier-material configurations to support obstacle-group construction
Unresolved issues	• Issues that the plan does not adequately address

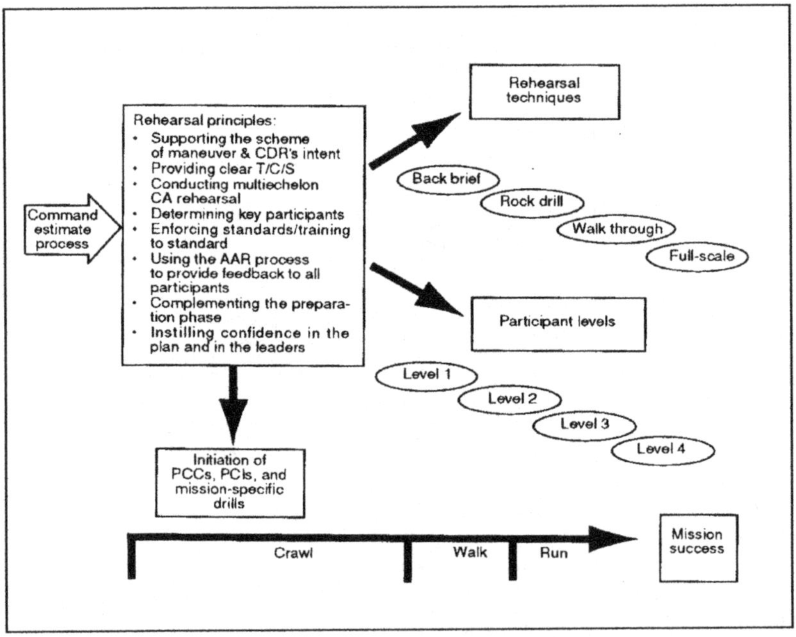

Figure 2-2. Rehearsal planning

PARTICIPANT LEVELS

The participant level details exactly who in the unit is required to attend the rehearsal. A CA-team rehearsal may have four distinct levels. The platoon leader may participate with or without his subordinates in the CA rehearsal. The three levels that apply to a platoon rehearsal follow:

- Level 1 - Squad leaders and platoon HQ.
- Level 2 - Squad leader, team leader, special teams, and vehicle commanders.
- Level 3 - Entire unit.

REHEARSAL TECHNIQUES

Rehearsal techniques fall into four different categories that follow the "crawl/walk/run" training concept. These categories are—

- Back brief.
- Rock drill.
- Walk through.
- Full-scale.

Each type of rehearsal reflects an increase in mission realism and a corresponding increase in rehearsal payoff. Each technique also increases the realism of the enemy, terrain, subunit actions, and time/distance relationships.

Back Brief

The back-brief rehearsal is an event that occurs after an operation order (OPORD) brief. Subordinate leaders repeat to the commander what he expects them to do and why. They identify all specified and implied tasks, determine their mission-essential tasks, and give their restated mission. During the back brief, they address the—

- Commander's intent.
- Concept of the operation.
- Scheme of maneuver.
- Timing to complete tasks.

During a typical back brief, each leader uses a map or a sand table and explains his mission. The back brief is the quickest of all the rehearsal techniques. It is a leader's tool and is typically the first rehearsal for the unit.

Rock Drill

A rock-drill rehearsal is the acting out of friendly and enemy actions based on the scheme of maneuver and the situation and event templates. Participants rehearse their actions by moving something that represents themselves or their unit, such as rocks or sticks. While acting out the plan, participants talk through their missions, critical tasks, actions, and decisions. Since all participants are simultaneously acting out their part of the scheme of maneuver, leaders can identify problems and disconnects in synchronization more clearly in a rock drill.

Walk Through

A walk-through rehearsal is the acting out of the scheme of maneuver using mounted or dismounted movement based on the assets employed during the attack. Participants should also communicate with the same type of equipment they will use during the action. Participants rehearse by—

- Maneuvering their vehicles (mounted movement) or themselves (dismounted movement).
- Reporting critical actions.
- Making required decisions.

Since the participants are in a more realistic environment, they rehearse the finer aspects of synchronization, C^2, and subunit actions. Aggressive portrayals of enemy actions and reactions are critical in walk-through rehearsals. This threat portrayal is the backbone of increasing the rehearsal realism. This rehearsal technique is more difficult to orchestrate than the back brief or rock drill; however, it is the optimum balance between resource constraints and realism. Successfully mastering a walk-through rehearsal should be the minimum goal for all units.

Full-Scale

During a full-scale rehearsal, participants use real-time mounted and dismounted movement over the actual or similar terrain. Typically, it is conducted at Level 3; however, a mix of participant levels is possible. At least one subunit must participate at Level 3 for a full-scale rehearsal. This type of rehearsal is obviously the most resource intensive, but it provides the most realistic training environment for the unit. It is often used to rehearse the operation plan (OPLAN) or OPORD when time is not an immediate constraint. One must plan extensively to execute full-scale rehearsals correctly and in a manner that does not waste the soldiers' time.

INITIATION OF PCCs, PCIs, AND MISSION-SPECIFIC DRILLS

The subordinate leader conducts PCCs to determine his unit's readiness to execute its assigned tactical mission. These inspections are informal. They are the leader's inspection of mission-critical tasks or areas of special interest to the commander. PCCs should cover—

- Vehicles.
- Weapons systems.

- Soldier's equipment, to include his weapon.
- Mission-critical equipment, such as explosives, detonators, and blasting machines.

The unit commander or leader conducts PCIs to determine the unit's readiness to execute its assigned tactical mission. These inspections may be formal or informal. Formal PCIs are the commander's meticulous and time-consuming inspections of all areas within the unit. They are seldom conducted during combat operations or assembly-area occupation. Informal PCIs are the commander's inspections of particular areas or activities of special interest or concern to him. He may designate the executive officer (XO) or first sergeant (1SG) to perform the PCI. PCIs should cover—

- Vehicles
- Major weapons systems.
- Soldiers.
- Communications equipment.

Mission-specific drills are rehearsals of special combat drills that subordinate units conduct. Special drills may include—

- Breach drills.
- Marking drills.
- Situational obstacles.
- Reserve demolitions.

CHAPTER 3
Mobility and Survivability

Mobility and survivability are the primary tasks of the engineer on the battlefield. The platoon leader requires a detailed understanding of this BOS to succeed. This chapter provides the tactics, techniques, and procedures (TTP) needed for the platoon to conduct mobility, countermobility, and survivability.

MOBILITY

Mobility enables the commander to maneuver tactical units into positions of advantage over the enemy. In the attack, engineers reduce enemy obstacles and fortifications that inhibit maneuver.

Expanded engineer reconnaissance capability is needed to identify routes, existing obstacles, and minefields. Open areas that are level provide good sites for aircraft landing strips; however, in most cases, the soil must be stabilized.

The application of the breaching fundamentals (suppress, obscure, secure, and reduce {SOSR}) and the organization of the force in terms of support, breach, and assault forces are standard. However, open areas may offer greater opportunity to bypass enemy obstacles because of the greater range of mobility afforded by the terrain. Exercise caution when choosing to bypass enemy obstacles, since the bypass may lead the force to the enemy's engagement area (EA). Additionally, expect the enemy to make extensive use of rapid mine-laying techniques that include SCATMINE delivery systems, as well as more conventional methods of mine emplacement.

OBSTACLE BREACHING

The platoon leader must decide where the best breach location is. Never breach where the avenue of approach crosses the obstacle. The enemy is sure to have preplanned artillery there. Pick a location away from obvious breaching sites to reduce the threat of artillery. The platoon leader should place himself where he can see the breach site and effectively command and control his platoon. He must be able to signal his breaching reserve (generally a squad) if he sees the breach failing. Observation of the breach site also reduces reporting requirements on the squad leaders to breach and mark the breach site.

Typically, there is a time lag as the maneuver force suppresses the enemy and obscures the breach site for the breach. The platoon leader should also consider dismounting an element from his lead squad to reconnoiter the best breach location during this tactical pause. This will assist the platoon leader in choosing the best breach site.

Obstacle breaching must be a CA effort. Units are more successful if they conduct—

- Obstacle reconnaissance.
- CA rehearsals.
- Breach planning.

For more information on obstacle breaching, see *FM 90-13-1*.

Breaching Fundamentals

When breaching against a defending enemy, the following fundamentals must be applied to ensure success:

- Suppress. Neutralize or destroy the enemy's weapons with our own fires. (The support force should have a 3:1 advantage over the enemy's weapons overwatching the obstacle.) Consider the effects and characteristics of the enemy's weapons. Ensure the correct focus of direct and indirect fires and the timely shift or lift of fires. Consider what ammunition is required.
- Obscure. Hamper the enemy's observation and target acquisition through the use of smoke (indirect, on board, pyrotechnic), terrain, limited visibility, and stealth.
- Secure. Eliminate enemy interference with obstacle reduction and lane usage. Occupy the obstacle with a friendly force and through the use of effective fires.
- Reduce. Create lanes through, over, or around an obstacle. Ensure that a company/team has at least one lane through an obstacle and that a battalion TF has two lanes. Consider what additional requirements are needed to guide follow-on forces through the obstacle or to mark and reduce the obstacle.

Types of Breaching Operations

When planning a breaching operation, the engineer leader should consider bypassing the obstacle as his first option. However, the enemy's intent may be to force the unit to turn or channelize into a fire sac. This is why a good reconnaissance of a possible bypass is essential. The unit conducting a bypass must have maximum security during the bypass. If bypass operations are not possible, consider the following options:

In-stride Breach. A unit conducts an in-stride breach when it can be successfully executed under the complete C^2 of a task-organized subordinate unit or when the situation is unclear and rapid movement is critical. Subordinate units serve as both the breach and assault forces. The subordinate commanders synchronize the breaching fundamentals. Doctrinally, TFs and above conduct in-stride breaching.

Deliberate Breach. A unit conducts a deliberate breach when the force-allocation ratios for support, breach, and assault forces are beyond the capability of a task-organized subordinate unit. One or more subordinate units are specifically tasked to perform the role of support, breach, and assault forces. Synchronization is critical and the principle of mass drives task organization.

Assault Breach. A unit conducts an assault breach during the penetration and destruction phases of actions on the objective to destroy an enemy that has had time to emplace protective obstacles. For the assault breach, a rule of thumb is one lane per assaulting platoon. Assault lanes are generally only footpaths until the objective is secured. Mounted assault lanes require the same effort as tactical vehicle lanes.

Covert Breach. A unit conducts a covert breach during limited visibility when surprise is essential. The covert breach relies on stealth; quiet, manual lane-reduction techniques; and dismounted maneuver. It can be employed to breach tactical obstacles or protective obstacles (more difficult). It requires the same combat ratios as other breaches for suppression and security, and these points should be considered during the planning process.

Breach Planning

The following are considered when developing a breaching plan (see *Figures 3-1 and 3-2, pages 3-4 and 3-5*):

- Reverse planning begins with actions on the objective.
- Actions on the objective drive the size of the assault force and determine the number and location of lanes to be breached.
- Lane requirements and the type of obstacle drive the allocation of mobility assets to the breach force.
- Ability of the enemy's infantry to interfere with the breach determines whether to secure the breaching site by force or by fire.
- Ability of the enemy to mass fires at the breaching site determines the amount of suppression required and the size of the support force.

The platoon leader must plan for actions after the breach is complete. Generally, the platoon is responsible for beaching, marking, and providing guides for the lane.

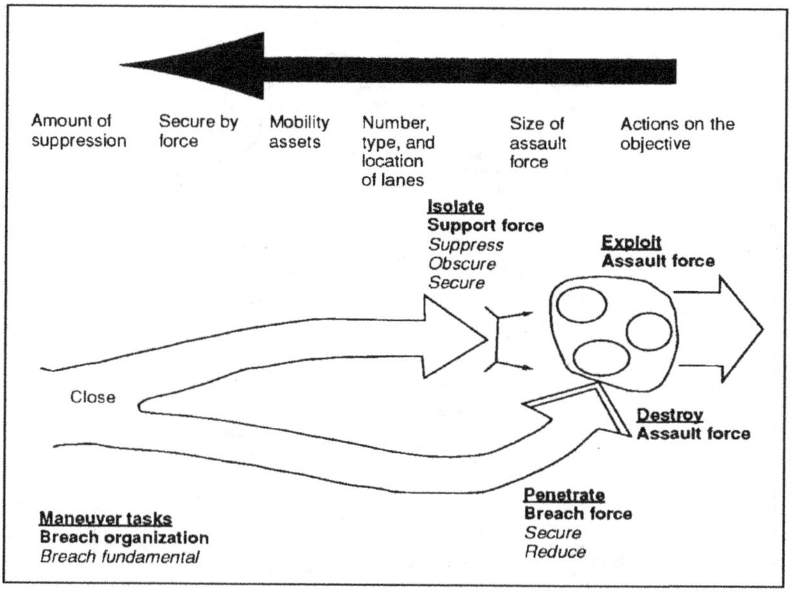

Figure 3-1. Breach planning

Normally, the platoon remains at the breach location and continues to improve the breached lane. The platoon must be prepared to reopen the lane if the enemy closes it with artillery-delivered mines. The platoon leader should also plan for vehicles that are stopped in the lane. The platoon may be forced to create an additional lane around the stopped vehicle or extract it from the breached lane.

Breach TF Organization

The breach TF is organized into three elements:

- Support force—eliminates enemy interference with the breach, isolates the objective area, and destroys enemy weapons that are able to fire on the breach force.
- Breach force—creates lanes in and through the obstacle. While creating the lane through the obstacle, it must also locally secure the breaching site and pass the assault force through the breach.

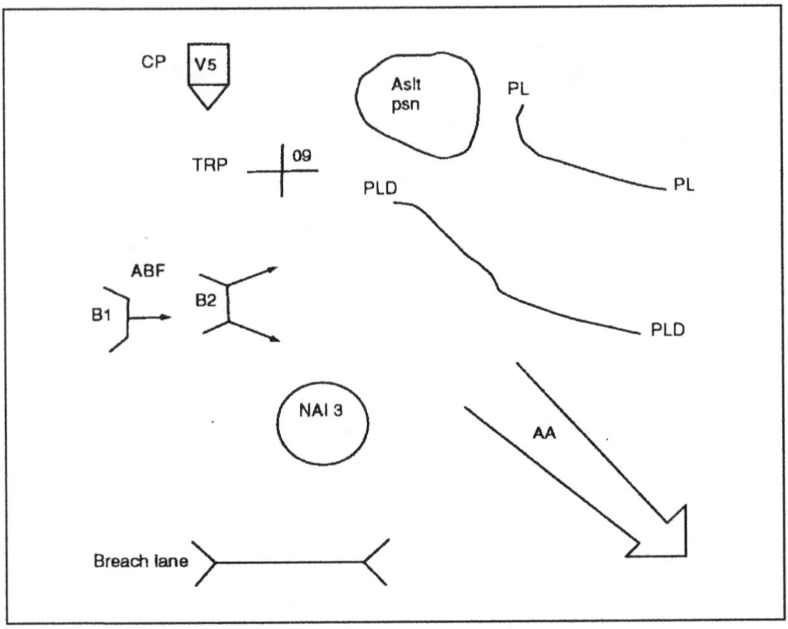

Figure 3-2. Breach control measures

- Assault force—destroys or dislodges the enemy force on the far side of the obstacle and seizes the initial foothold on the objective.

Obscurants

To obscure friendly activities and movement, units can use artillery- or mortar-delivered smoke and/or white phosphorous (WP) rounds, generators, smoke pots and M203 grenades, chemical units, and burning diesel. Obscurants—

- Should be preplanned and coordinated with the company commander and the supported TF.
- Can be used for deception.
- Can be combined with supporting fires.
- Can be used over large areas.

LANE MARKING

The critical components to any lane-marking system are the lane-marking patterns and devices. Lane-marking devices are not standardized Army-wide. *Figure 3-3* shows examples of lane-marking devices. *Table 3-1, page 3-8,* can be used as a guide to choose lane-marking devices. Lane-marking patterns have been standardized Army-wide. Standard breach lane-marking patterns are explained in *FM 90-13-1, Appendix E*. The following are lane markers and their use:

- Entrance markers—indicate the start and the width of a reduced lane. They must be visually different than handrail markers. Entrance markers are placed a minimum of 4.5 meters apart (1 meter for dismounted).
- Handrail markers—define the lane path and indicate the limits of the lane width. Mark, as a minimum, the left handrail.
- Exit markers—indicate the far-side limit of reduced lanes. Exit markers must be visually different from handrail markers but may be the same as entrance markers.
- Entrance-funnel markers—augment the entrance markers. They assist the small-unit commander in guiding the lead unit of his combat column formation.
- Final-approach marker—is a highly visible marker that augments the visual signature of entrance-funnel markers. It provides the assault-force commander with a highly visible reference point toward which to maneuver his formation.
- Far-recognition markers—are highly visible markers located between the final-approach marker and the friendly unit. They are primarily used when passing battalion-sized forces through a lane where distance, visibility, or terrain does not allow the passing force direct observation of the final-approach marker. When possible, far-recognition markers should be different from the final- approach marker.
- Traffic-control post (TCP) or guides—are a two-man team with communications means that assists the commander in controlling the movement of forces. When possible, military police (MP) should man TCPs. However, the breach force should plan to man TCPs until relieved.

The standard levels of lane marking for breach lanes and bypasses are initial, intermediate, and full (see *Figure 3-4, page 3-9*). The following are lane-marking patterns and their use:

- Initial lane-marking pattern—is emplaced by the breach force immediately after the lane is reduced and proofed. It is a signal to the assault force that the lane is ready for traffic to pass through.

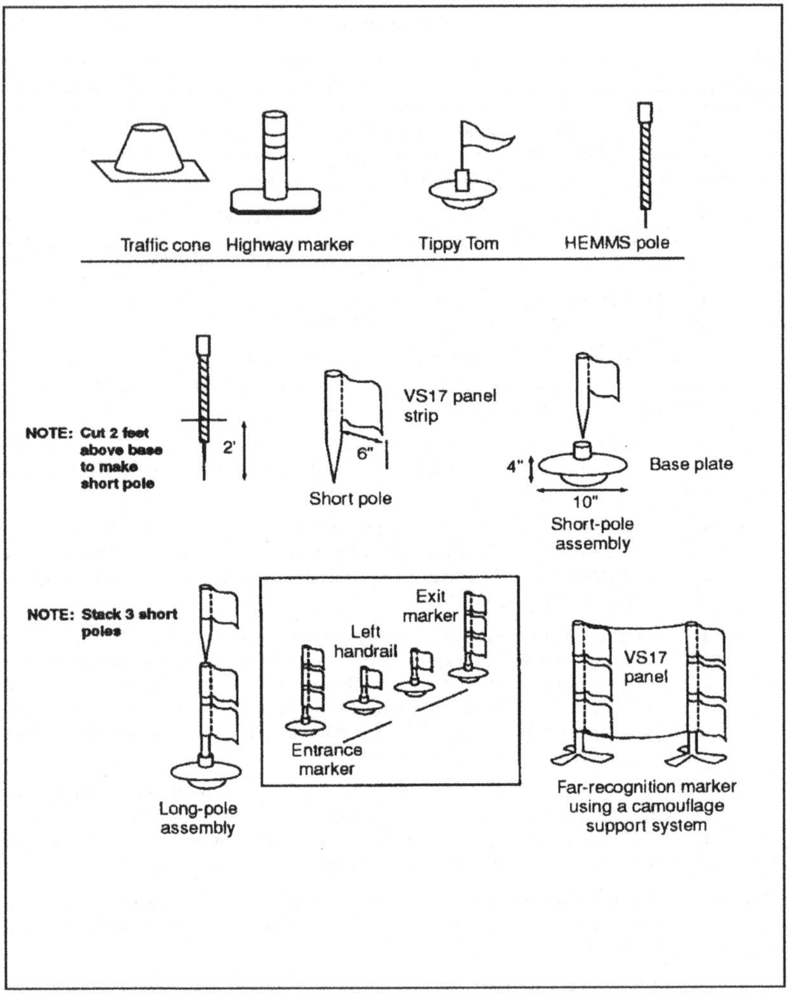

Figure 3-3. Lane-marking devices

Table 3-1. Breach lane marking

Markers	Mounted Forces	Dismounted Forces
Handrail and funnel	They— • Can be seen by the vehicle commander and driver buttoned up from 50 meters. • Are quick to emplace, minimizing soldier exposure.	They— • Are visible to a prone dismounted soldier from 50 meters. • Are lightweight and easy to emplace.
Entrance and exit	They— • Can be seen from 100 meters. • Are visually different from handrail and funnel markers. • Are easy to emplace and are man portable.	They— • Are visible from 50 meters. • Are visually different from handrail and funnel markers. • Are easy to emplace and are man portable.
Final approach and far recognition	They— • Can be seen from 500 meters. • Are visually different from each other. • Can be altered to facilitate traffic control through multiple lanes.	They— • Are visible from 100 meters. • Are visually different from each other. • Can be altered to facilitate traffic control through multiple lanes.

- Intermediate lane-marking pattern—is used for the commitment of larger combat forces who are unable to directly observe the breach or the rearward passage of sustainment traffic. It builds on the initial lane-marking pattern by adding right handrail markers, exit-funnel markers, far-recognition markers, and a far-side final-approach marker.
- Full lane-marking pattern—is usually not part of an initial breaching operation. It is a lane that will support uninterrupted two-way traffic.

Table 3-2, page 3-10, provides information as to who marks the various types of lanes, when they are to mark the lanes, and what markers they are to use. Lane marking is normally synchronized during the CA rehearsal to ensure that all key leaders understand the marking procedure and standard being employed.

COUNTERMOBILITY

Countermobility operations are conducted to augment natural terrain with obstacle systems according to the commander's concept. This adds depth to the battle in

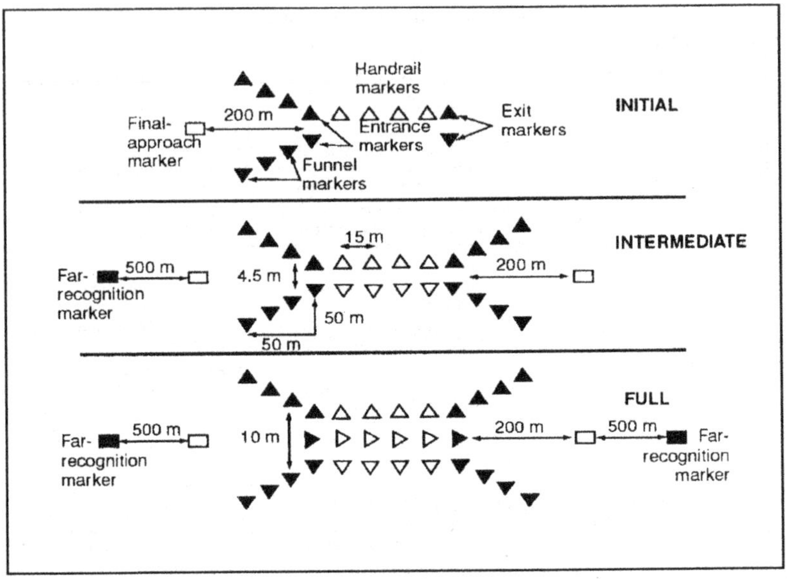

Figure 3-4. Lane-marking patterns

space and time by attacking the enemy's ability to maneuver its forces. With the enemy's movement impeded (disrupted, turned, fixed, or blocked), it is vulnerable to our forces. Engineers advise the commander on the best means to reinforce natural obstacles (terrain) and to emplace tactical obstacles that support his plan. Minefields, wire, antitank ditches (ATDs) or berms, road craters, bridge demolition, and all other traditional obstacles may be effective. For expedient obstacles, units should consider employing local materials.

ATD or berms require extensive preparation but are very effective when properly prepared. Many areas have irrigation ditches that can be used tactically. However, in sandy areas, ditches can easily be filled in. Obstacles should be planned to protect the unit's flanks during offensive operations.

All obstacles can be situational obstacles. Situational obstacles are obstacle resources held in reserve. They have a "be-prepared" mission but not an "on-order mission". Situational obstacles should be considered during planning to give the maneuver commander a flexible response to enemy or friendly maneuver.

Table 3-2. Lane-marking level—unit responsible, trigger events, and lane markers

Breach Type	Initial	Intermediate	Full
	Unit Responsible		
Deliberate	TF breach force	TF breach force	Brigade
Covert	TF breach force	TF breach force	Brigade
In-stride	Breach company/ team	TF mobility reserve	Brigade
Assault	Assault platoon	TF assault force	NA
	Trigger Events		
	When— • Lanes are reduced • Passing platoon- or company-sized forces	When passing— • Battalion- or company-sized forces • Forces that can-not see the lane • TF combat trains	When— • Passing brigade- or battalion-sized forces • Situation requires uninterrupted sus-tainment traffic
	Lane Markers		
	Entrance/exit Left handrail Entrance funnel Final approach	The following mark-ers are added: • Right handrail. • Exit funnel. • Far-side final approach. • Far recognition. • Guides/TCPs.	Lane width is expanded to 10 meters. Existing markers are adjusted. Far-side-recognition markers and guides/TCPs are added.

OBSTACLE PLANNING

To assist the TF commander in the defense, the TF engineer must integrate obstacle planning throughout the TF sector and coordinate with adjacent TFs according to the overall engineer plan.

Each echelon has its own obstacle-control measures for directing the location of obstacles on the ground. The division commander and his staff use obstacle zones to control obstacles to support future operations. Within these zones, the brigades plan obstacle belts, and the battalion TF plans obstacle groups. The engineer will then emplace each obstacle by doctrinal obstacle norms to support the commander.

Protective Obstacles

They are designed to repel the enemy's assault by emplacing them around a unit's defensive position (a survivability task).

Tactical Obstacles

They are designed and emplaced to (see *Figure 3-5*)—

- Disrupt, which—
  - Breaks up C^2.
  - Frustrates the enemy's timing.
  - Interrupts formations or attacks the enemy when they are massing into formations.
  - Coerces the enemy into our lines of fire.

Obstacle-Effect Graphic	Application	Examples Conveying Intent	Resource Factor[1]
Disrupt	Short arrow indicates where enemy is attacked by obstacles. Long arrows indicate where bypass is allowed and attacked by fires.	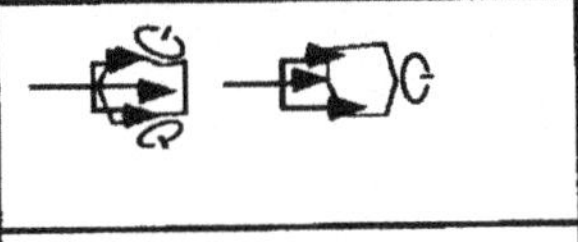	0.5
Turn	Heel of arrow is anchor point. Direction of arrow indicates desired direction of turn.		1.2
Fix	Irregular part of arrow indicates where enemy advance is slowed by obstacles.	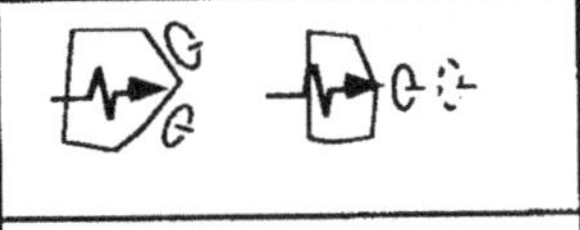	1.0
Block	The ends of the vertical line indicate the limit of enemy advance. The ends of the vertical line also indicate where obstacles tie in to severely restricted terrain.		2.4

[1] See *FM 90-7, Appendix C*, for an explanation of how to use the resource factor.

Direction of Enemy Attack

Figure 3-5. Obstacle-effect graphics

- Fix, which—
  - Slows the enemy's advance within a specified area.
  - Causes the enemy to breach repeatedly.
- Turn, which manipulates enemy maneuver in a desired direction.
- Block, which—
  - Stops the enemy along a particular avenue of approach.
  - Defeats the enemy's breaching effort.

Figure 3-6 shows how to employ conventional row minefields to obtain the desired obstacle effects and the planning figures for standard minefields by obstacle effect.

Situational Obstacles

Situational obstacles are "triggered" by an event or situation. Once the general location of the obstacle has been determined, it must be refined to better enhance its effect on the enemy. When planning, preparing, and executing situational obstacles (see *FM 90-7*)—

- Identify the need.
- Plan for appropriate resources.
- Plan the obstacle.
- Integrate the obstacle with friendly fires.
- Identify obstacle execution triggers.
- Withhold execution of the obstacle until it is needed.

Situational obstacles can be used to—

- Attack an enemy's vulnerability.
- Exploit success.
- Separate follow-on enemy forces.
- Provide flank protection.

Tables 3-3 and 3-4, pages 3-14 and 3-15, provide some characteristics of conventional and scatterable mines and are useful in planning minefields. When emplacing standard disrupt, fix, turn, and block minefields, see *Figures 3-7 through 3-12, pages 3-16 through 3-21, and Tables 3-5 through 3-8, pages 3-22 and 3-23,* for information on the employment of SCATMINE systems. FASCAM systems include the air/ground Volcano, Modular Pack Mine System (MOPMS), Flipper, area denial artillery munition (ADAM), and remote antiarmor mine (RAAM).

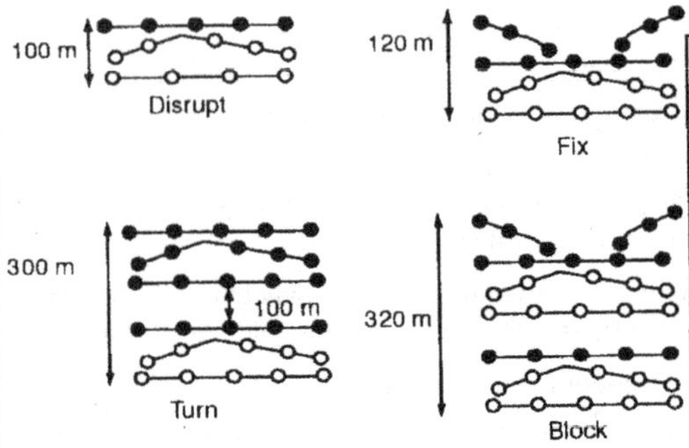

Mines spaced 6 m apart
Rows spaced 50 m apart
Minefield frontage
Disrupt: 250 m
Turn: 500 m
Fix: 250 m
Block: 500 m
o Track-width fuzed mine
● Full-width fuzed mine

AP mine employment

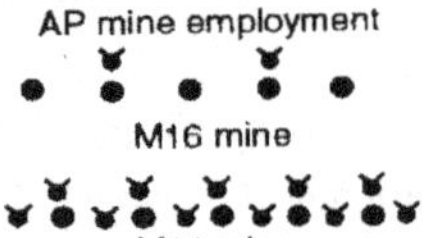

AP mines are—

- Used in block minefields only.
- Used in any two rows of full-width AT mines.
- Patterned based on AP mine type.

	Disrupt	Turn	Fix	Block
Frontage (m)	250	500	250	500
Depth (m)	100	300	120	320
AT full (# rows)	1	4	1	4
AT track (# rows)	2	2	2	2
AP (# rows)	0	0	0	2
IOE (yes or no)	No	No	Yes	Yes
AHD (yes or no)	Yes	No	No	Yes
Plt hrs reqd	1.5	3.5	1.5	5.0
Track-width AT mines	84	168	84	168
Full-width AT mines	42	336	63	378
AP mines M16	0	0	0	84
AP mines M14	0	0	0	500
Density AT	0.5	1.0	0.6	1.1
Density AP M16/M14	0	0	0	0.17/1.0

Figure 3-6. Minefield norms

Table 3-3. Conventional mines

Mine	Arming Time	Fuzing	Warhead	AHD	Mine Weight	Mines per 5-Ton Dump	Remarks
M14	NA	Pressure	Blast	No	3.3 oz	6,480	
M16A1	NA	Pressure Trip wire	Bounding fragment	Yes	8 lb	672	
M18A1	NA		Directional fragment	No	3.5 lb	1,782	
M86 (PDM)	50 sec		Bounding fragment	100%	1 lb	—	Self-destruct after 4 hours
M15	NA	Pressure	Blast	Yes	30 lb	90	
M15 w/ M624	NA	Tilt rod	Blast	Yes	30 lb	90	
M21	NA	Tilt rod	Shape charge	Yes	17 lb	192	

OBSTACLE SITING

When determining where to emplace obstacles, the platoon leader should consider the following:

- Supporting the maneuver commander's intent.
- Supporting the CATK plan.
- Supporting the defensive plan.
- Coordinating obstacle locations with the maneuver element, the fire-support officer (FSO), the Operations and Training Officer (US Army) (S3), and the weapons-systems commander.
- Prioritizing the engineer effort.
- Planning the maneuver unit's responsibility.
- Using observation and fields of fire, cover and concealment, obstacles, key terrain, and avenues of approach (OCOKA).

Company/Team Commanders' Coordination

The platoon leader must ensure that key obstacles are covered by indirect fire. These obstacles become priority targets for the artillery. When coordinating with company/team commanders, the platoon leader should consider the following:

- Determining the obstacle intent.
- Determining the company's/team's tactical purpose.

Table 3-4. Scatterable mines

Mine	Delivery System	Arming Time	Fuzing	Warhead	AHD	Self-Destruct Time	Mine Weight	Mines per 5-Ton Dump
M73	155-mm arty (RAAM)	45 sec 2 min	Magnetic	M-S plate	20%	48 hr	3.8 lb	NA
M70	155-mm arty (RAAM)	45 sec 2 min	Magnetic	M-S plate	20%	4 hr	3.8 lb	NA
M75	GEMSS Flipper	45 min	Magnetic	M-S plate	20%	5 days 15 days	3.8 lb	1,600
BLU 91/B	USAF (Gator)	2 min	Magnetic	M-S plate	No	4 hr 48 hr 15 days	3.8 lb	NA
M76	MOPMS	2 min	Magnetic	M-S plate	No	4 hr (recycle 3 times)	3.8 lb	30 modules (510 mines)
Volcano	Ground/air	2 min	Magnetic	M-S plate	No	4 hr 48 hr	3.8 lb	160 canisters (800 mine)
M72	155-mm arty (ADAM)	45 sec 2 min	Trip wire	Bounding fragment	20%	15 days 48 hr	1.2 lb	36 per M692 projectile
M67	155-mm arty (ADAM)	45 sec 2 min	Trip wire	Bounding fragment	20%	4 hr	1.2 lb	38 per M731 projectile
M74	GEMSS Flipper	45 min	Trip wire	Blast fragment	20%	5 days 15 days	3.2 lb	1,600
BLU 92/B	USAF (Gator)	2 min	Trip wire	Blast fragment	100%	4 hr 48 hr 15 days	3.2 lb	NA
M77	MOPMS	2 min	Trip wire	Blast fragment	100%	4 hr (recycle up to 15 days)	3.2 lb	30 modules (120 mines)
Volcano	Ground/air	2 min	Trip wire	Blast fragment	100%	4 hr 48 hr 15 days	3.2 lb	160 canister (160 mines)

- Ensuring that the obstacles the maneuver company/team commanders emplace on the battlefield serve as a combat multiplier and enhance the direct-fire plan.
- Providing forward security during obstacle emplacement.
- Conducting obstacle turnover with the maneuver company/team and ensuring that indirect-fire support is provided after the obstacle is finished.
- Rehearsing lane-closure plans.

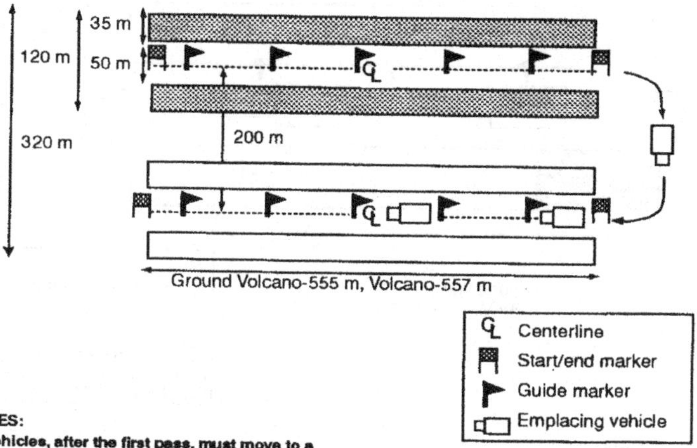

NOTES:

- Vehicles, after the first pass, must move to a safe distance for 2 minutes to allow mines failing the arming sequence to self-destruct.
- Guide markers are offset from centerline.
- Start and end markers should be used for air Volcano only.
- Turn and block minefields require the dispensing of 80 canisters (40 each side) on each pass, 160 canisters total.
- One full load of ground or air Volcano emplaces one turn or block minefield.

Figure 3-7. Volcano turn and block minefields

Direct-Fire Integration

When integrating obstacles with direct fire, the platoon leader should consider the following:

- Coordinating with the maneuver element overwatching the obstacle.
- Sighting obstacles with weapons.
- Coordinating with the maneuver element for—
  - Forward security.

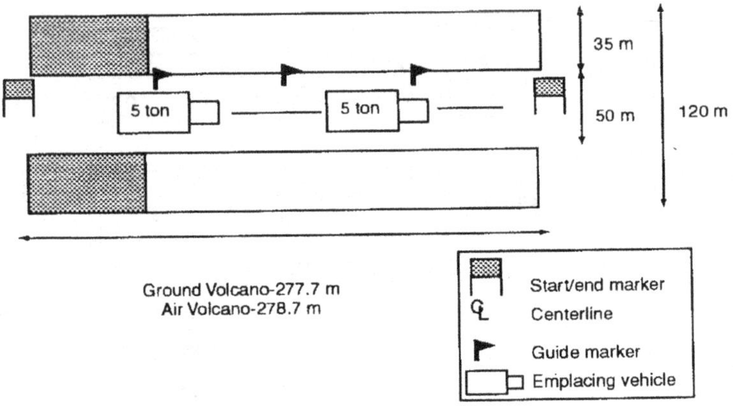

Notes:

- **Guide markers are offset to the left of the centerline to guide the driver and allow the vehicle to remain on the centerline.**
- **Start and end markers should be used for air Volcanos only. Guide markers are not required.**
- **Disrupt and fix minefields require the dispensing of 40 canisters (20 each side).**
- **One full load of air or ground Volcano provides four disrupt or fix minefields.**

Figure 3-8. Volcano disrupt and fix minefields

- Logistics support.
- Obstacle-turnover requirements.

Indirect-Fire Integration

The platoon leader must ensure that key obstacles are covered by indirect fire. These obstacles become priority targets for the artillery. The platoon leader coordinates indirect fire with the company FIST and company/team commander in heavy units. Indirect fire must be coordinated with the FSO and the TF commander if units are light or if the platoon leader is the TF engineer. For more information on

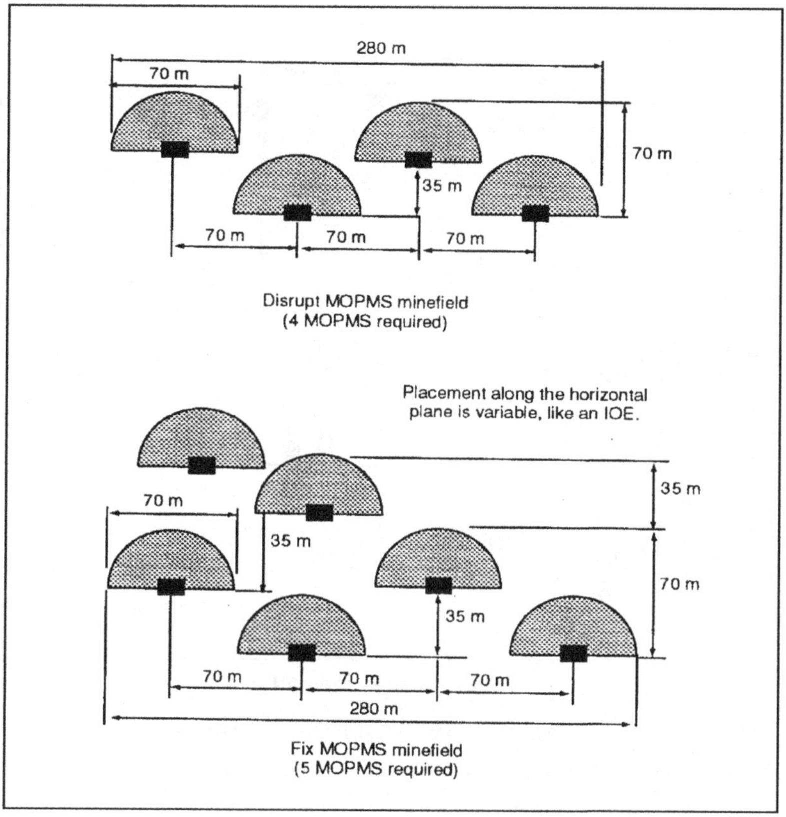

Figure 3-9. MOPMS disrupt and fix minefields

integrating obstacles with indirect fire, see *FM 90-7*. When integrating obstacles with indirect fire, the platoon leader should consider the following:

- Ensuring that there are target reference points (TRPs) on every obstacle.
- Preplanning ADAM and RAAM fires.
- Calling for fire nets (engineers).

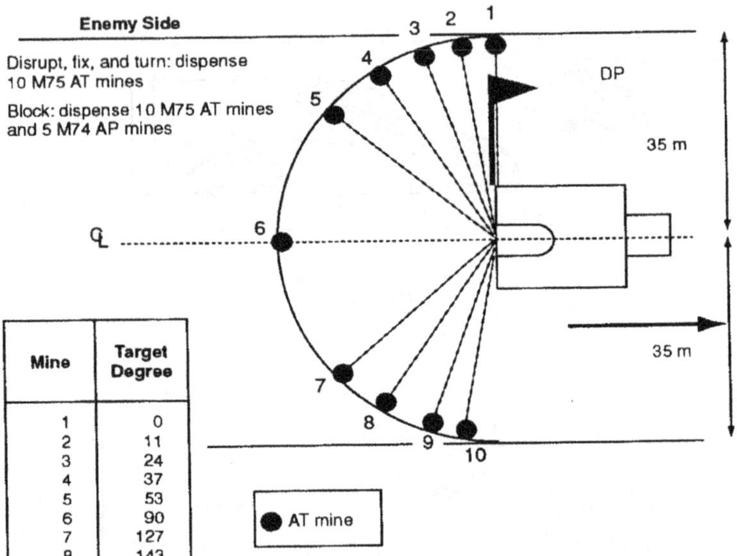

Mine	Target Degree
1	0
2	11
3	24
4	37
5	53
6	90
7	127
8	143
9	156
10	168

Figure 3-10. Flipper TTP

OBSTACLE CAMOUFLAGE (IF REQUIRED)

When camouflaging obstacles, the following should be considered:

- Ensuring that the obstacles emplaced support the intent (sometimes you want the enemy to see the obstacles).
- Using reverse slope siting.
- Burying mines.
- Using night and covert emplacement.
- Minimizing dust signatures from the excavation.

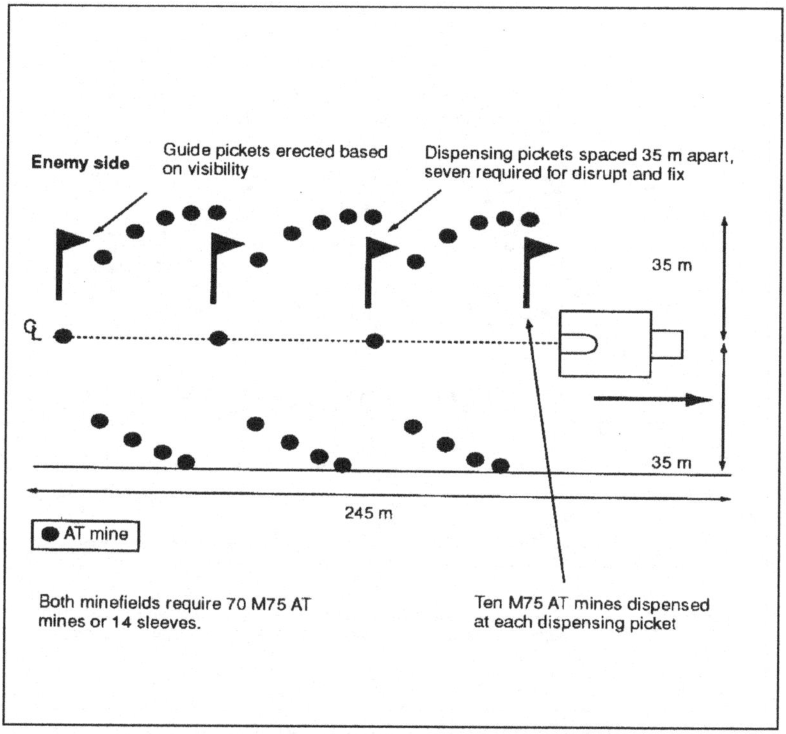

Figure 3-11. Flipper fix and disrupt minefields

OBSTACLE PROTECTION

The following should be considered for the protection of obstacles:

- Denying the enemy covert breaching opportunity.
- Denying the enemy reconnaissance
- Using observation posts (OPs) and counterreconnaissance.
- Displacing forces during low visibility to cover obstacles.

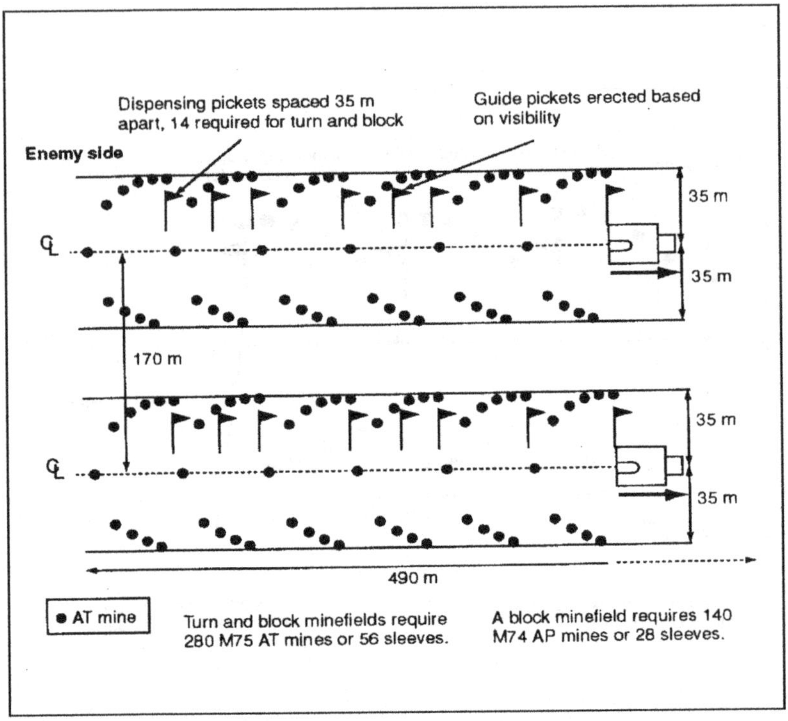

Figure 3-12. Flipper turn and block minefields

- Identifying and destroying the enemy's obstacle breaching equipment and C^2 vehicles.
- Covering the obstacle with direct and indirect fires.
- Patrolling obstacles at night.

SURVIVABILITY

Preparation of fortifications may be difficult. Fortifications in sandy soils often require revetments. It may be impossible to dig in rocky areas, arctic regions, or locations with a high or perched water table. To counter these problems, you may

Table 3-5. Flipper planning factors

Type of Minefield	Depth (Meters)	Front (Meters)	Number of Strips	DPs per Strip	M75 Mines per DP	M74 Mines per DP	Total M75 Mines	Total M74 Mines
Disrupt	70	245	1	7	10	0	70	0
Fix	70	245	1	7	10	0	70	0
Turn	240	490	2	14	10	0	280	0
Block	240	490	2	14	10	5	280	140

Table 3-6. Volcano planning factors

Type of Minefield	Depth (Meters)	Front Ground/Air (Meters)	Number of Strips	Canisters per Strip	Total Canisters	Minefields per Load
Disrupt	120	277/278	1	40 (20 each side)	40	4
Fix	120	277/278	1	40 (20 each side)	40	4
Turn	320	555/557	2	80 (40 each side)	160	1
Block	320	555/557	2	80 (40 each side)	160	1

Table 3-7. RAAM/ADAM planning factors

Type of Minefield	RAAM		ADAM		Width (Meters)	Depth (Meters)
	Area	Linear	Area	Linear		
Disrupt	0.001	0.2	0.0005	0.1	200	200
Turn	0.002	0.9	0.001	0.4	400	400
Fix	0.002	0.4	0.0005	0.1	200	200
Block	0.004	1.6	0.002	0.8	400	400

Table 3-8. Mine dump planning factors

Number of Personnel	Quantity of Mines
2-man team (2 minutes per mine)	25 mines per hour
Squad (8 soldiers)	100 mines per hour
Platoon	300 mines per hour; 3,000 mines per day
Company	6,000 mines per day
NOTE: For planning, soldiers work 10 hours per day.	

have to build up emplacements with existing materials or depressions. *FMs 5-34 and 5-103* provide the standards for dismounted and vehicular field fortifications.

Camouflage is very effective when properly employed. However, you must carefully select patterns and techniques to apply to the local environment. All equipment should have camouflage nets. Other survivability techniques include dispersion and frequent moves.

Direct-fire weapons are more effective from dug-in positions. Logistics areas (brigade support area/division support area (BSA/DSA)) require survivability support. These sites are large, relatively static, and difficult to camouflage. As a result, they are vulnerable to enemy interdiction. Emphasis should be placed on ammunition and fuel dumps.

Digging assets, such as bulldozers, small emplacement excavators (SEEs), or armored combat earthmovers, M9 (ACEs), should be tasked to provide survivability support to these sites. Particular emphasis should be placed on hardening ammunition and fuel-storage locations.

With long-range observation and fields of fire, defensive positions are very vulnerable to offensive fire. This fact, coupled with a lack of natural obstacles, may lead the commander to concentrate the bulk of his engineer effort on survivability positions. Survivability positions enhance the ability of all direct-fire elements to survive indirect-fire preparations and the direct fire from attacking enemy units.

Care should be taken when digging foxholes and tank hide positions in sandy or noncohesive soils since the walls have a tendency to cave in.

PLANNING

Survivability positions are key to surviving any battle on today's modern battlefield. This is especially true where there is a lack of vegetation and natural hiding positions. The TF engineer must know the following information to start planning:

- Number of positions: primary, alternate, and supplementary.
- Type of position: hull-down, turret-down, or hasty.
- Weapons and vehicle priorities.
- Battle-position and EA priorities.
- Unit priorities.

Once the TF engineer determines the number of positions required, then the number of positions that can actually be put in is determined. The engineer unit's capability is based on the following factors:

- Ten-hour days for soldiers.
- Fifteen-hour days for equipment.
- Time available (days).
- Number of blade teams (BTs) available. ACEs, dozers, combat engineer vehicles (CEVs), and bucket loaders can be used to make blade teams. A blade team must include at least one dozer or ACE.
- Work rates.

For work rates, use the planning factor developed for your area of operation. If work rates are unknown, use the following as a guide:

- One hull-down position (HDP) takes 1.5 BT hours.
- One turret-down position (TDP) takes 3.5 BT hours.
- Fifty meters of ATD takes 1 BT hour.

A variety of variables may increase or decrease the time required to dig a vehicle fighting position or ATD. These variables include, but are not limited to, local soils and terrain, operator experience, additional maintenance down time, and the tactical situation. The difference of blade widths may also affect work rates. The M9 ACE has a shorter blade width than the D7 dozer and requires a minimum of two passes to cut the same width.

Survivability assets should be massed on the commander's number one priority. This focuses the survivability effort and allows the TF engineer to monitor construction progress. It also reduces the C^2 burden on the engineer platoon and facilitates rapid transition to changing priorities.

The following equations can be used to determine the number of HDPs and TDPs and the length of ATDs that can be emplaced:

HDP = (# BTs) x (1 HDP/work rate) x (# days) x (15 hrs/day)

TDP = (# BTs) x (1 TDP/work rate) x (# days) x (15 hrs/day)

Meters of ATD = (# BTs) x (ATD m/work rate) x (# days) x (15 hrs/day)

POSITIONS

Coordination between the engineer/equipment operator and the maneuver element is critical to ensure the proper placement of positions. Without effective coordination, much valuable time is wasted redigging positions. The maneuver element, preferably the gunner or tank commander (TC), must tell the engineer/equipment operator the following:

- Type of position (deliberate or hasty).
- Gun target line.
- Sector of fire.
- TRP.

In heavy units, initial vehicle-fighting-position construction starts with the M9 ACE. The ACE has different digging capabilities than the dozer. In many cases, divisional ACEs start with hasty positions, which are then upgraded to improved positions as time allows. Also, the ACEs may start work before corps-level dozers arrive to augment the construction effort. In this situation, the ACEs would dig the hasty positions and the dozers would improve them. The platoon must plan for this augmentation to efficiently integrate corps assets into the construction effort. In heavy units, the company commander must integrate corps assets; however, in light units, the platoon leader must plan for and integrate corps assets.

Figure 3-13, page 3-26, shows the different vehicle fighting positions. Some terrain areas may not accommodate a hull or turret defilade. The use of a modified hasty vehicle-position design or simple marking techniques incorporated with terrain mottling may be required.

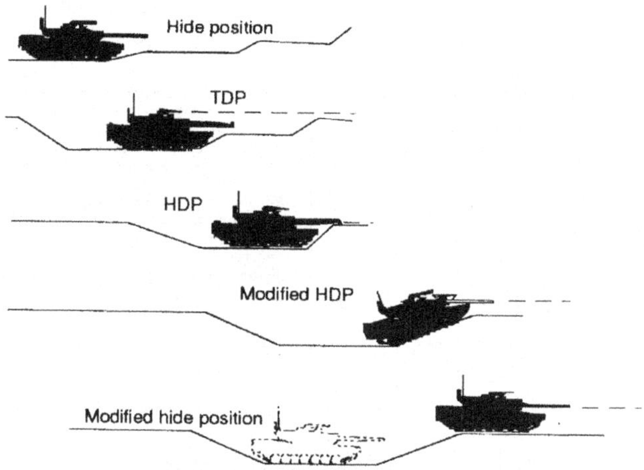

Figure 3-13. Vehicle survivability positions

Each soldier, not just engineers, must also emplace individual and crew fighting positions. Engineers have some limited assets, such as the SEE, that can assist the maneuver commander in accomplishing this mission. Fighting positions are—

- Located to effectively engage the enemy and mutually support each other.
- Designed to allow weapons to be properly fired.
- Constructed to provide adequate cover (see *Table 3-9*).
- Concealed from the enemy.

NUCLEAR, BIOLOGICAL, CHEMICAL (NBC)

Chemical and biological weapons are many countries' preferred weapons of mass destruction. Many of our potential adversaries have an NBC capability. *Figure 3-14 and Tables 3-10 and 3-11, page 3-28,* detail the NBC threat and protective procedures.

Table 3-9. Survivability

Camouflage	Defensive Positions
Soldiers should consider the following when camouflaging: • Breaking up the shape of the object. • Shading the object with natural shadows. • Concealing the shine of the object. • Disrupting the silhouette of the object. • Using folds, gullies, and wadies. • Using reverse-slope positions. • Scaring two to three positions for each vehicle or position. • Using irregular-shaped positions • Building positions a minimum of twice the size of the concealed object. • Exercising noise and light discipline. • Using camouflage for deception.	Soldiers should know and consider the following facts when constructing defensive positions: • Field expedient construction materials, such as corrugated metal, coverts, scrap metal, 40-gallon oil drums, sandbags, and sand-filled ammo boxes, can be used. • Small-arms fire penetrates loose sand 39 inches and sandbag parapets 24 inches. • A minimum of 18 to 24 inches of overhead cover is required to protect against indirect fire. • Revetments are required in sandy soils. • Design details are covered in *FMs 5-34 and 5-103.*

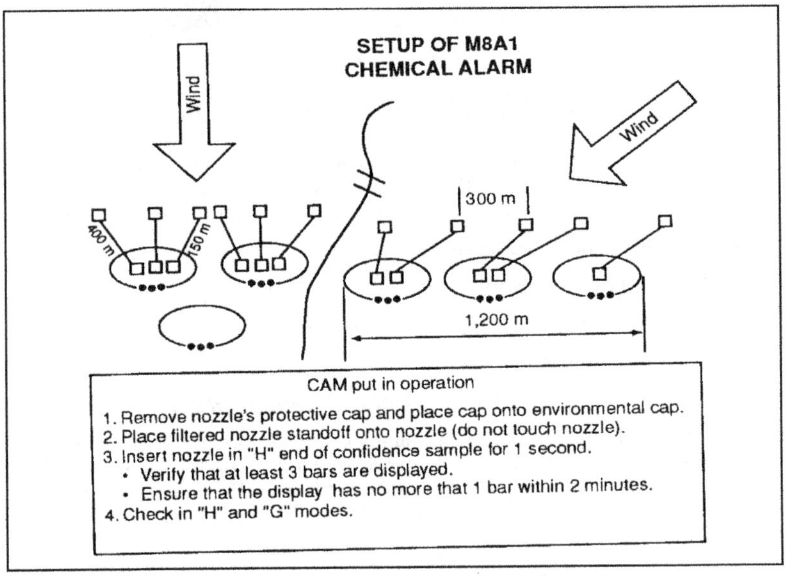

Figure 3-14. M8A1 chemical alarm

Table 3-10. Unmasking procedures

Unmasking with M256 Series Kit	Unmasking without M256 Series Kit	MOPP Gear Exchange
This procedure takes 10 minutes. See *FM 3-4* for more information. The leader— • Checks M256 results to ensure that they are negative. • Selects one or two soldiers. The soldiers— • Move to a shady place. • Unmask for 5 minutes. • Clear and reseal their masks. The leader— • Observes the soldiers for 10 minutes. • Gives the all-clear signal if the soldiers show no symptoms. • Watches for delayed symptoms. • Has first aid available.	This procedure takes 25 minutes. See *FM 3-4* for more information. The leader— • Checks the M8 paper for liquid contamination. • Selects one or two soldiers. The soldiers— • Move to a shady place. • Break the seal of their mask, with their eyes open, for 15 seconds. • Clear and reseal their masks. The leader observes the soldiers for 10 minutes. If no symptoms occur, the soldiers— • Unmask for 5 minutes. • Clear and reseal their masks. The leader— • Observes the soldiers for 10 minutes. • Gives the all-clear signal if the soldiers show no symptoms. • Watches the soldiers for delayed symptoms. • Has first aid available	Soldiers conduct MOPP gear exchange as a two-man team. See *FM 3-5* for more information. Soldiers— • Decon their individual gear. • Prepare their mask for decon and their overgarment for removal. • Decon their mask hood. • Remove their overgarment and overshoes. • Remove their gloves. • Put on their new overgarment. • Put on their new overshoes and gloves. • Secure their hood. • Secure their gear.

Table 3-11. NBC threat, detection, and protection

Agent Type	Agent Class	Agent Effect	M8A	M256	CAM	M8	M9	Protection Required	Skin Decon	Equip Decon
AC	NP	Blood	N	Y	N	N	N	Mask only	N	N
GA	NP	Nerve	Y	Y	Y	Y	Y	Mask and overgarment	Y	N
GB	NP	Nerve	Y	Y	Y	N	N	Mask and overgarment	Y	N
GD	P	Nerve	Y	Y	Y	Yellow	Y	Mask and overgarment	Y	Y
VX	P	Nerve	N	Y	N	Green	Y	Mask and overgarment	Y	Y
HL	P	Blister	N	Y	Y	Red	Y	Mask and overgarment	Y	Y
HD	P	Blister	N	Y	Y	Red	Y	Mask and overgarment	Y	Y
CG	NP	Choking	N	N	N	N	N	Mask only	N	N

CHAPTER 4
Fire Support

The platoon gets FS in the forms of indirect fires, smoke, and obstacle emplacement. FS protects the platoon while it is executing its mission. The platoon may provide survivability to FS units. The platoon leader must—

- Keep the TOC apprised of all element locations at all times. The FSO gets information on engineer unit locations from the TOC operations section (there are no FISTs with engineers).
- Work closely with the FSO or the FIST to plan and report SCATMINEs, cover friendly obstacles with fire, and provide suppressive fires and smoke on locations of expected enemy obstacles.
- Keep track of the amount of artillery FASCAM allocated to the TF and still available for employment.
- Get the frequencies, call sign, and procedures needed to call for indirect fires for the platoon. Most call for fires will be done by adjusting fires from a target or known point. See *FM 5-34* for call-for-fire procedures. Two examples of call-for-fire procedures follow:
  - From your location, you will adjust fire by distance and direction from the TRP to the target. Give the distance in meters and the direction as you see it. If the target is to the left or right of the TRP as you see it, call LEFT distance in meters or RIGHT distance in meters. If the target is behind the TRP as you see it, call ADD distance in meters and if the target is in front of the TRP, call DROP distance in meters.
  - During breaching operations, you may have to call for smoke. Ensure that the FSO or the FIST knows the location and the duration of your smoked target. The call-for-fire procedures are similar to those discussed above.
- Determine if the maneuver commander plans to provide survivability positions for artillery assets and what the priority is for those positions.
- Ensure that obstacles are synchronized with the indirect-fire plan.

CHAPTER 5
Air Defense

The platoon leader and sergeant should understand how AD weapons can support the platoon and how the platoon can support AD units. The platoon leader should know where the AD umbrella is in relationship to his area of operations and what kind of support he can expect. The platoon can provide its own AD measures, either passive or active. Passive AD measures are used to keep soldiers and equipment undetected by enemy air assets. Active AD measures are used to engage enemy aircraft. The following are active AD measures:

- Using the football-field method for high-performance aircraft (aim 200 meters in front of the aircraft).
- Using the football-field method for helicopters (aim 50 meters in front of the helicopter).
- Aiming weapons slightly above the nose of aircraft that is approaching head on.
- Firing all weapons at the reference point
- Ensuring that the AD-weapons control status is given. Status could be—
  - Hold—weapons are fired only in self-defense or in response to a formal order.
  - Tight—weapons are fired only at aircraft positively identified as hostile.
  - Free—weapons are fired at any aircraft not positively identified as friendly.
- Ensuring that AD warning conditions are given. Conditions could be—
  - White—an air attack is not probable.
  - Yellow—an air attack is probable.
  - Red—an air attack is imminent or in progress.
- Determining if the maneuver commander plans to provide survivability positions for AD assets and what the priority is for those positions.
- Ensuring that the platoon is able to receive AD early warning. When the AD warning condition changes to red, the warning of incoming enemy aircraft and their direction is transmitted over command or operations and intelligence networks. The platoon leader should ensure that the platoon is monitoring the appropriate networks.

CHAPTER 6
Combat Service Support

The platoon's CSS is determined by its supply requirements. The platoon leader and sergeant decide what CSS the platoon needs for an operation and initiate a request. Their request is forwarded to the 1SG of either their parent engineer company or supported maneuver unit, depending on their command and support relationships. For more information on command and support relationships, see *Chapter 2*. The platoon's CSS is normally delivered in a scheduled logistical package (LOGPAC).

RESUPPLY OPERATIONS

A LOGPAC is a resupply element that is formed on the basis of a unit's logistic requirements. Normally, it consists of a petroleum, oils, and lubricants (POL) truck; an ammunition truck (Class IV/Class V supplies); and a supply truck that carries rations, repair parts, and other requested items and pulls a water trailer. LOGPACs are—

- Assembled in the BSA and then moved forward under the control of the support-platoon leader to a logistics release point (LRP).
- Picked up by the 1SG at the LRP and moved forward to a secure area behind the company's position to feed, fuel, and resupply the company.
- Returned by the 1SG to the LRP where the support-platoon leader takes control of it and moves it back to the BSA.

LOGPACs are usually distributed by one of the following methods:

- Service-station method.
- Tailgate method.
- Modification of both.

In the service-station method, the 1SG establishes the LOGPAC in a centrally located and secure site. Successively, each platoon moves to the LOGPAC site. Each squad or crew passes through the various stations before the final inspection and returns to the work sites. Normally, the stations include maintenance support, fuel (POL), food, water, mail, administrative support, medics, and other CSS assets as required by the SOP and the mission, enemy, terrain, troops, and time available (METT-T). *Figure 6-1* shows the service-station method.

The tailgate method differs from the service-station method in two major aspects (see *Figure 6-2*). In the tailgate method, the LOGPACs—

- Are brought to each platoon's general location by the 1SG.
- Are usually not as comprehensive as the service-station LOGPACs because they are more mobile and spend more time in transit.

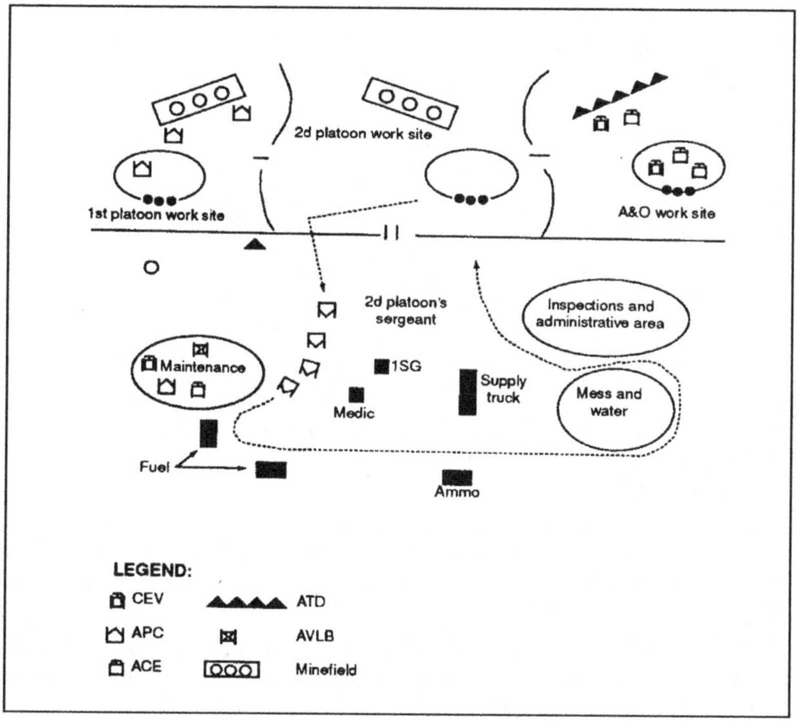

Figure 6-1. Service-station method

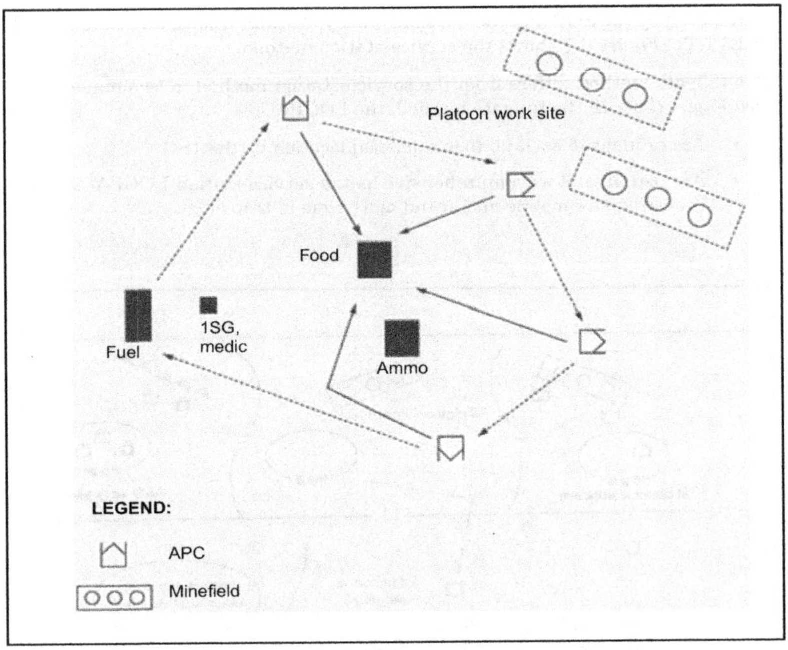

Figure 6-2. Tailgate method

The tailgate method takes more time for the 1SG but is the least interfering of the platoon's work effort. The platoon does not have to spend time moving to the LOGPAC.

Platoons often get resupplied from the maneuver company's/team's LOGPAC using the service-station or tailgate method. For security reasons, maneuver units normally provide LOGPACs to individual vehicles. Therefore, the squad rather than the platoon gets resupplied.

Even when being logistically supported by maneuver elements, platoons sometimes need engineer-specific resupply. Logistics-support responsibilities for command and support relationships are shown in *Table 6-1, page 6-4*.

Table 6-1. Command and support relationships

Relationship	Command and Task Organize	Assign Tasks	Provide Logistics Support	Reorganize as Infantry
Attached	Supported unit	Supported unit	Supported unit	Supported unit
OPCON	Supported unit	Supported unit	Parent unit	Supported unit
DS	Parent unit	Supported unit	Parent unit	Parent unit
GS	Parent unit	Parent unit	Parent unit	Parent unit

SUPPLY CLASSES

Normally, there are 10 classes of supplies; however, during defensive operations, Class IV/Class V supplies require special engineer considerations. Some supply classes require special considerations in resupplying the engineers. Supplies can either be pushed or pulled through the system. Supplies are pushed by predicted requirements and pulled by the needs of the user via a request for resupply.

CLASS I

Class I consists of subsistence and gratuitous health and welfare items. The platoon allocates Class I supplies based on unit strength. Class I supplies are normally pushed. The following are the amounts of water required for a soldier per day for—

- Drinking - 6 gallons.
- Personal hygiene - 2.5 gallons.
- Food preparation - 1.0 gallon.
- Vehicle radiators - 25 percent of the radiator's capacity.

CLASS II

Class II consists of such items as clothing, individual equipment, tentage, hand tools, administrative and housekeeping supplies, chemical-defense and decontamination items, and wet- and cold-weather contingency items. The company allocates Class II supplies based on need. It is normally pulled.

CLASS III

Class III consists of POL, including petroleum fuels, lubricants, hydraulic and insulating oils, preservatives, liquids and gases, bulk chemical products, coolants, deicer and antifreeze compounds, components and additives of petroleum and chemical products, and coal. Supply amounts are based on either offensive or defensive operations. When the TF is preparing to defend, additional Class III supplies are required for all equipment. ACEs require fuel every 8 hours when being heavily used. Additional fuel and package products, such as hydraulic fluid, should be maintained in the combat trains for this purpose. Forward fueling at the work sites should be a standard procedure. Class III supplies are usually pushed.

CLASS IV

Class IV consists of construction materials, including all fortification and barrier materials. These are items for which allowances are not prescribed. Class IV supplies are pulled.

CLASS V

Class V consists of all types of ammunition, including radiological and special weapons, bombs, explosives, mines, fuzes, detonators, pyrotechnics, missiles, rockets, propellants, and other associated items. Engineers need some unique Class V supplies (for example, cratering charges and line charges), which require special coordination with the Supply Officer (US Army) (S4) to get them into the TF's Class V system. Class V supplies are pulled.

CLASS VI

Class VI consists of all personal-demand items, such as candy, cigarettes, soap, and cameras (nonmilitary sales items). Sundry packs are also Class VI items. The platoon allocates Class VI supplies based on need. Class VI supplies are normally pulled.

CLASS VII

Class VII consists of major end items, such as launchers, tanks, mobile machine shops, vehicles, and organizational tool sets. Class VII supplies are demand driven.

CLASS VIII

Class VIII consists of medical material, including repair parts peculiar to medical equipment. Class VIII supplies are normally pulled.

CLASS IX

Class IX consists of repair parts and components, including kits, assemblies, and subassemblies (repairable and unrepairable) that are required for maintenance support of all equipment. The platoon's stock of repair parts is based on vehicle density, the authorized stockage list (ASL), the high-demand prescribed load list (PLL), and specific engineer-equipment requirements. Repair parts for items that the TF does not stock (for example, the M9 ACE or dozer) will have to come from the engineers. The platoon generates requests for repair parts by accurately completing *DA Form 2404*. When the maintenance contact team gets the form, it determines what repair parts are needed and the level of repair. The mechanics may bring the repair parts with them or the crew/operator may get them from the LOGPAC. Class V supplies are pulled.

CLASS X

Class X consists of materials to support nonmilitary programs, such as agriculture and economic development. Class V supplies are pulled.

SPECIAL CLASS IV/CLASS V SUPPLIES

These classes of supplies form the majority of the materials with which engineers construct the obstacles and fortifications required for the defense.

There are two types of Class IV/Class V loads: mission and basic. Mission loads consist of those materials required for a specific mission (for example, a standard fix minefield). Basic loads consist of those materials that the platoon carries to protect itself. The basic load can be used for missions to save time; however, it is to be replenished from the materials in the mission load. Basic loads are pulled and mission loads are pushed.

Mission loads are a TF responsibility regardless of the command and support relationships specified for engineers. The quantity of Class IV/Class V supplies (mines) normally stretches or exceeds the transportation assets of the TF. Palletized standard loads help solve the planning and distribution problem. Class IV/Class V supply points should be selected by their proximity to major obstacles.

Class IV/Class V resupply for the defense is one of the most demanding logistics operations the TF must carry out and requires all the assets that can be made available. A total cooperative effort by the TF, including engineers, is required if the defense is to be adequately resourced.

In the offense, TF planners anticipate Class IV/Class V mission loads for even a hasty defense on the objective. Prepackaging in the BSA speeds movement forward when it is needed. In addition to the Class V supplies that come forward as a part of the LOGPAC, the TF will maintain emergency Class V resupply in the combat trains.

MAINTENANCE PROCEDURES

When the TF is providing logistics support to the engineers, it will also provide organizational maintenance support to them. This causes no particular problem for common equipment. No engineer equipment repairers exist in the maneuver unit's maintenance sections lower than the brigade level except those at the engineer battalion. It may be unrealistic to expect the maneuver unit to provide all logistics support. The platoon may have to coordinate for maintenance assistance from its parent unit. When the engineers are providing logistics support, they will also provide organizational maintenance support.

Doctrinally, recovery is the responsibility of the owning unit. However, it is virtually impossible for engineers working in the TF area to recover their own vehicles. In the defense, the TF will have to provide recovery support at least back to the unit maintenance collection point (UMCP) located near the combat trains. If repairs cannot be made at the UMCP and the time and situation permit, the engineer's maintenance-support team (MST) can come forward and recover the down vehicle. The TF could provide additional recovery back to the BSA for repairs. The engineer battalion maintenance officer (BMO) or technician must ensure that each BSA has direct-support (DS) mechanics capable of repairing engineer equipment.

In the offense, the TF recovers the vehicle to the main supply route (MSR). Depending on the situation, the TF either turns vehicle recovery over to the engineers or recovery assets take the vehicle to the BSA. It must be clear in the TF SOP or order which MST has recovery responsibility for the engineer vehicles. The priority the TF commander places on the engineer assets determines the priority they are recovered in.

Engineer companies should have maintenance contact teams that are familiar with engineer heavy equipment to routinely visit work sites. The communications repairman and armorer can accompany this contact team to help anticipate problems and provide support before the mission is jeopardized.

MEDICAL SUPPORT

Medical support and evacuation must be coordinated for each operation and normally come from the nearest company/team (see *Table 6-2, page 6-8)*. The platoon's aidman or combat lifesavers should be required to coordinate with the TF's medics and ensure that the medical requirements of the platoon are planned for and met. In the event of a mass casualty incident, the platoon may have to designate a platoon vehicle as a litter vehicle.

Table 6-2. Command and support relationships and logistics support

Logistics Support	Class III	Class IV/V	Class VIII Medical	Maintenance	Emergency Resupply
Parent unit (OPCON, DS, GS)	Company LOGPAC	Basic loads are resupplied from the LOGPAC. Mission loads are supplied by the supported unit.	Parent unit	Parent unit performs maintenance on all equipment.	Parent unit
Company/team (attached)	Company/team LOGPAC	Basic loads are resupplied from the LOGPAC. Mission loads are supplied by the supported unit.	Company/team	Company/team MST performs maintenance on common equipment. Engineer MST performs maintenance on engineer-specific equipment.	Company/team
TF (attached)	TF support platoon	Basic loads are resupplied from the LOGPAC. Mission loads are supplied by the supported unit.	TF medical platoon	TF maintenance platoon performs maintenance on common equipment. Engineer MST performs maintenance on engineer-specific equipment.	TF support platoon

The platoon leader must—

- Understand the TF's evacuation and medical support plans.
- Understand the engineer company's evacuation and medical support plans.
- Know the locations of the—
  - Battalion aid station.
  - Jump aid station.
  - Personnel transfer points.
  - Casualty collection points.

CHAPTER 7
Command and Control

The platoon leader uses the C^2 processes to ensure that his platoon accomplishes its missions. This chapter provides the TTP needed to command and control the platoon and make sound tactical decisions.

COMMAND AND SUPPORT RELATIONSHIPS

Command relationships can be operational control (OPCON) or attached. Support relationships can be DS or general support (GS). Both supporting and support parties must clearly understand what each of these mean (see *Table 6-2, page 6-8*). The following should be considered when using command and support relationships:

- When OPCON is used, the maneuver element—
  - Commands and controls the platoon.
  - Has the authority to place squads OPCON to companies/teams even though it is not considered advisable to do so with a mechanized platoon. Additionally, the TF cannot attach them to the company.
- When attached is used, the maneuver element—
  - Commands and controls the platoon except for limitations imposed by the attachment order.
  - Has responsibility for the platoon's logistics support except for personnel transfer and promotion.
- When DS is used—
  - The platoon receives its missions from the maneuver element.
  - The engineer company and the engineer battalion support and command the platoon.
  - The platoon works on the tasks that the maneuver element assigns.
  - The TF does not have the authority to split the platoon up and attach squads to companies/teams.
  - Logistics support for the platoon comes from engineer higher HQ.

- When GS is used, subordinate commanders must request support from the engineer commander on a task-by-task basis, and the senior commander sets priorities and assigns tasks to the GS unit. GS is seldom used at TF level. When an engineer commander wants to retain control over the engineers, they are retained in GS of maneuver forces.

The TF always provides the materials necessary to support engineer operations in its sector for offensive and defensive missions. This is a key logistics issue that must be understood at the start of any C^2 discussion. Even though the engineers determine requirements and use the material, supplying it remains a TF responsibility regardless of the command or support relationship (see *FM 5-100*).

COMMUNICATIONS

Engineers must operate on the following frequency modulated networks:

- TF.
- Platoon internal.
- Engineer company.
- Maneuver company/team.

Engineers have a two-net capability to support their communication requirements. In the DS or GS relationship, the engineer platoon should operate the first network on the engineer command network. The second network, the platoon's internal network, should be used to control the platoon's assets. In the attached or OPCON relationship, the engineer platoon should operate the first network on the engineer command network or supported maneuver company/team command network based on the task organization. The platoon leader uses his platoon's internal network to enter other networks for coordination and situation reporting. Before the platoon leader leaves his platoon's internal network, he must first hand over control of the network to the platoon sergeant to control the platoon's assets.

The platoon submits the following reports:

- Routine reports:
  - Commander's situation report (SITREP)—identifies changes to a unit's tactical situation.
  - Sensitive-items report—gives the results of a serial-number check on all sensitive items.
  - Supply-status report—informs the commander and staff of a subordinate unit's supply status.
  - Personnel-status report—is used to maintain the accountability and status of assigned and attached personnel and generate requests for replacements.

  - Maintenance-status report—provides information on the operational status of equipment and is used to generate requests for maintenance and parts.
- Event-driven reports:
  - Spot report (SPOTREP)—identifies any known or suspected enemy activity.
  - Reconnaissance report—provides technical and tactical engineer data on a point/area target.
  - Countermobility report—gives the status of all obstacles.
  - Mobility report—gives the status and conditions of lanes, routes, bypasses, crossing sites, air lines of communication, as required, and ground lines of communication. It also includes status changes and the upgrade/downgrade of previously reported trafficability of locations/sites.
  - Survivability report—provides the detailed status of the unit's survivability level (for example, the number, type, and protective level (hull or turret) of positions).
  - Class IV/Class V special report—is used to keep track of critical engineer classes of supplies. For example, a platoon establishes a mine dump and keeps track of the types and quantities of mines available and the amounts issued.
  - Flash SITREP—is used to notify the next higher HQ when a subordinate unit becomes decisively engaged or has critical shortages that prevent mission accomplishment.
  - NBC report—is used to report NBC events.
  - Shelling report (SHELREP)—provides information on enemy indirect fire.
  - Enemy prisoner of war (EPW) report—is used to account for prisoners captured.
  - SCATMINE report—is used to notify affected units of the emplacement of SCATMINEs.
  - Meaconing, intrusion, jamming, and interference (MIJI) report—is used to notify the Intelligence Officer (US Army) (S2) when the reception of radio signals is hindered, confused, or distorted by any external source or when instructions are received from a station that cannot authenticate.
  - Splash report—is used to report downed friendly aircraft.
  - Casualty report—identifies battle and nonbattle casualties to the parent unit and alerts medical-treatment facilities of incoming casualties.
  - Casualty-feeder report and witness-statement report—provide detailed eyewitness information on a casualty. They are used to account for casualties and provide administrative information for use in notifications and awards.

ENGINEER INTEGRATION

The company commander of a mechanized engineer company normally associated with a battalion/TF is also the staff engineer and advisor to the TF commander. Likewise, in a light TF, the staff engineer advisor to the TF commander is the light engineer platoon leader. The staff engineer integrates engineers into the TF's planning process. He coordinates with other BOS to ensure engineer synchronization. During the preparation of the battle, the staff engineer ensures that soldiers and equipment are in the correct location and on time for linkups and rehearsals. He then executes the engineer mission to support the operation. The platoon is the lowest level engineer unit that can still effectively accomplish independent tasks. The platoon usually operates under the control of an engineer company; however, it can be attached to a maneuver company/team.

The platoon leader must keep both the engineer company and maneuver company/team commanders informed on critical tasks. He must continuously forward and update tactical and technical information upward and laterally by all available means.

In heavy forces, the TF normally receives an engineer company in support, and the company commander is the TF engineer. In light forces and in some special situations in heavy forces, the TF only receives an engineer platoon, and the platoon leader is the TF engineer. When the platoon leader is acting as the TF engineer, he should consider the following information:

- Orders should include when and where the TOC is to be established. If the TOC is to be moved before the platoon arrives, this should also be included.
- Orders should identify an assembly area where the platoon can be consolidated.
- The entire platoon must **not** go to the TOC.

PLATOON LEADER'S RESPONSIBILITIES

Once the platoon leader arrives at the TOC, he should—

- Check in with the S3 or company commander and report the platoon's status and location and any problems.
- Provide the status of equipment and the number of soldiers to the operations noncommissioned officer (NCO).
- Obtain the TF's and the team's signal operation instructions (SOI) and enter the maneuver force's network.

- Ensure that the communications equipment is set for the correct variable for secure operations and is compatible with the supported unit's communications equipment.
- Check with the operations sergeant for any changes to routine or required reports and SOP items.
- Be familiar with the TF SOP. **This is not the time to start studying the SOP.**
- Provide the status of obstacles being constructed.
- Check on format requirements for the periodic report. He should also check to see that TOC personnel are familiar with the reports being sent and know how to post them to their charts or overlays.
- Determine if a specific individual at the TOC has been designated as an expediter or overwatcher for engineers in the sector requiring assistance.
- Provide any terrain, route, or other information acquired from the reconnaissance unit or personnel enroute to the S2 or company/team commander.
- Obtain such information as the current enemy situation, weather and light data, and terrain details, to include topographic products, from the S2 or company/team commander.
- Provide any priority intelligence requirements (PIR) or information requirements (IR) needed for upcoming operations to the S2 or company/team commander.
- Stop at the combat trains and coordinate for classes of supplies and services when joining the maneuver force.
- Obtain frequencies and call signs for the administrative and logistics net control stations (NCSs).
- Provide a platoon roster and identify any critical shortages to the Adjutant (US Army) (S1) or operations sergeant.
- Review the TF SOP for reporting and evacuating casualties.
- Coordinate with the S4 or company/team 1SG and review the platoon's logistics status. The channel used for operational logistics for the platoon varies based on the command and support relationships.

NOTE: The platoon leader should remember that if a problem arises which will prevent the accomplishment of the mission or if normal channels are not working fast enough and logistics become an operational issue, the S3 or commander should be apprised of the problem.

As a minimum, the platoon leader must be present when the—

- Maneuver force receives the WO.
- Commander conducts reconnaissance.
- Commander makes and announces a decision and explains the intent.
- OPORD is received.

MANEUVER FORCE'S CONTROL

Normally, the platoon is kept under the engineer company's control and works for the battalion/TF commander. A major consideration in planning for the use of engineers is to maintain platoon integrity, keeping the entire platoon under the platoon leader's control. When maneuver forces are moving on two axes during mechanized offensive operations, the platoon will not place a part of the platoon on each axis. During the planning process, the engineer planner must weigh the main effort and determine which axis is critical to the mission and requires engineer assets. Placing an engineer NCO with the scout platoon when it is performing reconnaissance can help provide more specific information on the obstacles that the scouts find. Obstacle breaching can thus be preplanned. See *Appendix A* for basic formations, movement techniques, and hand-and-arm signals and *Appendix B* for checklists.

ENGINEER ATTACHMENTS

If the platoon has been augmented with AVLBs, they normally come in pairs. This is for both maintenance and tactical reasons. The AVLBs should be placed with the element that needs a gap-crossing capability. In the deliberate attack, AVLBs can be integrated individually into attack formations. Employment as a pair allows the squad leader to control them. If the AVLBs are not immediately required, they should be returned to the company's control. Curbs on the AVLB must be removed if M1 tanks with rollers are to cross the AVLB, and the roller's dog bone must be raised. If a CEV is attached to the platoon, it normally moves with the platoon.

Heavy equipment and wheeled vehicles should move with the platoon only on tactical road marches over good roads. During a movement to contact or over rougher terrain, METT-T determines where and what heavy equipment moves in the formation.

APPENDIX A

Basic Formations, Movement Techniques, and Hand-and-Arm Signals

The platoon uses a variety of mounted and dismounted formations and movement techniques to maneuver on the battlefield. This appendix gives examples of many of the basic formations, movement techniques, and hand-and-arm signals the platoon leader could use. It is not designed to be all encompassing. For more information on these topics, see *FMs 7-8, 5-34, and 71-1.*

MOUNTED MOVEMENT TECHNIQUES

The mounted engineer platoon must be proficient in moving with its maneuver counterpart. In the following paragraphs, formations, movement techniques, and actions taken during movement for the mounted engineer platoon are discussed.

WEDGE

The engineer platoon almost always maneuvers as part of another larger formation. Normally, this will either be the parent engineer company or a maneuver company/team. *Figure A-1, page A-2,* shows the formation that the platoon is most likely to use. The platoon wedge provides the most defensible formation with the easiest C^2. Ordinarily, the platoon follows a maneuver platoon if task-organized to a maneuver company/team. The platoon might lead if it is part of the engineer company formation. In either case, the wedge is the best formation to use if enemy contact is likely.

The platoon frequently receives assets from the assault and obstacle (A&O) platoon. These could include the AVLB, CEV, or armored vehicle-launched mine-clearing line charge (AVLM). These additional vehicles can strain the C^2 capability of the platoon. *Figure A-2, page A-2,* shows the technique of "tethering". The AVLB, CEV, and AVLM are tethered to individual engineer squads. Each squad leader has the responsibility of commanding and controlling an A&O platoon asset. This improves the C^2 of the platoon and provides each A&O asset with a security element as it moves across the battlefield.

Figure A-3, page A-3, shows a simple wedge formation. Note where the key leaders in the platoon are. The wedge formation uses the wingman concept. The platoon sergeant and leader are both in APCs and each has a wingman. This technique

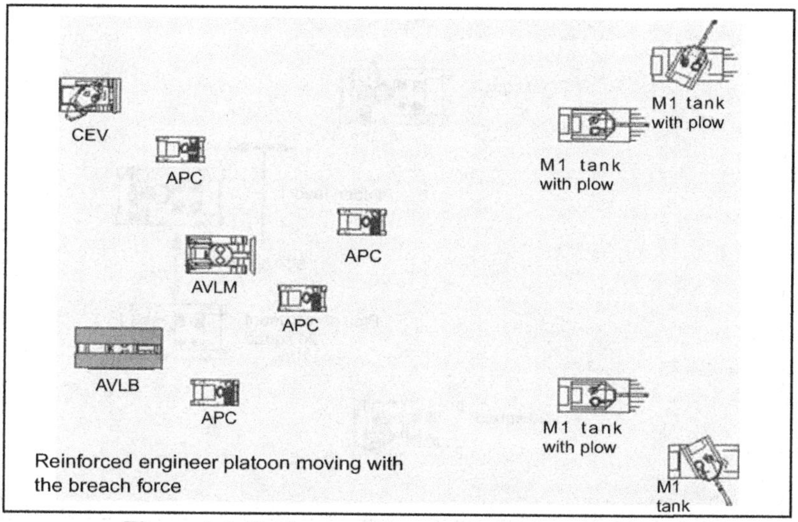

Figure A-1. Engineer platoon with the breach force

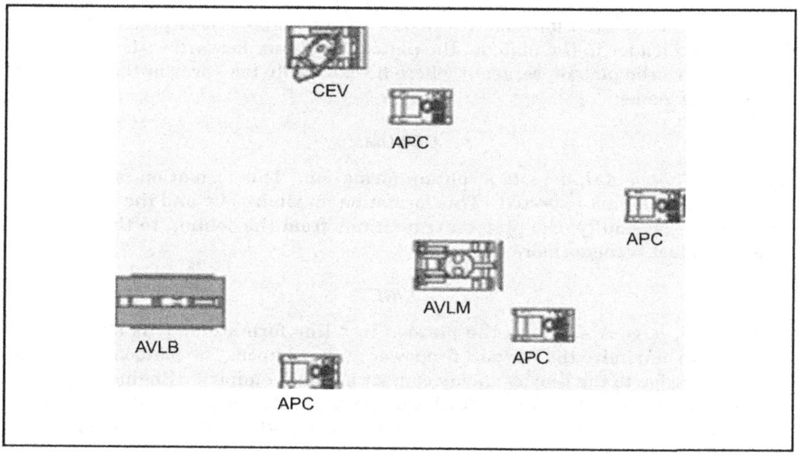

Figure A-2. Modified mounted wedge formation

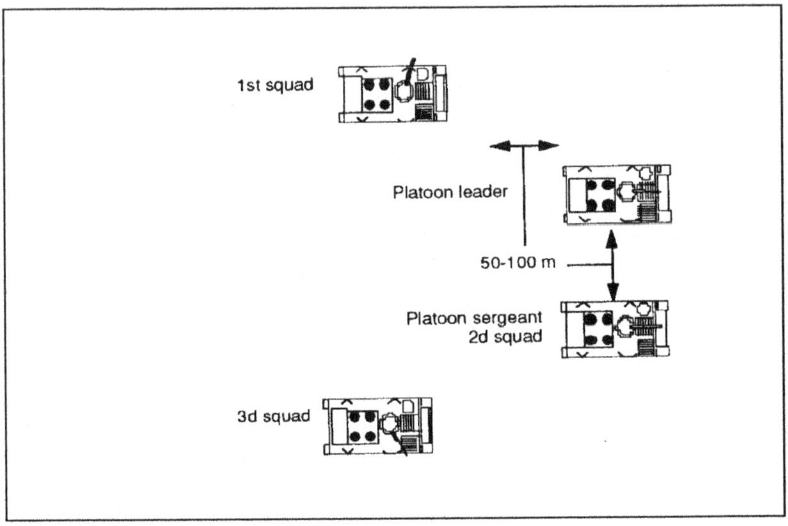

Figure A-3. Mounted wedge formation

also simplifies C^2 and movement. It has the added benefit of placing the most experienced leader in the platoon, the platoon sergeant, forward with the platoon. It also places the platoon sergeant where he can easily take over in the absence of the platoon leader.

COLUMN

Figure A-4, page A-4, depicts a column formation. This formation is used when enemy contact in not expected. This formation maximizes C^2 and the speed of the formation. Normally, the platoon transitions from the column to the wedge as enemy contact becomes more likely.

LINE

Figure A-5, page A-4, shows the platoon in a line formation. This formation is designed to maximize the forward firepower of the platoon. The platoon transitions from the wedge to the line as enemy contact becomes eminent. Engineer platoons do not generally use this formation. However, if the platoon is tasked to suppress a dismounted threat while another platoon maneuvers, this formation works well.

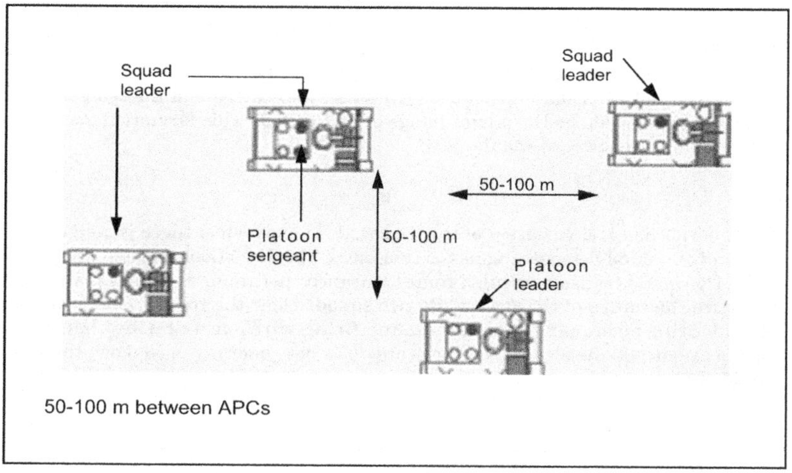

Figure A-4. Mounted column formation

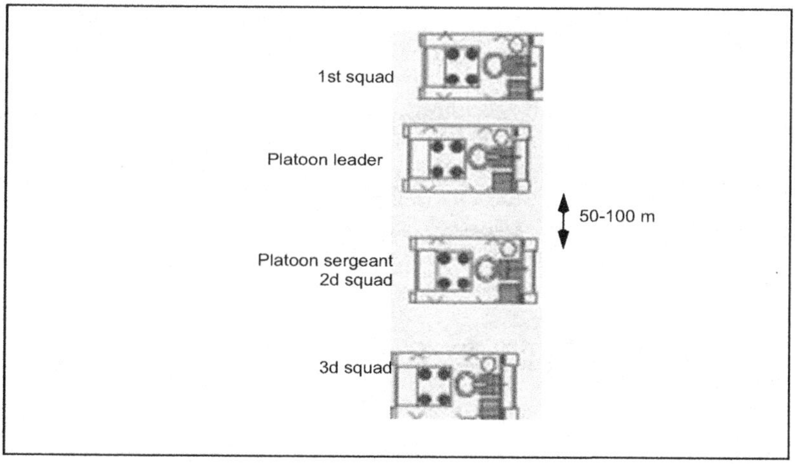

Figure A-5. Mounted line formation

ECHELON

The echelon formation is used when the platoon is on the flank of the company and the enemy threat is also from the flank. *Figure A-6* shows an echelon right formation. The echelon formation can also be used on the left flank of the company and, in that case, it would be the mirror image of *Figure A-6*. This formation maximizes the platoon's firepower to the flank.

V

The V-formation is a variation of the wedge. It is used when there is a significant threat of command-detonated mines or explosives. The formation shown in *Figure A-7, page A-6,* would be used during a route-clearance operation. It allows the platoon to secure the flanks of the road while two squads clear the route. This formation also identifies command-detonated-mine firing wires or ambushes before the enemy can attack the element on the route. It is not generally used when there is a significant enemy threat.

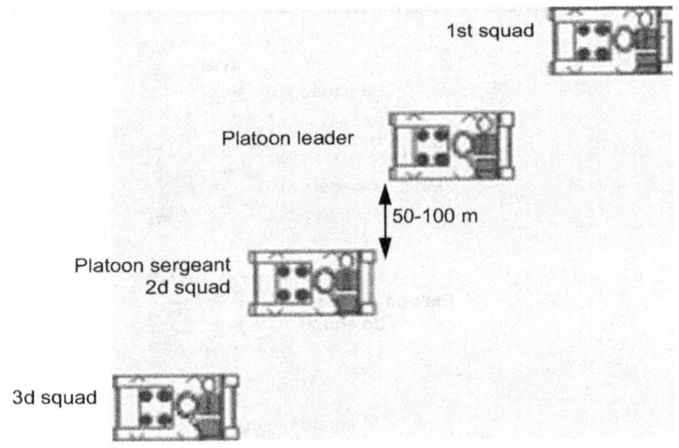

Figure A-6. Mounted echelon formation

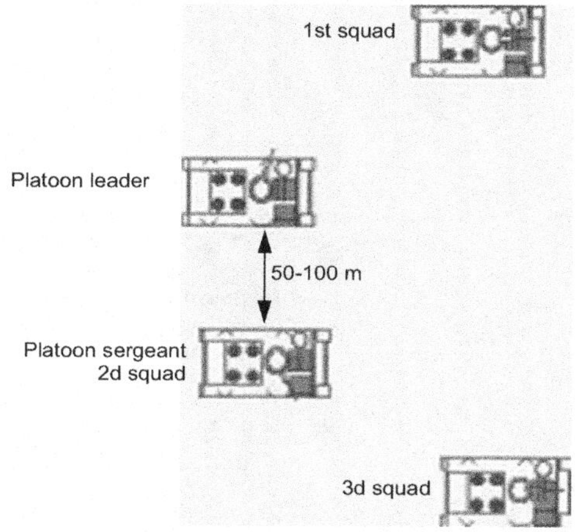

Figure A-7. Mounted V-formation

DISMOUNTED MOVEMENT TECHNIQUES

The dismounted engineer platoon must be proficient in moving with its maneuver counterpart. In the following paragraphs, formations, movement techniques, and actions taken during movement for the dismounted engineer squad and platoon are discussed.

SQUAD FORMATIONS

The squad formation is built from the fire-team wedge (see *Figure A-8*). The interval between soldiers in the wedge formation is normally 10 meters. The wedge expands and contracts depending on the terrain. When rough terrain, poor visibility, or other factors make control of the formation difficult, the squad uses the modified wedge. In this formation, the normal interval between soldiers is reduced so that all team members can still see their team leader and each team leader can see their squad leader. The sides of the wedge can contract to the point where the wedge resembles a single file. When moving in less rugged terrain, where control is easier, soldiers expand the formation or resume their original positions.

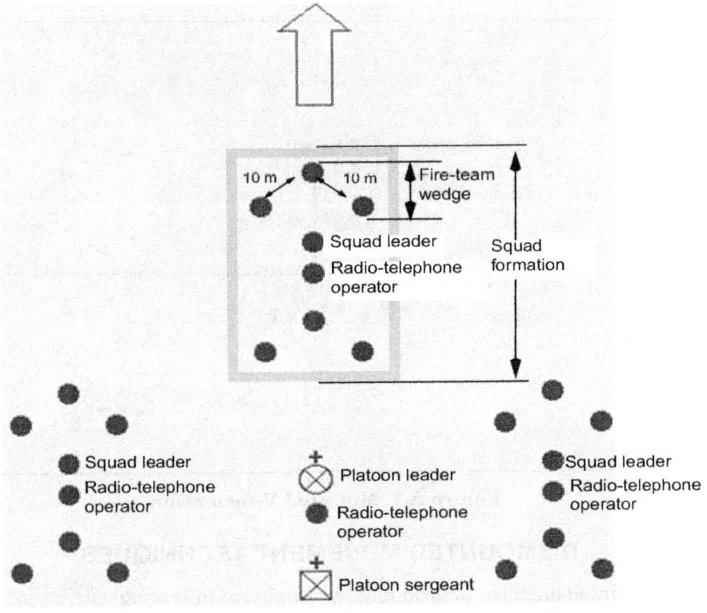

NOTE: Unlabeled positions are dismounted soldiers.

Figure A-8. Dismounted-platoon wedge formation

PLATOON FORMATIONS

Platoon formations are built from multiple squad formations. These formations include the platoon column, line, V, and wedge. The platoon leader selects the best formation based on his METT-T analysis. *Table A-1, page A-8*, shows a comparison of each formation and its use.

Table A-1. Comparison of platoon formations

Dismounted Movement Formations	Movement Use	Movement Characteristics			
		Control	Flexibility	Fire Capability/ Restriction	March Rate
Column	When the platoon performs primary movement formations	Good for maneuver (fire and movement)	Provides good dispersion laterally and in depth	Allows limited firepower to the front and rear, high volume to the flank	Good
Line	When the enemy situation is unknown and the leader wants all soldiers forward for maximum firepower to the front	Difficult	Is minimal	Allows maximum firepower to the front, little to the flanks and rear	Slow
V	When the enemy situation is vague, but contact is expected from the front	Difficult	Provides two squads up front for immediate firepower and one squad to the rear for movement upon contact from the flank	Allows an immediate heavy volume of firepower to the front or flanks	Slow
Wedge	When the enemy situation is vague, but contact is not expected	Difficult but better than the platoon V and platoon line	Enables the leader to make contact with the smallest element and still have two squads to maneuver	Provides a heavy volume of firepower to the front or flanks	Slow but faster than the platoon V
File	When visibility is poor due to terrain or light	Easiest	Is the most difficult formation from which to maneuver	Allows immediate fires to the flanks; prevents focused fires to the front and rear	Fastest

Wedge

The dismounted wedge is used when the enemy situation is vague and enemy contact is likely (see *Figure A-8, page A-7*). This formation allows a large volume of fire around the formation. Generally, at least one squad or element will be free to maneuver from the wedge after contact.

Column

The column formation is the platoon's primary movement formation (see *Figure A-9, page A-10*). It provides good dispersion both laterally and in depth and simplifies control. The lead squad is the base squad for fire control.

Line

The line formation allows the delivery of maximum fire to the front but little fire to the flanks (see *Figure A-10, page A-11*). This formation is hard to control and does not lend itself well to rapid movement. It is the basic platoon assault formation during an attack.

V

The V-formation has two squads up front to provide a heavy volume of fire on contact (see *Figure A-11, page A-12*). It also has one squad in the rear that can either overwatch or trail the other squads. This formation is hard to control, and movement is slow.

File

The file formation gives maximum control to leaders and is used for speed during movement (see *Figure A-12, page A-13*).

Traveling

The dismounted engineer platoon supporting a maneuver TF normally travels as part of one of the larger maneuver elements. When breaching assets are needed forward with the lead company, the engineer platoon travels behind the lead platoon for security (see *Figure A-13, page A-14*). If each maneuver company needs a breaching capability, the dismounted engineer platoon can be broken into squad-sized elements, where each squad travels with a maneuver company. In this case, the platoon HQ travels with either the battalion tactical operations center or the main effort for C^2.

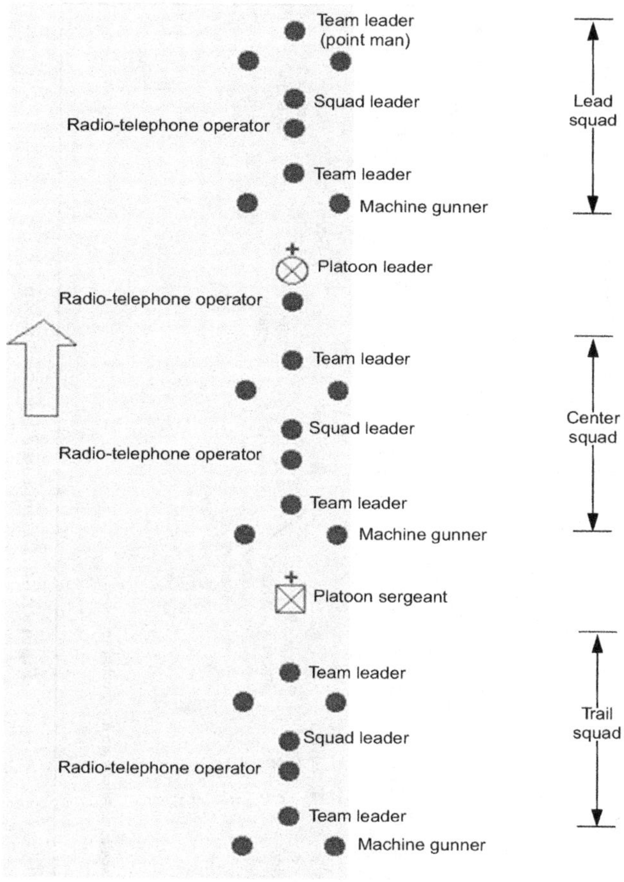

NOTE: Unlabeled positions are dismounted soldiers.

Figure A-9. Dismounted column formation

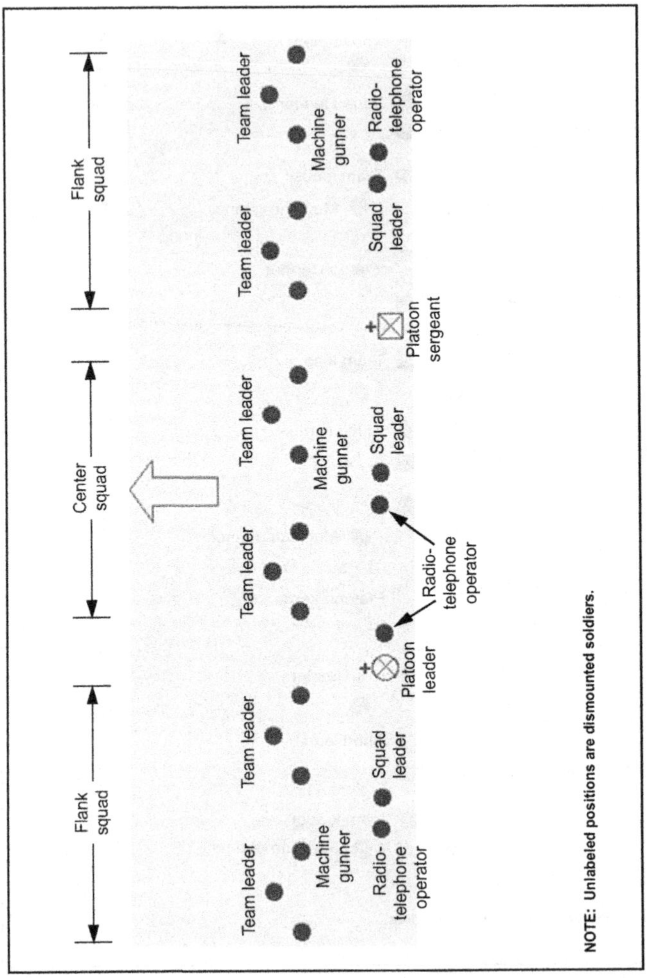

Figure A-10. Dismounted line formation

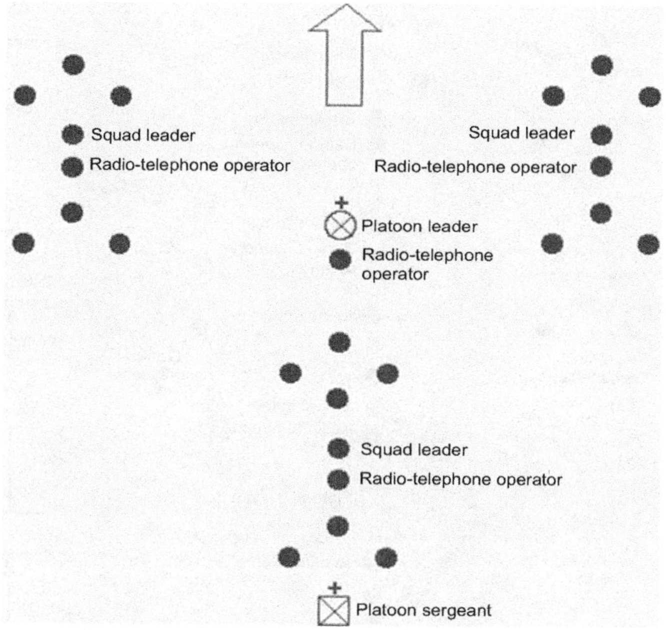

NOTE: Unlabeled positions are dismounted soldiers.

Figure A-11. Dismounted V-formation

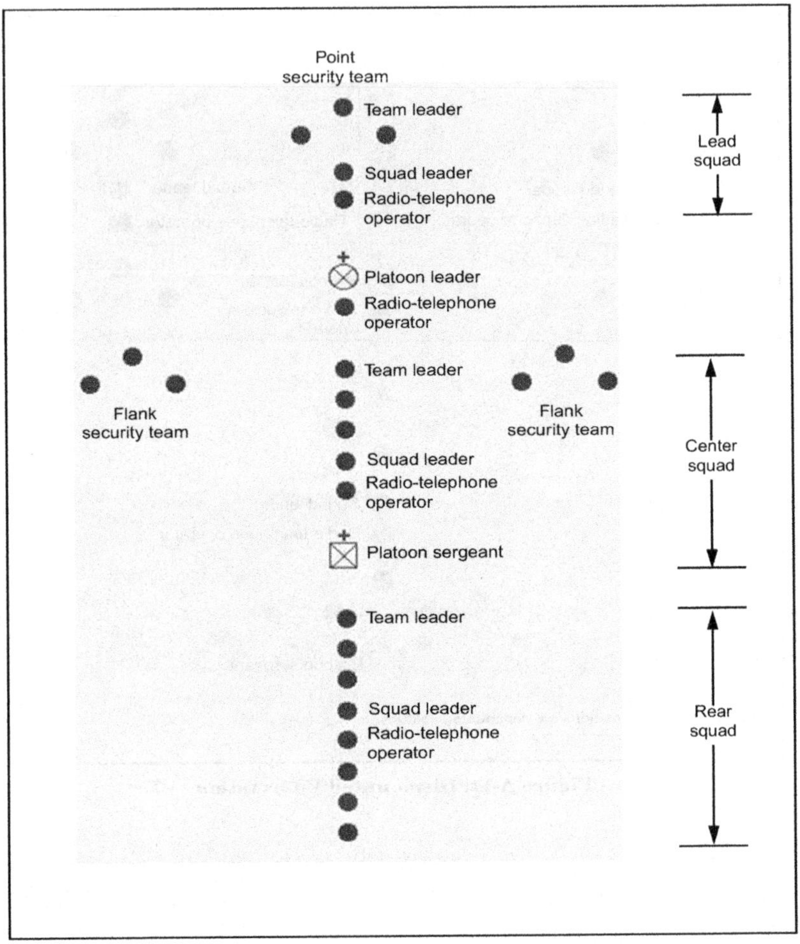

Figure A-12. Dismounted file formation

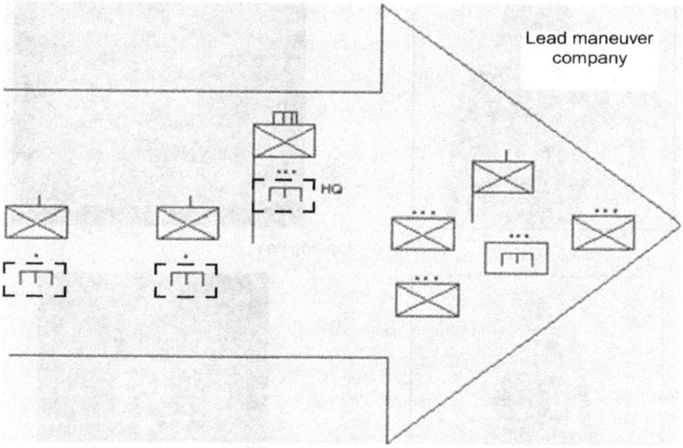

NOTE: The engineer platoon traveling with a TF could travel as part of the lead infantry company to provide immediate breaching capability, or the platoon could be broken down into squad-sized elements to travel with each maneuver company.

Figure A-13. Dismounted traveling formation

HAND-AND-ARM SIGNALS

During many operations, the platoon has to use hand-and-arm signals for C^2. *Figures A-14 and A-15, pages A-15 through A-20,* show many of the command hand-and-arm signals used when the platoon is mounted or dismounted. These signals are normally used when either radio listening silence is in effect or stealth is needed.

DAY	NIGHT

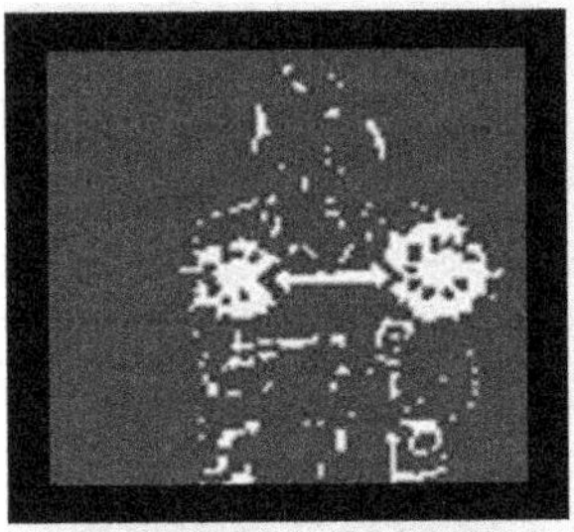

Stop or stop engines

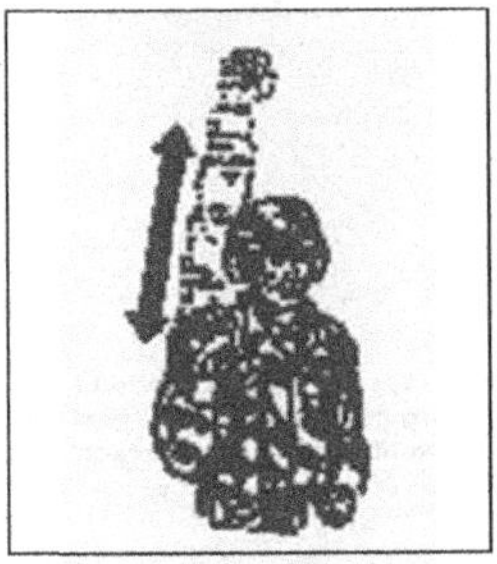
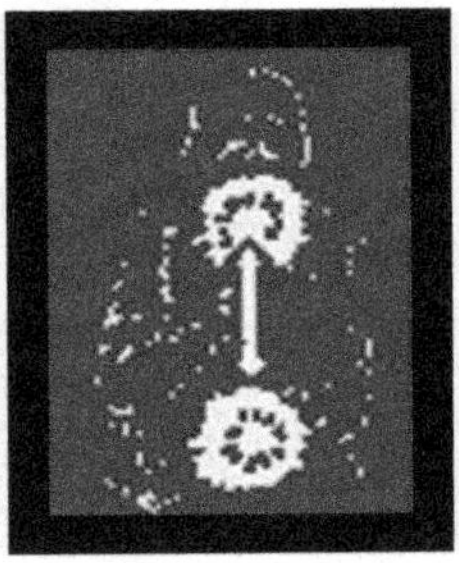

Move forward, move out, increase speed, or double-time

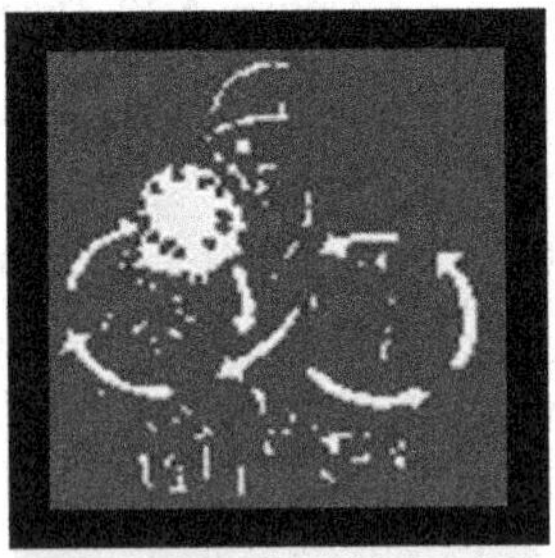

Start engine or prepare to move

Figure A-14. Hand-and-arm signals

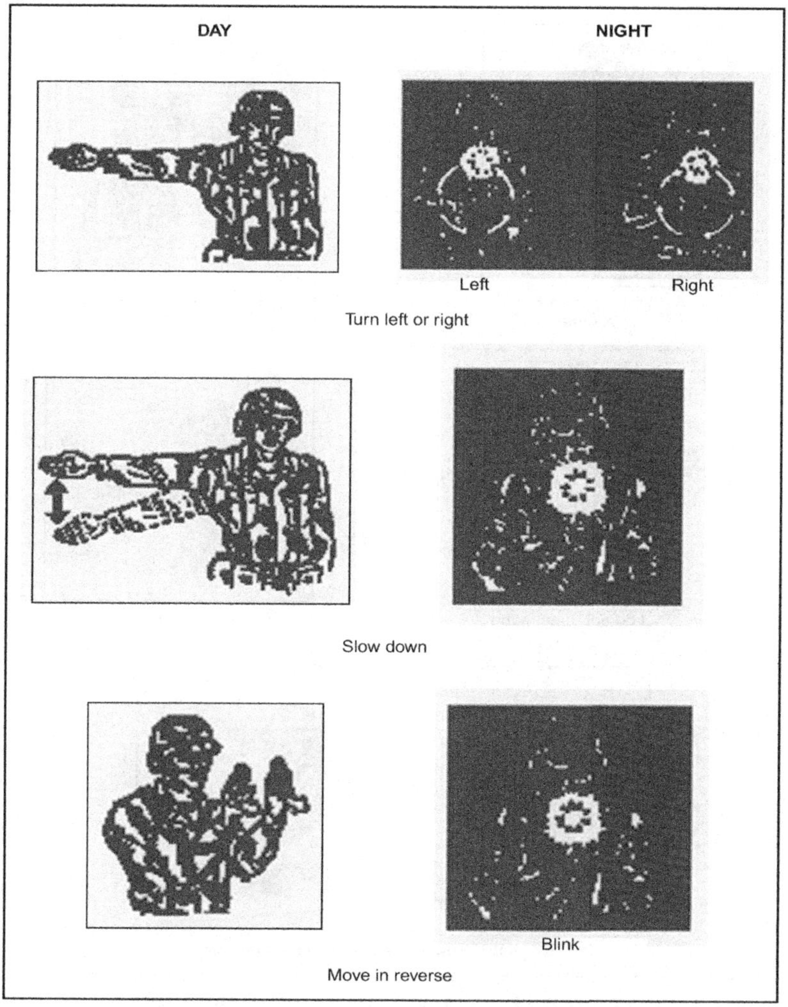

Figure A-14. Hand-and-arm signals (continued)

Disregard previous command

Attention

I do not understand

I am ready

Close the distance between vehicles and stop

Figure A-14. Hand-and-arm signals (continued)

Stop

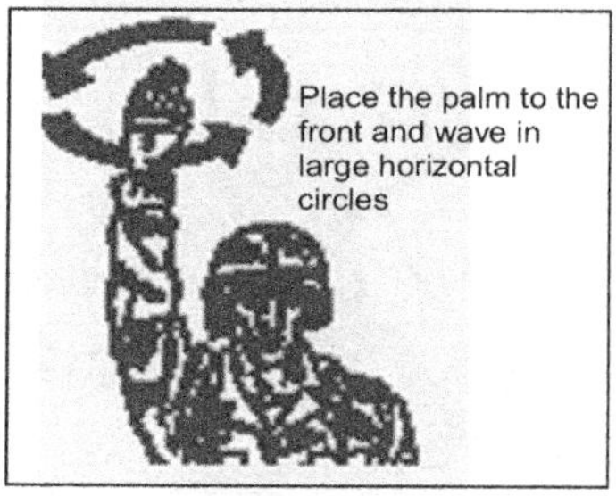

Assemble

Enemy is in sight

Commence firing

Cover our move

Cease firing

Figure A-14. Hand-and-arm signals (continued)

Move out

Enemy is in sight

Form a line

Figure A-14. Hand-and-arm signals (continued)

Mounted Movement Formations	Hand-and-Arm Signals	Formation Examples
V		
Echelon right (left)		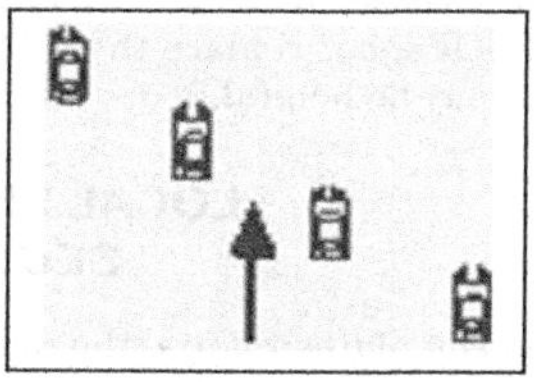
Wedge		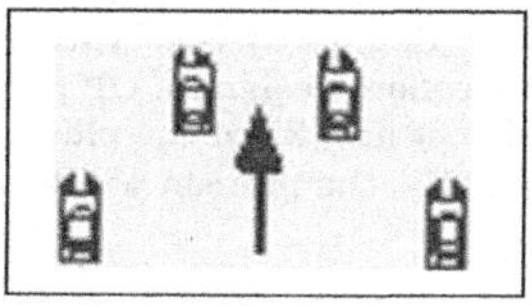
Line		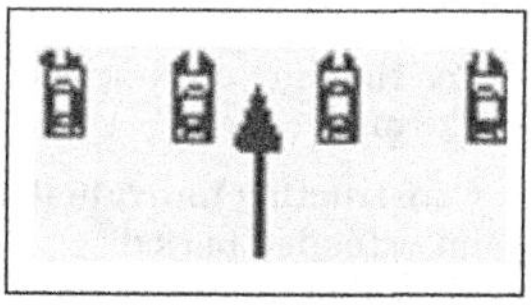
Column		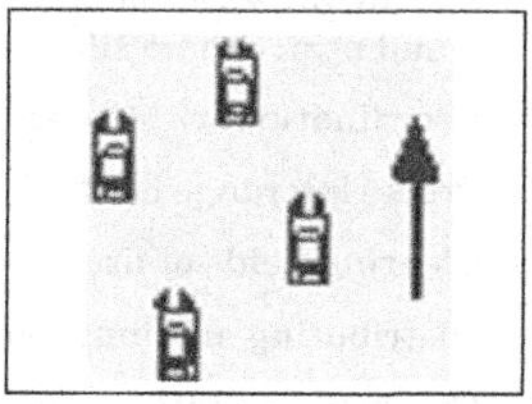

Figure A-15. Mounted movement formations and hand-and-arm signals

APPENDIX B

Checklists

This appendix gives several checklists that the platoon and its leaders will find useful. These checklists are not all encompassing and should be used to augment local SOPs, not replace them. Where units do not have standardized checklists, these may be helpful in developing local unit SOPs.

LOCAL SECURITY AND JOB-SITE SECURITY CHECKLISTS

When the platoon halts, the platoon leader establishes and maintains local security. He assigns each squad a sector of the perimeter to ensure mutual support and to cover all gaps by observation and fire. The platoon leader designates OPs and squad leaders designate OP personnel. OPs communicate with the platoon's command post and warn the platoon of the approaching enemy before it is attacked. Frequently, the platoon will be in a tactical halt while waiting for an operation to start.

When the platoon is halted, the platoon leader establishes a priority of work, to include—

- Positioning crew-served weapons, protective obstacles, and chemical alarms.
- Constructing individual and crew-served fighting positions, if the halt is for an extended period.
- Setting up wire communications between the squads and the platoon's command post. Radio silence should be maintained at the halts for OPSEC.
- Coordinating with adjacent units.
- Preparing range cards.
- Clearing fields of fire.
- Distributing ammunition, rations, water, supplies, and special equipment.
- Conducting PMCS on weapons, vehicles, and equipment.
- Inspecting the platoon's members and equipment.

- Rehearsing critical aspects for the upcoming mission.
- Test firing small-arms weapons, if the tactical situation permits.
- Conducting personal hygiene and field sanitation.
- Instituting a rest plan.
- Completing the work priorities as time permits.

Job-site-security actions are similar to those actions that are taken at a halt. However, job-site security is specifically designed to ensure that the platoon is able to work without being harassed.

Typically, the platoon will use this checklist when constructing obstacles and field fortifications. Before moving to the job site, the platoon leader informs everyone of the warning signals, code words, and pyrotechnics signals to be used. Upon arrival at the job site, the platoon—

- Occupies a job-site overwatch position.
- Dispatches reconnaissance/minesweeping/NBC teams to clear the job site.
- Establishes local security.
- Establishes escape routes; identifies possible enemy avenues of approach; and places OPs, chemical alarms, and crew-served- and antitank-weapons positions to cover enemy avenues of approach.
- Divides the job site into defensible sectors and assigns them to the squads.
- Establishes communications with the engineer company.
- Starts work.

PRECOMBAT CHECKLISTS

Each individual should have—

- Load-bearing equipment.
- A chemical protective mask with an M256 chemical-detection kit, an M13 decontamination kit, a waterproof bag for the mask, M8 chemical-detection paper, and three chemical-antidote injectors.
- A kevlar helmet with a cover and a band.
- An individual weapon with magazines and the zero posted in the pistol grip and a cleaning kit.
- A ruck sack and duffle bags packed with military clothing and table of allowance (TA)-50 equipment.

- Wire gauntlets.
- A fragmentation vest.

Each leader should have—

- A map and a compass.
- Signal operating instructions.
- His unit's SOP.
- *FMs 5-34 and 5-10* in their uniform cargo pockets.
- *FMs 20-32, 90-13-1, and 90-7* in their ruck sack.
- Other FMs, as required or needed.
- A watch and a solar-powered calculator.
- A notebook, pens, and pencils.
- Binoculars.
- Night-vision goggles.
- Casualty cards.
- A laminated range card.
- *DA Forms 1355 and 1355-1-R.*

All vehicles should have—

- Material, tied down and loaded according to the load plan.
- Fuel tanks, full.
- Preventive maintenance checks and services (PMCS), completed and dispatched.
- Packaged POL products, stowed.
- Water and extra fuel cans, full.
- Meals, ready-to-eat (MREs), stored.
- Basic-issue items (BIIs) and additional authorized list (ALL) items, inventoried, cleaned, and stowed.
- Spare parts and tires or track blocks, stowed.
- First-aid kits, inspected and stowed.
- A vehicle log book, the vehicle's TM, and *DA Forms 2404,* stowed.

- A camouflage screen system, stowed.
- Chemical-detection equipment, mounted.
- A tow cable with locking pins, stowed.
- VS17 marking panels for emergency marking.
- Ammunition evenly distributed and demolitions properly segregated and prepared for use.

UNIFORM-, DUFFLE-BAG-, AND RUCK-SACK-PACKING CHECKLISTS

A field uniform should include—

- A kevlar helmet with a cover and a band.
- A fragmentation vest.
- A pistol belt with ammunition pouches.
- A canteen with a cover and a cup, on the right hip.
- Ear plugs.
- A flashlight, on the right suspender strap (optional).
- A holster with a lanyard (if armed with a pistol), on the right side of the pistol belt.
- A bayonet, on the left side of the pistol belt.
- A compass, on the left side of the pistol belt.
- An M17A1 protective mask, on the left hip, and an M25 mask under the left arm.
- A battle dress uniform (BDU) with black boots.
- A field jacket.
- A chemical-decontamination kit connected to the protective mask.
- A nerve-agent antidote.
- A first-aid pouch, on the left suspender strap.

One duffle bag should contain—

- Three sets of BDUs.
- Three pairs of underwear.

- Three pairs of wool socks.
- Three T-shirts.
- One towel.
- Two laundry bags.
- One waterproof bag.
- One pair of boots with shoe polish, a brush, and a sewing kit.

The additional duffle bag should contain—

- Five tent pegs (10 for officers).
- Three tent poles (6 for officers).
- One tent rope (2 for officers).
- One sleeping bag.
- One pair of trouser suspenders.
- Two pairs of coveralls.
- One field jacket with a liner.
- One shelter half (2 for officers).
- One wool blanket.
- One pair of field pants.
- One pair of field pants with a liner.
- One cold-weather mask.
- One pair of extreme cold-weather boots.

The ruck sack should contain—

- Wet-weather pants and a parka.
- Overshoes.
- A weapons cleaning kit.
- Rations.
- A waterproof bag.
- A mission-oriented protective posture (MOPP) suit with gloves.
- Cold-weather mittens, winter use only.
- A parka with a liner and a hood, winter use only.

- A wool sweater, winter use only.
- A pile cap, winter use only.
- A scarf, winter use only.
- A towel and toilet articles.
- A pair of underwear.
- A T-shirt.
- Two pairs of wool socks.
- A BDU hat.
- An entrenching tool.

APPENDIX C

Breaching Case Studies

FM 90-13-1 and *Chapter 3* provide doctrinal overviews for breaching, breaching tenets, and breaching theory. The following case studies are designed to provide the platoon with instructive examples of dismounted and mounted breaching. Either case study could be applied to many different field applications. For example, the dismounted case study could be used to examine covert breaching, assault breaching, or deliberate breaching by light engineers. Likewise, the mounted case study could apply to both deliberate and in-stride breaching. These case studies are not designed to be prescriptive. They are designed to illustrate the thought process that the platoon leader must go through to successfully breach.

All breaching operations follow a similar pattern. They differ in technique. During breaching, the—

- Platoon leader determines the obstacle location and its composition.
- Platoon leader selects the breach site and lane locations.
- Platoon leader orders the breach to start, if the platoon is the primary reduction force.
- Squad prepares to breach, if the platoon is the backup reduction force.
- Squad breaches the obstacle.
- Squad marks the lane.
- Platoon leader reports the lane location to the company commander.

DISMOUNTED CASE STUDY

The company commander orders the dismounted platoon leader that is attached to him to breach an obstacle. The company is assaulting an enemy objective. The platoon leader moves to a location where he can see the obstacle and determines that it is a concertina fence and antitank minefield. He orders a squad forward to conduct reconnaissance and ascertain the best lane locations. The squad leader reports the locations and marks them with orange tape. The squad has determined that there are no antipersonnel mines in the minefield. The squad leader reports this to the platoon leader (see *Figure C-1, page C-2)*.

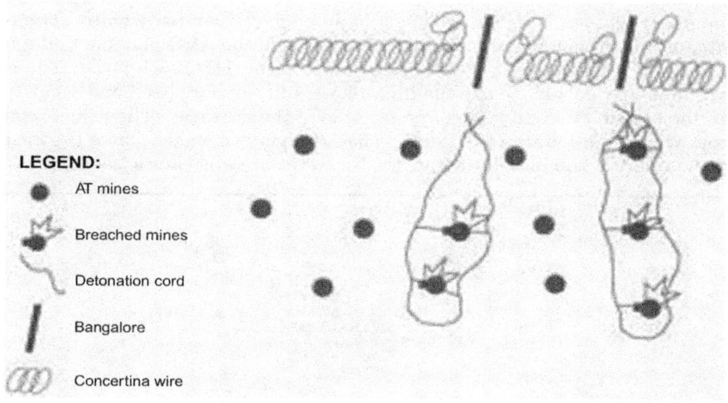

1 – The platoon leader determines the obstacle location and its composition.
2 – The platoon leader selects the breach site and lane locations.
3 – The squad breaches the concertina fence using cutters or Bangalores.
4 – The squad breaches the minefield using hand-emplaced explosives.
5 – The squad marks the lanes for the assault force.
6 – The platoon leader reports to the company commander that the lane is open.

Figure C-1. Dismounted case study

The platoon leader then orders the forward squad to remain at the obstacle as guides. He orders his second squad to come forward with two Bangalore sets. The first squad guides the second squad to the lane locations. Both squads breach the lanes in the concertina using the Bangalores and then move through the minefield placing charges to destroy antitank mines in the proposed lanes. The squads move to safety and initiate the explosives destroying the mines. Both squads provide security while the third squad comes forward to mark the lanes and guide the infantry company through the breach. The platoon leader keeps the third squad in reserve in case one of the squads fail or require additional help. However, in this scenario that was not necessary. The platoon leader observes the lanes and reports their completion to the infantry company commander who then starts his assault. The platoon leader then switches to the engineer company frequency and gives his company commander a situation report in reference to the lanes.

MOUNTED CASE STUDY

The engineer platoon is fighting as part of an engineer company/team. The company/team, which includes the engineer line platoon, the A&O platoon, and a tank platoon equipped with track-width mine plows, is the TF breach force. The company commander decides to use the tank plows, initially for speed and security, to create the breach site and secure the far side of the obstacle. Then the engineer platoon widens and marks the lanes. The A&O platoon leader places the far-lane recognition signals and initially orients the TF on the breached lanes (see *Figure C-2)*.

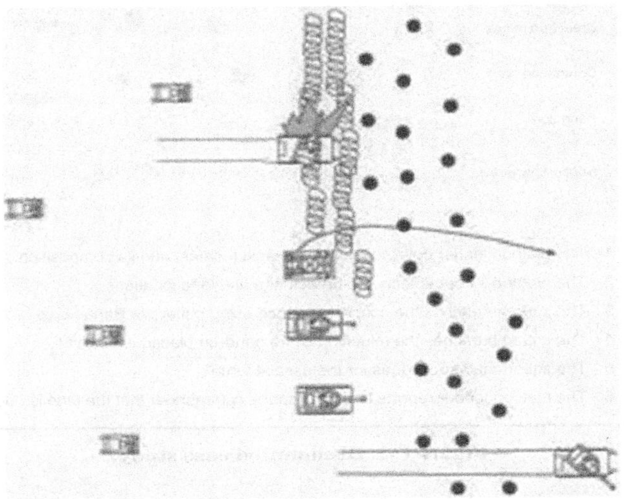

1 – The platoon leader determines the obstacle location and its composition.

2 – The platoon leader selects the breach site and lane locations.

3 – The tank crew breaches the obstacle with the tank plows.

4 – The platoon leader orders the AVLM forward to breach with the MICLIC when the enemy destroys a tank plow.

5 – The squad marks the lanes for the assault force.

6 – The platoon leader reports to the company commander that the lane is open.

Figure C-2. Mounted case study

In this scenario, the engineer company commander orders the engineer platoon forward to identify the obstacle location and its composition and potential lane locations. The platoon leader orders a dismounted element forward to conduct his reconnaissance. He reports his reconnaissance to the company commander. He states that the obstacle is a mixed minefield with antitank and antipersonnel mines, and a partial, multiple row of concertina fence is forward of the minefield. The platoon leader also reports two potential breach sites with an alternate site between them.

The engineer company commander orders the tank platoon to breach two lanes, based on the engineer platoon leaders' reconnaissance. He orders the engineer platoon to be prepared to back up the tank platoon. The northern tank equipped with a mine plow is destroyed by direct fire as it starts to breach. The engineer platoon leader observes this and orders his AVLM forward to breach with the MICLIC. The AVLM and the other tank plow successfully breach the obstacle.

The platoon leader orders two squads forward to mark each lane. These squads widen the lanes, per the initial plan, with hand-emplaced explosives after the assault force passes through. The platoon leader keeps one squad in reserve, with the AVLM, to counter any enemy attempts to close the lanes with artillery-delivered mines. He reports to the engineer company commander that the lanes are marked.

Glossary

&	and
+	plus
#	number
%	percent
1SG	first sergeant
1st	first
2d	second
3d	third
A&O	assault and obstacle
AA	axis of advance
AAR	after-action review
ABF	attack by fire
AC	hydrogen cyanide
ACE	armored combat earthmover, M9
AD	air defense
ADA	air-defense artillery
ADAM	area denial artillery munition
AHD	antihandling device
ALL	additional authorized list
ammo	ammunition

AP	antipersonnel
APC	armored personnel carrier
app	appendix
arty	artillery
ASL	authorized stockage list
aslt	assault
AT	antitank
AT-3, AT-5, AT-8	antitank missiles
ATD	antitank ditch
attn	attention
AVLB	armored vehicle-launched bridge
AVLM	armored vehicle-launched MICLIC
BDU	battle dress uniform
BII	basic-issue items
BMO	battalion maintenance officer
BMP	Soviet tracked amphibious infantry combat vehicle
BMP-1, BMP-2	Soviet infantry fighting vehicles
BOS	Battlefield Operating Systems
BSA	brigade support area
BT	blade team
C^2	command and control
CA	combined arms
CAM	chemical agent monitor
CATK	counterattack
CDR	commander

CEV	combat engineer vehicle
CG	phosgene
℄	centerline
COA	course of action
CP	checkpoint
CSS	combat service support
cu	cubic
DA	Department of the Army
decon	decontaminate/decontamination
DP	dispersing point
DS	direct support
DSA	division support area
EA	engagement area
EPW	enemy prisoner of war
FASCAM	family of scatterable mines
FEBA	forward edge of the battle area
FIST	fire-support team
FLOT	forward line of own troops
FM	field manual
FMFM	fleet Marine force manual
FS	fire support
FSO	fire-support officer
ft	foot (feet)

GA	tabun
GB	sarin
GD	soman
GEMSS	Ground-Emplaced Mine-Scattering System
GS	general support
HD	distilled mustard
HDP	hull-down position
HEMTT	heavy expanded mobility tactical truck
HL	mustard lewisite
HMMWV	high mobility multipurpose wheeled vehicle
HQ	headquarters
hr	hour(s)
IOE	irregular outer edge
IPB	intelligence preparation of the battlefield
IR	intelligence requirements
km	kilometer(s)
lb	pound(s)
LD	line of departure
LOGPAC	logistical package
LRA	local reproduction authorized
LRP	logistics release point
m	meter(s)

M-S	Miznay-Schardin
METT-T	mission, enemy, terrain, troops, and time available
MICLIC	mine-clearing line charge
MIJI	meaconing, intrusion, jamming, and interference
min	minute(s)
mm	millimeter(s)
MOPMS	Modular Pack Mine System
MOPP	mission-oriented protective posture
MP	military police
MRB	motorized rifle battalion
MRC	motorized rifle company
MRE	meal, ready-to-eat
MRP	motorized rifle platoon
MSR	main supply route
MST	maintenance support team
N	no
NA	not applicable
NAI	named area of interest
NBC	nuclear, biological, chemical
NCO	noncommissioned officer
NCS	net control station
No	number
NP	nonpersistent
OBSTINTEL	obstacle intelligence
OCOKA	observation and fields of fire, cover and concealment, obstacles, key terrain, and avenues of approach

OP	observation post
OPCON	operational control
OPLAN	operation plan
OPORD	operation order
OPSEC	operations security
ORP	objective rally point
oz	ounce(s)
P	persistent
PCC	precombat check
PCI	precombat inspection
PDM	pursuit-deterrent munition
PIR	priority intelligence requirements
PL	phase line
PLD	probable line of deployment
PLL	prescribed load list
plt	platoon
PMCS	preventive maintenance checks and services
POL	petroleum, oils, and lubricants
psn	position
R&S	reconnaissance and surveillance
RAAM	remote antiarmor mine
recon	reconnaissance
reqd	required
S&P	stake and platform
S1	Adjutant (US Army)

S2	Intelligence Officer (US Army)
S3	Operations and Training Officer (US Army)
S4	Supply Officer (US Army)
SCATMINE	scatterable mine
sec	second(s)
SEE	small emplacement excavator
SHELREP	shelling report
SITREP	situation report
SOI	signal operation instructions
SOP	standing operating procedure
SOSR	suppress, obscure, secure, reduce
SPOTREP	spot report
T/C/S	tasks, conditions, and standards
T72/80	Soviet tank
TA	table of allowance
TC	tank commander
TCP	traffic-control post
TDP	turret-down position
TF	task force
TLP	troop-leading procedures
TOC	tactical operations center
TRP	target reference point
TTP	tactics, techniques, and procedures
UMCP	unit maintenance collection point
US	United States

USAF	United States Air Force
VX	VX
w/	with
WO	warning order
WP	white phosphorous
wt	weight
XO	executive officer
Y	yes

Index

References

SOURCES USED

These are the sources quoted or paraphrased in this publication.

Army Publications

FM 3-4. *NBC Protection* (FMFM 11-9). 29 May 1992.

FM 3-5. *NBC Decontamination* (FMFM 11-10). 17 November 1993.

FM 5-34. *Engineer Field Data*. 14 September 1987.

FM 5-100. *Engineer Combat Operations*. 22 November 1988.

FM 5-103. *Survivability*. 10 June 1985.

FM 7-8. *Infantry Rifle Platoon and Squad*. 22 April 1992.

FM 20-32. *Mine/Countermine Operations*. 30 September 1992.

FM 71-100. *Division Operations*. 16 June 1990.

FM 90-7. *Combined Arms Obstacle Integration*. 29 September 1994.

FM 90-13-1. *Combined Arms Breaching Operations*. 28 February 1991.

FM 21-60. *Visual Signals*. 30 September 1987.

DOCUMENTS NEEDED

These documents must be available to the intended users of this publication.

Department of the Army Forms

DA Form 1155. *Witness Statement on Individual*. 1 June 1966.

DA Form 1156. *Casualty Feeder Report*. 1 June 1966.

DA Form 1355. *Minefield Record*. March 1987.

DA Form 1355-1-R. *Hasty Protective Minefield Record (LRA).* July 1975.

DA Form 2028. *Recommended Changes to Publications and Blank Forms.* 1 February 1974.

DA Form 2404. *Equipment Inspection and Maintenance Worksheet.* 1 April 1979.

Made in the USA
Monee, IL
05 November 2026

67810963R00184